THE ABC TRANSPORTERS OF HUMAN PHYSIOLOGY AND DISEASE

Genetics and Biochemistry of ATP Binding Cassette Transporters

THE ABC TRANSPORTERS OF HUMAN PHYSIOLOGY AND DISEASE

Genetics and Biochemistry of ATP Binding Cassette Transporters

Edited by

Kenneth J Linton
Queen Mary University of London, UK

I Barry Holland
Université Paris-Sud, France

World Scientific

NEW JERSEY • LONDON • SINGAPORE • BEIJING • SHANGHAI • HONG KONG • TAIPEI • CHENNAI

Published by

World Scientific Publishing Co. Pte. Ltd.
5 Toh Tuck Link, Singapore 596224
USA office: 27 Warren Street, Suite 401-402, Hackensack, NJ 07601
UK office: 57 Shelton Street, Covent Garden, London WC2H 9HE

British Library Cataloguing-in-Publication Data
A catalogue record for this book is available from the British Library.

THE ABC TRANSPORTERS OF HUMAN PHYSIOLOGY AND DISEASE
Genetics and Biochemistry of ATP Binding Cassette Transporters

The front cover shows a cartoon based on the crystal structure of a related drug efflux ABC transporter (Sav1866) from *Staphylococcus aureus*. The relevance of this structure for our understanding of how human ABC transporters work is described in Chapter 1. The back cover shows an unrooted phylogenetic tree of the human ABC transporter family.

All 48 full length proteins were aligned in Clustal W and phylogenetic analysis was performed in Seaview.

ISBN-13 978-981-4280-06-8
ISBN-10 981-4280-06-2

Typeset by Stallion Press
Email: enquiries@stallionpress.com

Printed by FuIsland Offset Printing (S) Pte Ltd Singapore

Contents

Contributors

Lydia Aguilar-Bryan, MD, PhD
Principal Scientist
Pacific Northwest Diabetes Research Institute
720 Broadway
Seattle, WA 98122, USA

Eric W. F. W. Alton, MD, FRCP, FmedSci
Department of Gene Therapy
Imperial College London
Manresa Road
London SW3 6LR, UK

Irwin M. Arias, MD
Senior Scientist
Head Unit on Cellular Polarity
Cell Biology and Metabolism Program
National Institute of Child Health and Human Development
National Institutes of Health
Bethesda, MD 20892, USA

Patrick Aubourg, MD
Faculty of Pharmaceutical and Biological Sciences
Inserm U745 and University Paris Descartes
4 avenue de l'Observatoire
75279, Paris cedex 06, France

Yuya Azuma, PhD
Laboratory of Cellular Biochemistry
Division of Applied Life Sciences
Graduate School of Agriculture, Kyoto University
Kitasirakawa, Sakyo-ku
Kyoto 606-8502, Japan

Joseph Bryan, PhD
Principal Scientist
Pacific Northwest Diabetes Research Institute
720 Broadway
Seattle, WA 98122, USA

Jean A. Chmielewski, PhD
Department of Chemistry
Purdue University
560 Oval Dr.
West Lafayette, IN 47907-2084, USA

Peter H. Dixon, PhD, BSc
Maternal and Fetal Disease Group
Institute of Reproductive and Developmental Biology
Department of Surgery and Cancer
Imperial College Hammersmith Campus
Du Cane Road
London W12 0NN, UK

Dana M. Emmert, PhD
Department of Chemistry
Purdue University
560 Oval Dr.
West Lafayette, IN 47907-2084, USA

Martin F. Fromm, MD
Institute of Experimental and
Clinical Pharmacology and Toxicology
Department of Clinical Pharmacology and Clinical Toxicology
Friedrich-Alexander Universität Erlangen-Nürnberg
Fahrstrasse 17
91054 Erlangen, Germany

Yu Fukuda, PhD
Department of Pharmaceutical Sciences
St. Jude Children's Research Hospital
262 Danny Thomas Place — MS 313
Memphis, TN 38105, USA

Martina Gentzsch, PhD
Cystic Fibrosis Research Center and
Department of Cell and Developmental Biology
University of North Carolina, Chapel Hill, North Carolina
6103 Thurston-Bowles Bldg., CB# 7248
Chapel Hill, NC 27599-7248, USA

Hartmut Glaeser, PhD
Institute of Experimental and
Clinical Pharmacology and Toxicology
Department of Clinical Pharmacology and Clinical Toxicology
Friedrich-Alexander Universität Erlangen-Nürnberg
Fahrstrasse 17
91054 Erlangen, Germany

Uta Griesenbach, PhD
Department of Gene Therapy
Imperial College London
Manresa Road
London SW3 6LR, UK

Hiroshi Hashiguchi, MD, PhD
Postdoctoral fellow
Pacific Northwest Diabetes Research Institute
720 Broadway
Seattle, WA 98122, USA

Tamás Hegedüs, PhD
Cystic Fibrosis Research Center and
Department of Biochemistry and Biophysics
University of North Carolina, Chapel Hill, North Carolina
6103 Thurston-Bowles Bldg., CB# 7248
Chapel Hill, NC 27599-7248, USA

Masako Hozoji-Inada, PhD
Institute for Integrated Cell-Material Sciences (iCeMS)
Kyoto University
Kitasirakawa, Sakyo-ku
Kyoto 606-8502, Japan

Christine A. Hrycyna, PhD
Department of Chemistry
Purdue University
560 Oval Dr.
West Lafayette, IN 47907-2084, USA

Qiujie Jiang, PhD
Assistant Professor
Department of Dermatology and Cutaneous Biology
Jefferson Medical College
Thomas Jefferson University
233 S. 10th Street, BLSB 430
Philadelphia, PA 19107, USA

Jörg König, PhD
Institute of Experimental and
Clinical Pharmacology and Toxicology
Department of Clinical Pharmacology and Clinical Toxicology
Friedrich-Alexander Universität Erlangen-Nürnberg
Fahrstrasse 17
91054 Erlangen, Germany

Qiaoli Li, PhD
Assistant Professor
Department of Dermatology and Cutaneous Biology
Jefferson Medical College
Thomas Jefferson University
233 S. 10th Street, BLSB 430
Philadelphia, PA 19107, USA

Kenneth J. Linton, PhD
Blizard Institute of Cell and Molecular Science
Barts and The London School of Medicine and Dentistry
Queen Mary, University of London
4 Newark Street, Whitechapel
London E1 2AT, UK

John Lynch, PhD
Department of Pharmaceutical Sciences
St. Jude Children's Research Hospital
262 Danny Thomas Place — MS 313
Memphis, TN 38105, USA

Laurie L. Molday, BSc
Department of Biochemistry and Molecular Biology
Centre for Macular Research
University of British Columbia
2350 Health Sciences Centre
Vancouver, B.C. V6T 1Z3, Canada

Robert S. Molday, PhD, FRSC
Department of Biochemistry and Molecular Biology
Centre for Macular Research
University of British Columbia
2350 Health Sciences Centre
Vancouver, B.C. V6T 1Z3, Canada

Kohjiro Nagao, PhD
Institute for Integrated Cell-Material Sciences (iCeMS)
Kyoto University
Kitasirakawa, Sakyo-ku
Kyoto 606-8502, Japan

Yumiko Nakamura, MD, PhD
Postdoctoral Fellow
Pacific Northwest Diabetes Research Institute
720 Broadway
Seattle, WA 98122, USA

Daphne Nikles, PhD
Institute of Biochemistry
Biocenter, Goethe-University Frankfurt
Max-von-Laue Str. 9
D-60438 Frankfurt a.M., Germany

David Ortiz
Graduate Student in the Program of Structural and Computational Biology and Molecular Biophysics at Baylor College of Medicine
Pacific Northwest Diabetes Research Institute
720 Broadway
Seattle, WA 98122, USA

Ronald P. J. Oude Elferink, PhD
Tytgat Institute for Liver and Intestinal Research
Academic Medical Center, S1-162
Meibergdreef 69-71
1105 BK Amsterdam, The Netherlands

Coen C. Paulusma, PhD
Tytgat Institute for Liver and Intestinal Research
Academic Medical Center, S1-162
Meibergdreef 69-71
1105 BK Amsterdam, The Netherlands

Marcos M. Pires, PhD
Department of Chemistry
Purdue University
560 Oval Dr.
West Lafayette, IN 47907-2084, USA

Heidrun Potschka, DVM
Institute of Pharmacology, Toxicology, and Pharmacy
Ludwig-Maximilians-University Munich
Koeniginstr. 16
80539 Munich, Germany

Aurora Pujol, MD, PhD
Institut d'Investigació Biomèdica de Bellvitge
Gran Via de l'Hospitalet, 199
08907 L'Hospitalet de Llobregat
Barcelona, Spain

John R. Riordan, PhD
Cystic Fibrosis Research Center and
Department of Biochemistry and Biophysics
University of North Carolina, Chapel Hill, North Carolina
6103 Thurston-Bowles Bldg., CB# 7248
Chapel Hill, NC 27599-7248, USA

Balazs Sarkadi, MD, PhD, DSc
Membrane Research Group
Hungarian Academy of Sciences
Dioszegi 64, 1113 Budapest, Hungary

Lutz Schmitt, PhD
Institute of Biochemistry
Heinrich Heine University Düsseldorf
Universitätsstr. 1
40 225 Düsseldorf, Germany

John D. Schuetz, PhD
Member, Vice Chair
Department of Pharmaceutical Sciences
St. Jude Children's Research Hospital
262 Danny Thomas Place — MS 313
Memphis, TN 38105, USA

Gergely Szakacs, MD, PhD
Institute of Enzymology
Hungarian Academy of Sciences
Karolina 29, 1113 Budapest, Hungary

Robert Tampé, PhD
Institute of Biochemistry
Biocenter, Goethe-University Frankfurt
Max-von-Laue Str. 9
D-60438 Frankfurt a.M., Germany

Kazumitsu Ueda, PhD
Institute for Integrated Cell-Material Sciences (iCeMS) and
Laboratory of Cellular Biochemistry
Division of Applied Life Sciences
Graduate School of Agriculture, Kyoto University
Kitasirakawa, Sakyo-ku
Kyoto 606-8502, Japan

Jouni Uitto, MD, PhD
Professor and Chair
Department of Dermatology and Cutaneous Biology
Jefferson Medical College and
Director of Jefferson Institute of Molecular Medicine
Thomas Jefferson University
233 S. 10th Street, BLSB 450
Philadelphia, PA 19107, USA

Dagny Ulrich, PhD
Department of Pharmaceutical Sciences
St. Jude Children's Research Hospital
262 Danny Thomas Place — MS 313
Memphis, TN 38105, USA

Saskia W. C. van Mil, PhD
Department of Metabolic and Endocrine Diseases and
Netherland's Metabolomics Center
Universitair Medisch Centrum Utrecht
Room KC02.069.0, Lundlaan 6
3584 EA Utrecht, The Netherlands

Alan S. Verkman, MD, PhD
Department of Medicine and Physiology
University of California, San Francisco
1246 Health Sciences East Tower, Box-521
San Francisco, CA 94143, USA

Catherine Williamson, MD, FRCP
Professor of Obstetric Medicine
Institute of Reproductive and Developmental Biology
Department of Surgery and Cancer
Imperial College Hammersmith Campus
Du Cane Road
London W12 0NN, UK

Joseph K. Zolnerciks, PhD
Blizard Institute of Cell and Molecular Science
Barts and The London School of Medicine and Dentistry
Queen Mary, University of London
4 Newark Street, Whitechapel
London E1 2AT, UK

Preface

ATP binding cassette (ABC) proteins are versatile machines that convert chemical energy from the hydrolysis of ATP into mechanical energy. These domains can be used to drive diverse cellular processes such as chromatin organization, DNA repair and mRNA trafficking, but the vast majority power the flow of solutes across cellular membranes.

ATP binding cassette transport proteins are integral membrane proteins that catabolize ATP to drive the uptake (prokaryota only) or efflux (prokayota and eukaryota) of solutes across membranes. Their ubiquity and abundance (many different forms are found in all species) is indicative of their importance to the maintenance of a healthy cell, organ and organism with an impact both directly and indirectly on human health. Chloroquine resistance in *Plasmodium* sp., antibiotic efflux from bacterial producers and some pathogenic bacteria, and fungal-pathogen resistance in wheat, remind us that you do not have to stray far from a human centric view of the biosphere to see the influence of this superfamily of transporters.

This volume focuses specifically on the direct influence imparted by the complement of more than 40 ABC transporters encoded by the human genome. Increasingly, studies from laboratory bench to clinical care reveal that human ABC transporters are crucial to both normal physiology and pathology. These include important roles, for example, in haem, lipid and fatty acid metabolism, bile flow, drug disposition and ion flux, which encompass nearly every organ, tissue and cell type in the body.

Their malfunction or dysregulation is linked to more than 30 diseases and conditions, and it seems likely that more will be discovered as disease aetiologies, or the precise function of some of the transporters, are clarified.

This is a particularly exciting time for those in the ABC transporter research field. The recent FEBS special meeting on ABC transporters in Innsbruck in 2010 attracted 300 delegates committed to the analysis of the structure, biochemistry and genetics of diverse ABC transporters from many species. The field is certainly broad but the seemingly highly conserved fundamentals of ABC transporter mode of action ensures that those of us who work in the realm of biomedical sciences can learn from the advances in distant disciplines; from the structure and function of bacterial transporters, to the role of plant transporters in lipid or wax metabolism. Although we now have a reasonable idea of the general principles driving the action of these transporters, we still lack details of crucial aspects of the mechanism, including the nature of endogenous allocrites (the transported "substrates"), how these proteins fold and how the expression of individual ABC transporter genes are regulated. With greater understanding of the basic science and knowledge of disease aetiology, this will afford the development of diagnostic and prognostic tools, and consequently more informed clinical intervention. Concerning the latter, several recent examples herald much promise: gene therapy for the neurodegenerative disease, X-linked adrenoleukodystrophy; the development and use of small molecules to switch on the chaperones needed to facilitate maturation and function of folding mutants of the cystic fibrosis transmembrane regulator. One of the most difficult challenges remains the design of specific drugs for clinical use that can successfully modulate the action of the multidrug efflux ABC transporters, without serious side effects. More distant objectives include drug therapy developed on personalized lines, and the targeting of new drugs to circumvent the action of the resident ABC transporters. This will allow penetration of specific barriers to treat, for example, brain tumors, epilepsy or psychotic

disease. If such aims are to be achieved, this is certainly not the time to restrict research activity in this field. On the contrary we would very much encourage young scientists to apply their skills to the study of these fascinating proteins not least in the clinical environment.

Kenneth Linton
Barry Holland

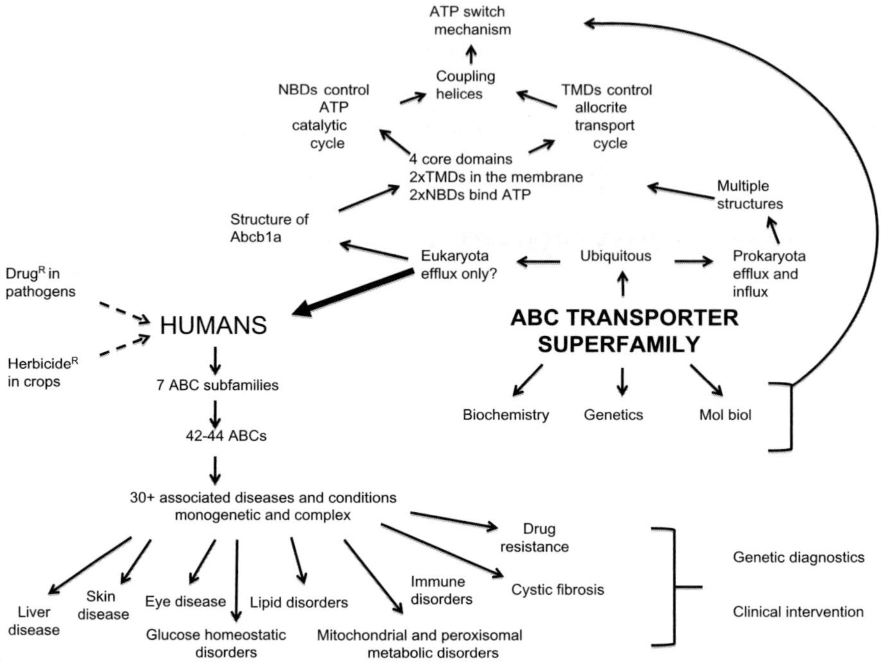
ATP switch mechanism
Coupling helices
NBDs control ATP catalytic cycle
TMDs control allocrite transport cycle
4 core domains
2xTMDs in the membrane
2xNBDs bind ATP
Structure of Abcb1a
Multiple structures
Eukaryota efflux only?
Ubiquitous
Prokaryota efflux and influx
Drug^R in pathogens
HUMANS
Herbicide^R in crops
ABC TRANSPORTER SUPERFAMILY
7 ABC subfamilies
Biochemistry
Genetics
Mol biol
42-44 ABCs
30+ associated diseases and conditions monogenetic and complex
Drug resistance
Genetic diagnostics
Immune disorders
Cystic fibrosis
Liver disease
Skin disease
Eye disease
Lipid disorders
Clinical intervention
Glucose homeostatic disorders
Mitochondrial and peroxisomal metabolic disorders

Chapter 1

General Introduction, Structure and Likely Mechanism of Action of ABC Transport Proteins

Kenneth J. Linton, Joseph K. Zolnerciks and Lutz Schmitt

1. Introduction

ATP binding cassette (ABC) proteins are versatile machines that convert chemical energy from the hydrolysis of ATP into mechanical energy. These domains can be used to drive diverse cellular processes such as chromatin organization, DNA repair and mRNA trafficking but the vast majority are coupled to integral membrane proteins where they power the flow of solutes across cellular membranes.

These so called ABC transporters are ubiquitous and are likely to be found in the membranes of all cells across all species. Most are fairly specific for a small subset of solutes, which range from small ions and peptides, to lipids, drugs and proteins. Generally it is the transmembrane domains (TMDs) that determine the specificity and many species therefore encode numerous subtly different ABC domains that are coupled, specifically, with distinct TMDs to generate transporters for distinct solutes. This paralogous expansion has resulted in a protein superfamily with surprising ubiquity and abundance (Table 1), typified by *E. coli*, which devotes almost 5% of its genome to encode its complement of ABC transporters.

Table 1. Ubiquity and abundance of ABC transporters.

Organism	ABC proteins
Archaeoglobus fulgidus	33
Escherichia coli	79
Saccharomyces cerevisiae	28
Caenorhabditis elegans	58
Drosophila melanogaster	51
Arabidopsis thaliana	129
Homo sapiens	48

The number of genes encoding proteins with ABC domains is given; as described in the text, and because of the modular nature of the transporter complexes, the actual number of ABC transporters per species will always be lower.

Although the human genome encodes 48 genes for ABC proteins, this corresponds to only 42–44 transporters, as some of these ABC domains are known to encode half transporters apparently functioning as heterodimers (see Section 3 below). Many of the human ABC transporters have important roles in normal physiology and consequently their dysregulation or malfunction underlies a number of important diseases. This volume describes the current understanding of the biology of the human complement of ABC transporters as it pertains to normal human physiology and disease. In this first chapter, an overview of the human proteins is provided and, drawing on data from the wider superfamily, particularly the bacterial ABC transporters for which there is now a rich literature in structure determination, the mechanistic features of a generic ABC transporter are described.

2. The Human Complement of ABC Transporters

The human genome contains 48 genes encoding for closely related ABC nucleotide binding domains (NBDs) that are likely to be involved in transport of solutes across cellular membranes. This excludes a number of pseudogenes that are unlikely to be expressed and also the more distantly related motor domains involved in non-transport processes like chromatin organisation. The actual number of transporters is smaller as some genes encode half transporters that heterodimerise to form the functional transporter. Phylogenetic analysis of the protein sequences of the human complement of ABC transporters has identified seven distinct sub-families A to G (Table 2). The NBDs are the most highly conserved domains, but the TMDs, when analysed separately, fall into the same groups as their cognate NBDs. This suggests two waves of paralogous expansion within higher vertebrates; an initial expansion into the seven groups followed by gene amplification and diversification (rather than a modular form of expansion with an NBD from one subfamily picking up a TMD from another to generate a novel transporter).

3. The Core Architecture of an ABC Transporter Comprises Two TMDs and Two NBDs

Although the familial name was not coined until much later, Higgins *et al.*, were the first to describe an ABC transporter when the permease for the uptake of histidine into *Salmonella typhimurium* was cloned and sequenced in 1982.[1] Bacteria often encode proteins that are part of the same complex in an operon under the control of a single promoter. The "his" operon contained four genes, encoding a single nucleotide binding domain (HisP), two integral membrane domains (HisQ and HisM) and a periplasmic soluble domain (HisJ). Although these were proposed to form a complex at the membrane, this was not formally established until 1991,[2] when the stoichiometry was also determined to comprise HisQ, HisM, 2xHisP, i.e., heterodimeric TMDs combined with homodimeric NBDs. The fifth component, HisJ, a high affinity binding protein for histidine, is secreted into the periplasmic space of these Gram negative bacteria and is required to deliver histidine to the transporter for uptake. These

Table 2. The human ABC transporters.

Protein	Common alias	Domains	Chromosome	Disease	Function	Chapter
ABCA1	ABC1, HDLDT1	T-N-T-N	9q31	Tangier disease, FHA, Scott syndrome	Cholesterol efflux	2
ABCA2	ABC2	T-N-T-N	9q34			
ABCA3	ABC3	T-N-T-N	16p13.3	PILD		
ABCA4	STGD1, ABCR	T-N-T-N	1p22	RP, MD, Stargardt's disease		5
ABCA5		T-N-T-N	17q21-q25			
ABCA6		T-N-T-N	17q21			
ABCA7	ABCX	T-N-T-N	19p13.3			
ABCA8		T-N-T-N	17q24			
ABCA9		T-N-T-N	17q24			
ABCA10		T-N-T-N	17q24			
ABCA12		T-N-T-N	2q34	HI		9
ABCA13		T-N-T-N	7p12.3	Schizophrenia, bipolar disease, depression		
ABCB1	MDR1, PGY	T-N-T-N	7q21.12	MDR	Drug efflux	4, 14
ABCB2	TAP1	T-T-N-T-N	6p21.3	Bare lymphocyte syndrome	Peptide transport	6
ABCB3	TAP2	T-T-N-T-N	6p21.3			
ABCB4	MDR3, PGY3	T-N-T-N	7q21	PFIC3, ICP	PC efflux	3, 14

(*Continued*)

Table 2. (*Continued*).

Protein	Common alias	Domains	Chromosome	Disease	Function	Chapter
ABCB5		T-N-T-N	7p14			
ABCB6		T-N	2q36	porphyria		7
ABCB7		T-N	Xq13.3			
ABCB8		T-N	7q35-q36			7
ABCB9	TAPL	T-N	12q24		peptides	6
ABCB10		T-N	1q32			7
ABCB11	BSEP, PFIC2, SPGP	T-N-T-N	2q24	PFIC2, ICP	Bile salt efflux	3
ABCC1	MRP1	T-T-N-T-N	16p13.1	MDR	Drug efflux	8
ABCC2	MRP2, CMOAT	T-T-N-T-N	10q24	DJS	Drug efflux	8
ABCC3	MRP3	T-T-N-T-N	17q21	MDR	Drug efflux	8
ABCC4	MRP4	T-N-T-N	13q31	MDR	Drug efflux	8
ABCC5	MRP5	T-N-T-N	3q27	MDR	Drug efflux	8
ABCC6	MRP6, PXE, ARA	T-T-N-T-N	16p13.1	PXE		9
ABCC7	MRP7, CFTR	T-N-T-N	7q31-q32	Cystic fibrosis	Cl^- channel	10, 14
ABCC8	MRP8, SUR1	T-T-N-T-N	11p15.1	HHI	K^+ channel regulation	11
ABCC9	SUR2	T-T-N-T-N	12p12.1			11

(*Continued*)

Table 2. (*Continued*).

Protein	Common alias	Domains	Chromosome	Disease	Function	Chapter
ABCC10	MRP7	T-T-N-T-N	6p12.3			
ABCC11	MRP8	T-N-T-N	16q12			
ABCC12	MRP9	T-N-T-N	16q12.1			
ABCC13	PRED6	Pseudogene	21q11.2			
ABCD1	ALD, ALDP	T-N	Xq28	ALD, ZS		12
ABCD2	ALDL1, ALDR	T-N	12q11-q12			12
ABCD3	PXMP1, PMP70	T-N	1p22-p21			12
ABCD4	PXMP1L	T-N	14q24			12
ABCF1	ABC50	N-N	6p21.33			
ABCF2	ABC28, HUSSY-18	N-N	7q35-q36			
ABCF3		N-N	3q25.1-q25.2			
ABCG1	ABC8	N-T	21q22.3		Cholesterol efflux	
ABCG2	MXR, BCRP, ABCP	N-T	4q22-q23	MDR, gout	Drug, Uric acid efflux	4, 14
ABCG4	WHITE2	N-T	11q23			
ABCG5	STSL	N-T	2p21 }	Sitosterolaemia, HC	Sterol efflux	13
ABCG8	GBD4	N-T	2p21 }			

FHA, familial hypoalphalipoproteinemia; PILD, pediatric interstitial lung disease; PFIC, Progressive familial intrahepatic cholestasis; ICP, Intrahepatic cholestasis of pregnancy; MDR, multidrug resitance; RP, retinosa pigmentosa; MD, age-related macular dystrophy; DJS, Dubin Johnson syndrome; PXE, pseudoxanthoma elasticum; HI, Harlequin ichthyosis; HHI, hyperinsulinemic hypoglycemia of infancy; ALD, adrenoleukodystrophy; ZS, Zellweger syndrome; HC, hypercholesterolemia.

periplasmic components are restricted to the subset of ABC transporters involved in solute import into bacteria. All of the eukaryotic ABC transporters of known function appear to be exporters, i.e., the direction of transport is away from the compartment in which the NBDs reside. This includes both export into organelles and to the extracellular milieu. However, the core architecture of 2xTMDs combined with 2xNBDs (Fig. 1) is a consistent feature of the ABC transporter family and can be realized either as a protein complex of two to four proteins in which any of the domains can be fused in any combination, or as a single polypeptide with four domains. The first example of the latter was identified with the cloning in 1986 of the first mammalian ABC transporter, the drug-resistance P-glycoprotein (ABCB1), an exporter of an unusually broad range of solutes including many drugs of clinical importance.[3] Most of the human ABC transporters are similarly encoded on a single polypeptide, but there are also "half transporters" of two-domain polypeptides (one TMD plus one NBD fused in either order) which function either as a homodimer (for example ABCG2) or a heterodimer (for example TAP1 and TAP2). Some eukaryotic transporters also have additional domains, for example, CFTR has a "regulatory" or R domain that is important for function of the channel (see Chapter 10) and several have an additional TMD (often referred to as TMD0 providing an extra 4, 5 or 6 transmembrane helices). It is sometimes important for regulation, or for interaction with other proteins if the transporter functions as part of a macromolecular complex, although in most cases a function has not yet been described.

4. Structure Analysis and Its Relevance to Molecular Mechanisms

Structural studies have had a strong impact on the field of ABC transporter research in recent years. The first milestone arrived with the determination of the novel structural fold of the soluble NBD of the histidine permease in 1998 by X-ray

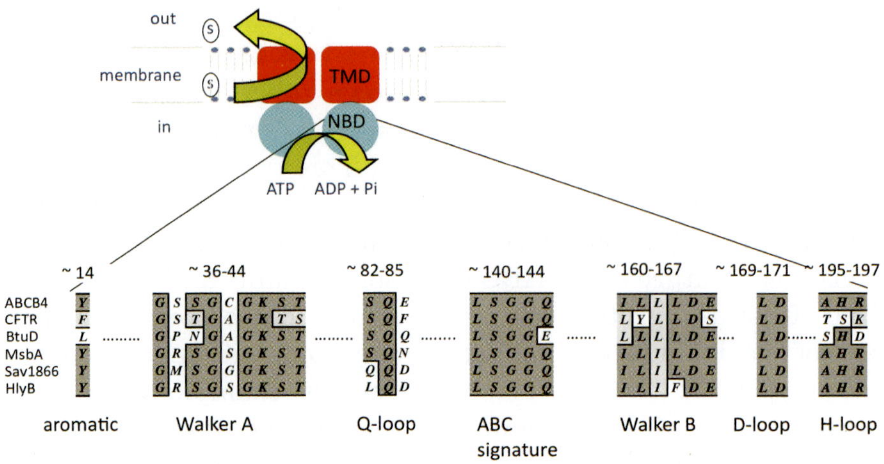

Fig. 1. The core domain organization of ABC transporters and key motifs. The four core domains of an ABC transporter comprise two TMDs within the membrane and two NBDs. The direction of allocrite (S) (the term we prefer for the transported solute, to distinguish from the true substrate of the transporter which is ATP) movement is shown for an exporter. The sequences of the TMDs vary widely throughout the ABC transporter family, but the NBDs are highly conserved. Alignment: ABCB4, the human hepatic phosphatidylcholine exporter NBD2; ABCC7, the human cystic fibrosis chloride channel NBD1; BtuD, the vitamin B12 uptake system from *Escherichia coli*; MsbA, the lipid A exporter from *Vibrio cholera*; Sav1866, a putative multidrug exporter from *Staphylococcus aureus* and HlyB, the haemolysin exporter from *Escherichia coli*.

crystallography (Fig. 2A).[4] In this paper, a crystallographic dimer (i.e. a dimer generated by the symmetry of the crystal lattice) was wrongly interpreted as physiologically meaningful. The nature of the true NBD dimer interface was not elucidated for several years, when bacterial homodimeric NBDs in various conformations of the ATP catalytic cycle were resolved at high resolution (Fig. 2B).[5–7] These provided insights into the roles of the conserved motifs, the allosteric coupling of the ATP binding sites and allowed the conformational changes that would likely ensue from the binding and hydrolysis of ATP to be modeled. The structure of the first full-size ABC transporter, the vitamin B12 importer from *E. coli*, with all four of the core domains

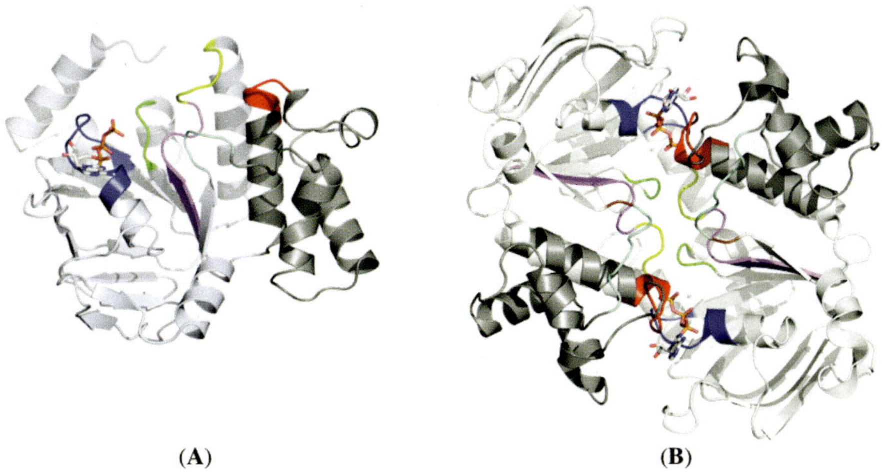

Fig. 2. The nucleotide binding pockets of an ABC transporter. (**A**) The conserved motifs of an NBD, modelled on the structure of HisP and viewed from the TMDs.[4] (**B**) Dimeric arrangement of a NBD dimer, modelled on HlyB.[6] The helical sub-domain is shown in dark grey and the F1-core in light grey. Conserved sequence motives are color-coded as follows: Walker A motif in blue and Walker B motif in magenta, Q-loop in cyan, ABC signature in red, D-loop in yellow and the H-loop in green. Bound ATP is shown in stick representation.

present was solved at medium resolution in 2002.[8] We now have medium resolution structural data for eight ABC transporter complexes (including some in multiple conformations). Seven of these are bacterial ABC transporters; five importers in which the conserved NBDs are coupled to TMDs from three classes with distinct protein folds (Fig. 3),[9] and two exporters, the lipid A transporter MsbA from *E. coli*[10] and the multidrug transporter Sav1866 from *Staphylococcus aureus*.[11] These share a common fold throughout their NBDs and TMDs and are also homologous with the more distantly related multidrug efflux pump Abcb1a from mouse, whose structure was determined in 2009 (Fig. 4).[12] These exporters exhibit some features that are not apparent in the importers. For example, the longer intracellular loops of the TMDs, their more extensive contact with the NBDs (exporters have four coupling helices connecting the

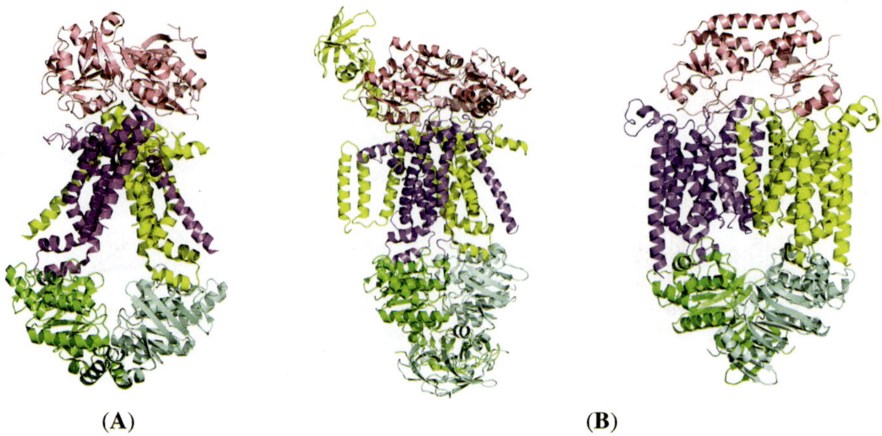

Fig. 3. The different classes of ABC importers.

(**A**) Class I represented by the structures of ModABC (left, pdb entry 2ONK) and the maltose importer (right, pdb entry 2R6G). The NBDs (ModC and MalK) are colored green and cyan, the TMDs (ModB and MalF/MalG) are colored yellow and magenta and the substrate-binding proteins (ModA and MalE) in salmon. (**B**) BtuCDF (pdb entry 2QJ9) as an example for class II of ABC importers. Color coding is identical to (A).

two TMDs with the two NBDs, while importers have only two), and the apparent wide separation of the NBDs in the absence of nucleotide. The latter, in particular, has been the source of some controversy in the field. However, perhaps the most significant difference between importers and exporters was the unexpected "domain swapping", in which each TMD in an exporter contacts both NBDs, i.e., the first coupling helix from each TMD contacts the cis (directly adjacent) NBD while the second coupling helix contacts the trans (diametrically apposed) NBD. Such trans-interaction has not been observed in ABC importers where only cis-interactions are present.

In the 1960s, Oleg Jardetzky[13] proposed that membrane transporters in general work using a ligand binding site that changed in affinity concomitant with alternating access from one side of the membrane to the other. Thus, an exporter would have

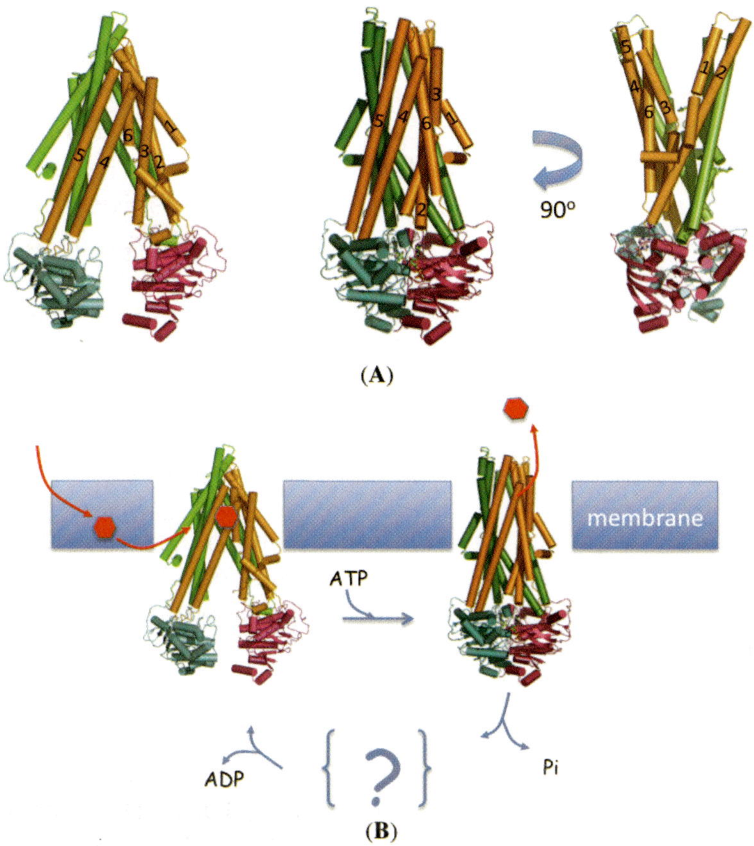

Fig. 4. Structural and mechanism of a mammalian ABC transporter. (**A**) Structure of Abcb1a (left) and Sav1866 (centre and right), with helices from Abcb1a TMD1, and the equivalent helices from Sav1866, numbered 1–6 to show the "domain swapping", and the inward- and outward- facing cavities of the two models, respectively. (**B**) The ATP switch model (see text for more details) describes an apo structure of a human ABC exporter with an inward facing high affinity site for allocrite binding. Allocrite binds to the high affinity site in a large cavity formed by the TMDs, most likely co-ordinated by hydrogen bonds, van der Waals and hydrophobic interactions. The hydrophobic binding cavity feeds via several helices to the main conduits for energy transduction, the coupling helices and their associated TMHs. Allocrite binding causes transmission of a conformational change that allows closure of the NBDs around two ATPs to form the closed NBD:NBD dimer. This represents the ATP switch between the open and closed conformations.

an inward facing high affinity site for ligand that would switch to an outward facing site with low affinity. In a primary active transporter (meaning one using a chemical energy source like ATP, rather than an ion gradient, to power directional transport), this alternating access model for ligand efflux must be coupled to an ATP catalytic cycle. In the following sections, the structural elements common to all ABC transporters, and the unusual features apparently unique to the ABC exporters, will be considered in the context of the biochemical data, as we ponder how these transporters work.

4.1 *The nucleotide binding domains and the ATP catalytic cycle*

The primary sequence conservation of the NBDs defines the ABC transporter superfamily. This is most apparent in the seven motifs described by the alignment shown in Fig. 1. These motifs

Fig. 4. (*Continued*) Mechanistic details of the closure of the gap separating the NBDs are obscure but must involve tilting of the TMHs to a more perpendicular aspect (with respect to the membrane). It may also involve repositioning of the Q-loop to engage the γ-phosphate of the ATP bound by the β-roll sub-domain, which in turn would bring the α-helical sub-domain forward to complete the second ATP binding pocket and provide a mechanism for allosteric coupling of the two pockets. That the Q-loop is also the likely site of energy transduction with coupling helix 2 from the TMDs suggests that this motif is the hub of the transporter, linking both ATP binding pockets to each other, and to the allocrite binding sites of the TMDs. Formation of the NBD:NBD dimer reorganises the TMHs to form the outwardly open conformation described by the nucleotide-bound Sav1866 structure with the re-association of the TMHs into different bundles of helices (akin to the repositioning of the message faces in the children's origami game of "fortune teller"). These movements result in an outward facing binding site that, for the multidrug resistance transporter ABCB1 (Abcb1a), has low affinity for allocrite. Allocrite disengages ABCB1 in this conformation (for other transporters it remains possible that a further step is required before allocrite is released). ATP hydrolysis, phosphate release and then ADP release, restores the transporter to the initial state for the next cycle of efflux.

are not invariant in sequence and some residues considered to have a key role in one transporter may even be absent in another. For example, the human multidrug resistance pump ABCB1 is unable to efflux drugs if the glutamate (E) in the Walker B motif is replaced with a glutamine, yet this residue is naturally absent from NBD1 of the CFTR (ABCC7). Such idiosyncrasies may make it difficult to describe a detailed model of the transport mechanism that is consistent with all the available data on each of ABC transporters studied. However, what we aim to derive here is a generic model which retains sufficient flexibility to allow for subtle refinement to account for these differences.

The overall conservation of the motifs within the NBDs throughout the wider superfamily implies conservation of the molecular mechanism of ATP binding and hydrolysis. This allows us to draw on the plethora of data derived from analyses of different systems; high resolution structures of bacterial NBDs, biochemical and biophysical analyses of NBDs *in vitro,* as well as detailed studies on the eukaryotic transporters in order to derive a consistent mechanism.

4.1.1 *Lessons from structure analysis of the NBDs*

The original structure of HisP indicated that the NBDs of ABC transporters had a unique protein fold.[4] In the crystallized monomer two subdomains were apparent, the larger sub-domain formed an anti-parallel β-roll which brought the Walker A and B motifs together with the stacking aromatic (or A-loop as coined by Ambudkar[14]) and H-loop (see also Fig. 2). The D-loop was somewhat distant but on the same face of the NBD and in close proximity to the smaller α–helical subdomain that contained the ABC signature motif linked to the core domain by the Q-loop. The isolated NBD of HisP and other bacterial ABC transporters retain the ability to hydrolyse ATP dependent on transient dimerisation.[15] However, the nature of the dimer interface only became known following crystallisation of mutant NBDs that were unable to hydrolyse the bound ATP. From these studies it was clear that the nucleotide binding pockets were formed at the interface of the dimer with both nucleotides co-ordinated by residues from both NBDs. Two sets of key motifs are brought together at the dimer interface to enclose the two ATPs. Thus, within the β-roll sub-domain, the aromatic ring of the A-loop residue stacks with the adenine ring of the nucleotide, the Walker A motif provides most of the hydrogen bonds with the three phosphate groups, either directly, or via the metal ion cofactor (usually magnesium). The Walker B, through its highly conserved glutamate, is also seen to contact the bound ATP via a water molecule that is appropriately positioned for nucleophilic attack on the γ-phosphate. There remains some controversy as to the exact mechanism of catalysis with both substrate-assisted[16] and general base[7] catalysis proposed for different transporters. This also influences the interpretation of the role of the H-loop histidine, which, in crystals of different mutants of different NBDs, is seen to stabilise the position of the Walker B glutamate and/or contact the γ-phosphate directly. While the precise details of catalysis remain equivocal it is clear that the highly conserved residues in these motifs form the nucleotide binding pockets and have a strong influence on the catalytic cycle.

The Q-loop glutamine also makes contact with the γ-phosphate indirectly via water and/or the bound metal ion cofactor. The Q-loop appears to occupy a pivotal position for both allosteric coupling of the two nucleotide binding pockets and allosteric coupling between these and the allocrite binding sites in the TMDs. The Q-loop glutamine is at the base of a groove on the surface of the NBD into which sits one of the "coupling helices" from the TMD (providing a link to the allocrite binding sites, see below). The Q-loop is also the direct link between the β-roll sub-domain and α–helical sub-domain that contains the ABC signature motif. The ABC signature forms several hydrogen bonds with the γ-phosphate group and one with the ribose moiety of the bound ATP.

That leaves the D-loop, for which a more prosaic role is suggested from these structures of dimeric NBDs. The D-loop extends from the Walker B motif to culminate in the highly conserved aspartate at the head of an α-helix juxtaposed between the β-roll and α–helical sub-domains. In the high resolution structural model of the NBD from the *E. coli* haemolysin exporter HlyB, the D-loop aspartate hydrogen bonds with the Walker A motif and the H-loop of the apposing NBD, providing two of several interactions across the dimer interface.[17] The function of this direct interaction between the two NBDs in the transport cycle of ABC transporters remains enigmatic.

4.1.2 *Allosteric coupling of the two nucleotide binding pockets*

Biochemical characterisation of the first four-domain single-polypeptide ABC transporters identified (human ABCB1 and its rodent homologues), demonstrated that both nucleotide binding sites were catalytically active, at least in these transporters. If either was inhibited by mutation or trapping in a post hydrolytic state, then hydrolysis was prevented at the other site indicative of the allosteric coupling of the two sites.[17,18] Purified NBDs from bacterial transporters have also been shown to dimerise to hydrolyse ATP *in vitro*.[15] Notably, acquisition of high resolution structural data for bacterial NBDs in different conformations of the catalytic cycle have allowed the allosteric coupling of the two catalytic pockets to be modelled. In particular, the differences between the apo monomeric NBD and the ATP bound dimer, centre on the conformation of the Q-loop and the position of the α-helical sub-domain with respect to the β-roll core. As ATP binds to the β-roll core, the Q-loop glutamine and the entire helical subdomain rotates by 20° towards the bound ATP to engage the nucleotide and the metal ion cofactor as well as possibly the hydrolytic water that is also co-ordinated by the γ-phosphate of the bound ATP.

Such a rigid body rotation of the helical domain probably describes very well how free NBDs in solution bind ATP and dimerise transiently to form the closed NBD dimer. However, whether this accurately describes what occurs in the full transporter is a moot point. The influence of the TMDs on the conformational changes of the NBDs is missing, and is likely very important. This is already evident from studies on the maltose importer from *E. coli*. The NBD (MalK) of this importer remains homodimeric in solution by virtue of a fused ~150 amino acid regulatory domain which interdigitates to act as a hinge at the base of the dimer. The free dimer hydrolyses ATP in solution and opens and closes, tweezer-like to do so.[5] However, after reconstitution of the MalK dimer with its cognate TMDs (MalF and MalG), binding of ATP is no longer sufficient to promote formation of the closed NBD dimer, and ATP hydrolysis becomes tightly coupled to the presence of both the maltose binding protein and maltose.[18] Thus, the TMDs somehow regulate the ATPase activity of the NBDs.

4.2 *The transmembrane domains*

The TMDs form the translocation pathway across the membrane and often, but not always, specify the transported substrate (allocrite). Across the entire super-family of ABC transporters the TMDs are not homologous. Indeed it is now clear from structural analyses of members of different subfamilies of bacterial ABC transporters that the NBDs are coupled to non-homologous TMDs of disparate primary sequence and protein fold. This, in part, provides an explanation for the broad range of allocrites handled across the ABC transporter superfamily. The interaction between the NBDs and TMDs is usually very specific but recent evidence has emerged for a novel family of bacterial membrane transporters which appear to be capable of using ABC-ATPases and which mix and match the same NBD with different TMDs to generate transporters with different allocrite specificity.[19]

In contrast to the situation in bacteria, the core TMDs of all the human ABC transporters may well share the same topology with each domain traversing the membrane six times. This seems likely for four reasons. First, the three exporters (MsbA, Sav1866 and Abcb1a) for which we have structural data exhibit this topology. Second, phylogenetic analysis of the TMDs in isolation from their highly conserved NBDs identifies the same seven sub-families. Third, hydropathy analysis (calculation of the hydrophobicity of a sliding window of several amino acids used to identify regions of sufficient length and hydrophobicity to cross the membrane) convincingly suggests the presence of six hydrophobic transmembrane helices (TMHs) per TMD in at least one member from each family. Lastly, when the topology has been analysed experimentally, clear evidence of six helices per core TMD has been obtained. Thus, even though their primary sequences may have diverged, it is likely that all of the human ABC transporters have evolved from a single subfamily of bacterial ABC exporters to which the lipid A transporter, MsbA and the drug exporter Sav1866 belong.

4.2.1 *The 6 × 6 topology and domain swapping*

After several years of study by several groups, Loo and Clarke finally determined the membrane topology of human ABCB1 by introducing novel cysteines into the hydrophilic loops of an otherwise cysteine-free protein.[20] The topology was deciphered after asking whether the unique cysteines were exposed extracellularly and available to bind sulphydryl-specific reagent. Each TMD was found to traverse the membrane six times with both the amino and carboxy-termini of the domain in the cytoplasm, thus giving rise to the 6 × 6 topological model. Similar technology was used to demonstrate that ABCC1, ABCC7 (CFTR), and TAP1/TAP2 shared this topology despite sharing little primary sequence identity within the TMDs.[21–23] Structure determination of the, albeit distant, bacterial homologues Sav1866 and MsbA, and more recently mouse Abcb1a, have shown the commonality of the protein fold and the α-helical nature of the membrane spanning regions. In particular, Sav1866 and Abcb1a are trapped in distinct conformations and may offer intriguing insights into the molecular mechanism of export (see following sections and Chapter 4 for more details).

The structural studies show that the TMHs of these ABC exporters cross the membrane in two bundles (Fig. 4A). Surprisingly, the bundles of six helices do not equate with the TMD boundaries but comprise four helices from one TMD and two helices from the other and they differ in the two distinct conformations found for these proteins. Thus, in the apo structure of Abcb1a, TMH 1, 2, 3 and 6 cross the membrane in concert with TMH 10 and 11, and TMH 4 and 5 cross the membrane in a bundle with TMH 7, 8, 9 and 12. In the closed, nucleotide-bound conformation of Sav1866 (but maintaining the TMH numbering of Abcb1a), TMH 1 and 2 cross the membrane with TMH 9, 10, 11 and 12, while TMH 7 and 8 cross with TMH 3, 4, 5 and 6.

One of the consequences of this arrangement is that the TMHs are tilted significantly with respect to the plane of the membrane, and that the long intracellular loops between TMHs 2 and 3, and TMHs 4 and 5 are able to interface through the "coupling helices" with NBD 1 and 2, respectively (see following sections for more detail on the coupling helices). This unexpected interface observed in the crystallised exporters has also been confirmed with a functional ABCB1 in the cell membrane, again by cysteine crosslinking experiments.[24] Such "domain swapping" with respect to the TMD helices is not common to all the TMDs in the ABC transporter superfamily; crystal structures of bacterial ABC importers with TMDs that are unrelated to the human TMDs show no, or little, evidence of intermingling of their TMHs. If, as argued above, all human TMDs are homologous and share a common protein fold then domain swapping is likely to feature in all of the human ABC transporters. The possible mechanistic implications of this structural organisation, is discussed in the following sections.

4.2.2 *Allocrite binding to the TMDs*

The Abcb1a structures also offer insight into the multi-specificity of the transporter as two of the three crystal forms solved have inhibitors bound in a non-productive conformation.[12] Although not an allocrite (i.e. not a transport substrate), the stereo-isomers of these cyclic peptide inhibitors are bound in distinct but overlapping regions of a large central cavity (measuring 6000Å^3), formed by the TMDs in the plane of the membrane (see also Chapter 4). Ueda *et al.*, had already speculated that ABCB1 should contain a large binding pocket in order to accept the variety of structurally distinct allocrites, and that some of the remaining volume, especially with respect to the smaller allocrites, would be occupied by cholesterol that presumably enters the pocket directly from the lipid bilayer (discussed further in Chapters 2 and 4).[25] This offers not only an explanation for the surprisingly large size range of allocrites handled by these drug exporters but also, given the co-binding of cholesterol, provides an explanation for why very closely related transporters (such as ABCB4 and ABCB11; Chapter 3) can handle lipids and bile salts.

Taking a quite different approach, Tampé's group (see Chapter 6) gained insight into the nature of the peptide binding pocket of the TAP transporter (TAP1/TAP2) by determining the character of the peptides accepted by the transporter. Thus, it became clear that while the three amino-terminal residues and the carboxy terminal residue provided some specificity, there must also be a certain amount of flexibility (either in the pocket or the peptide) in order to accommodate the full range of 8–40 mers transported, although peptides 8–12 amino acids in length are favored. While the co-ordinating residues in the TMDs of TAP1/2 have not been identified, the simplest explanation, that the amino- and carboxy- terminal residues of the transported peptide are recognised by the different TMDs within a large and flexible cavity, would be consistent with the structural data obtained for Abcb1a.

4.2.3 *The energy transduction interface*

In the ABC exporters, the TMHs of the TMDs extend well beyond (approximately 20Å) the likely position of the head groups of the inner leaflet of the lipid bilayer. The second and fourth TMH of each TMD culminate in a tight turn, followed by a short helix and another tight turn before extending into the third and fifth helices of the intra-membranous domains. These inter-membranous loops, the tips of which form the respective short coupling helices 1 and 2, provide the most substantial contact with the NBDs (Fig. 5A). While there is insufficient resolution to identify, unambiguously, the key residues involved at the interface, coupling helix 1 of TMD1 comes into close proximity with the Walker A motif of NBD1. Coupling helix 2 of TMD1 fits into a groove formed between the β-roll and the α-helical sub-domains of NBD2. Thus, coupling helix 2 of each TMD, interfaces with the Q-loop of the diametrically apposed NBD. These interfaces are the sites of energy transduction that allow communication between the TMDs and NBDs (thus coupling the allocrite binding sites to the ATP binding pockets). In the bacterial importers, the TMHs are much shorter and the coupling to the NBDs is via a single coupling helix per domain (equivalent, structurally, to coupling helix 2 of the exporters, Fig. 5B). Thus, each TMD in the bacterial importers only has significant contact with one NBD, and in a further departure from the organisation found in the exporters, the TMD of the importer contacts the NBD immediately below it and not the one diametrically opposite. The coupling helices are not highly conserved in primary sequence (except in a subset of more closely related importers)[26] but there is conservation in the structural fold. The length of the TMHs and the need for four coupling helices in the exporters is discussed in following sections in relation to molecular mechanism. Interestingly, a comparison of the available crystal structures of isolated NBDs already suggested in 2003[27] that a structural diverse region was present in the helical subdomain of ABC-ATPase. This region also covers for example, the X-loop that was identified in the crystal structure of Sav1866[11] as being a hallmark for MDR-ABC pumps and important for NBD-TMD communication.

(A) (B)

Fig. 5. The coupling helices of ABC transporters.

(**A**) The coupling helices of an ABC exporter. The NBD is shown as a transparent solid (cyan). The two coupling helices from both TMDs (cyan and green) are highlighted in red. (**B**) The coupling helix of an ABC importer. The NBD is shown as a transparent solid (magenta). The single coupling helix from the TMD (green) is highlighted in red.

5. Conformational Changes During the Transport Cycle

One of the curiosities of the crystal structure of Abcb1a is the large degree of separation of the NBDs (Fig. 4A). It was unexpected that the two NBDs would ever lose contact, let alone separate by 30Å, and this finding remains controversial. It is possible that the TMDs (or perhaps the TMH bundles that cross the membrane together) are relatively flexible in their relation to one another, particularly in a detergent micelle from which the crystal was grown, and that they are pulled apart by the strong non-physiological crystal contacts made by the hydrophilic NBDs in the three dimensional crystal lattice. If this is the case and the crystal conformation is an artefact, it is not irreversible since the protein is still able to regain drug-stimulated ATPase activity on solubilisation from the crystal.

Is there any evidence from biochemical or other biophysical studies for conformational changes in ABC transporters, including such a separation of the NBDs? The TMDs of human ABCB1 can undergo large conformational changes in response to trapping with non-hydrolysable ATP analogues or with ADP and vanadate.[28] Unfortunately, these data, obtained by transmission electron microscopy of 2D crystals of ABCB1, were not able to address the question of conformational change at the level of the NBDs. More recently, however, Rosenberg *et al.*,[29] report that the conformation of ABCC1 in the apo state, exhibits notable separation of the NBDs in a 2D crystal array consistent with the Abcb1a 3D crystal structures. Nevertheless, while the crystal contacts in these 2D and 3D crystals are likely to be very different, and are analysed by different techniques, they are both derived from transporters solubilised in detergent.

ABC transporters tend not to function in detergent and it is possible that the detergent could be the cause of the wide separation of the NBDs in the apo Abcb1a and MsbA structures. Fortunately, there is a technique, electron paramagnetic resonance (epr), which allows the structure of a membrane protein to be probed while the protein is functional within the lipid bilayer of a proteoliposome. This technique is nicely described in a recent review by Hellmich and Glaubitz,[30] but essentially, pairs of paramagnetic groups, usually nitroxide groups with an unpaired electron, are used to label the protein of interest at introduced cysteines. The electrons are excited by microwaves and their interaction, akin to the interaction of two bar magnets in a classroom experiment, can be measured in a magnetic field. The strength of the interaction is directly proportional to the distance separating the groups. MsbA has been studied extensively by epr and, the evidence obtained is consistent with a large (water filled) cavity formed by the TMDs in the apo state facing towards what would be the cytoplasmic face of the cell membrane. Notably, there is also significant separation of the NBDs (55–60Å between the spin labels) which closes by ~30Å in the presence of ADP plus vanadate.[31] The data reporting the separation of the NBDs is derived from a single cysteine positioned towards the base of the NBD in this homodimeric transporter (i.e. in a region furthest from the membrane). We are therefore reliant, so far, on data from just one mutant. Nevertheless, it is consistent with the observations from the crystal structures.

6. The ATP Switch Hypothesis

The current data set suggests that ABC transporters switch between the conformation described by the Abcb1a structure in the absence of nucleotide, representing an inwardly open allocrite binding site and open, i.e., well separated, NBDs, and the conformation described by the Sav1866 structure, in which the allocrite binding site is open to the exofacial side of the membrane and the NBD:NBD interface is closed around the bound ATP (Fig. 4B). The inwardly open conformation has a large cavity facing the cytoplasm for those transporters that can handle more hydrophilic allocrites, such as TAP1/TAP2 and ABCC1 and to the inner leaflet of the membrane for transporters like ABCB1 that transport hydrophobic allocrites. Some of the allocrites transported by human ABC transporters are very large; ergosterol, paclitaxel and phosphatidyl choline range from 400 to 850 Da and require a large binding pocket, and the smaller molecular weight drugs may also bind to the cavity alongside cholesterol. For ABCB1 and similar transporters, their promiscuous nature rules out any lock and key model for the interaction of allocrite and transporter. We have already argued that all of the human ABC transporter family likely share similar "domain swapping" between the TMDs, with helices from one TMD associating with helices from the second TMD as they cross the membrane. This architecture may be necessary to physically open the large cavity to accommodate the larger size range of allocrites. This may necessitate complete separation of the NBDs and to achieve this, the TMDs may have evolved a second coupling helix (i.e. four per transporter) to maintain direct contact with both NBDs throughout the transport cycle. The

Sav1866 structure gives us a snapshot of the ATP-bound structure (the protein is actually co-crystallised with ADP, but the NBD:NBD interface closely resembles the ATP-bound closed dimers of the bacterial NBDs solved at higher resolution). The nucleotide-bound Sav1866 appears to be in a conformation in which the putative allocrite binding cavity is open to the outside of the cell. These observations therefore satisfy one of the conditions of the "alternating accessibility" model for transmembrane transport originally postulated by Jardetzki in the 1960s. If the inwardly open and closed conformations also equate with high and low affinities for allocrite then the second condition of the "alternating accessibility" model of an exporter will be satisfied. That this is indeed the case was definitively shown for hamster Abcb1 in 2001 when the inwardly open conformation (i.e. in the absence of ATP) was shown to have a high affinity for the drug allocrite vinblastine, and the inwardly closed conformation (i.e. in the presence of non-hydrolysable ATP) was shown to have the same binding capacity for the drug, but at much lower affinity.[32] ATP binding and formation of the closed NBD:NBD dimer should therefore be sufficient to efflux allocrite (at least for the multidrug resistance ABCB1). This is the crux of the ATP switch model proposed independently by two groups[33,34] and summarized in Fig. 4B). However, the *in vitro* affinity for ATP reported for many ABC transporters is apparently lower than normal cellular concentrations, meaning that ABCB1 in the cell membrane should either hydrolyse ATP in a futile cycle, or be locked in a conformation with a low affinity for allocrite! But, if the observations in the maltose transport system, that the NBDs can bind ATP, but the TMDs prevent NBD:NBD closed dimer formation in the absence of allocrite represents a general phenomenon, (thus preventing futile ATP hydrolysis or locking the transporter in a non-productive conformation) then this conundrum is circumvented.

The energy released by ATP hydrolysis, we presume, would be required to resolve the NBD:NBD dimer and reset the transporter for the next cycle. In ABCB1 this appears to require hydrolysis of both ATPs (as mutation of either of the binding pockets results in the non-functional transporter). In other ABC transporters, such as ABCC1 or TAP1/TAP2, one of the nucleotide binding pockets contains non-canonical motifs that allow ATP binding but a significantly reduced hydrolysis rate.[35,36] Whether hydrolysis of one or two ATPs is required in any ABC transporter may well depend on the nature of the NBD:NBD interface (i.e. how much energy is required to break the bonds formed during dimerisation), and the kinetics/efficiency of transporter function required for a healthy cell. Even in ABCB1, hydrolysis seems to be non-simultaneous,[37] and the protein can be trapped in a distinct conformation[28] when the liberated phosphate is replaced with vanadate to trap one ADP in the nucleotide binding pocket.[38] Thus, phosphate is released first following hydrolysis, then ADP to restore the transporter to the apo state described by the Abcb1a structural model.

7. Future Perspectives

This is a particularly exciting time for those in the ABC transporter field. The recent FEBS special meeting on ABC transporters in Innsbruck had 300 delegates who research the structure, biochemistry and genetics of diverse ABC transporters from different species. To understand the molecular mechanism of these proteins would have an immense impact on the field and significant progress has been made in recent years. To drive this forward would require higher resolution structures of full-length transporters trapped in different conformations, with and without substrate and allocrite, combined with biophysical analysis of the transport mechanism, most likely in the form of the promising epr studies. Only then, will the detail emerge and surely inspire biochemists to probe the transport cycle of these transporters further.

References

1. Higgins, C.F., Haag, P.D., Nikaido, K., Ardeshir, F., Garcia, G., and Ames, G.F. (1982) *Nature* **298**, 723.
2. Kerppola, R.E., Shyamala, V.K., Klebba, P., and Ames, G.F. (1991) *J. Biol. Chem.* **266**, 9857.
3. Chen, C.J., Chin, J.E., Ueda, K., Clark, D.P., Pastan, I., Gottesman, M.M., and Roninson, I.B. (1986) *Cell* **47**, 381.
4. Hung, L.W., Wang, I.X., Nikaido, K., Liu, P.Q., Ames, G.F., and Kim, S.H. (1998) *Nature* **396**, 703.
5. Chen, J., Lu, G., Lin, J., Davidson, A.L., and Quiocho, F.A. (2003) *Mol. Cell* **12**, 651.
6. Zaitseva, J., Jenewein, S., Oswald, C., Jumpertz, T., Holland, I.B., and Schmitt, L. (2005) *Biochem. Soc. Trans.* **33**, 990.
7. Smith, P.C., Karpowich, N., Millen, L., Moody, J.E., Rosen, J., Thomas, P.J., and Hunt, J.F. (2002) *Mol. Cell* **10**, 139.
8. Locher, K.P., Lee, A.T., and Rees, D.C. (2002) *Science* **296**, 1091.
9. Locher, K.P. (2009) *Philos. Trans. R Soc. Lond. B Biol. Sci.* **364**, 239.
10. Ward, A., Reyes, C.L., Yu, J., Roth, C.B., and Chang, G. (2007) *Proc. Natl. Acad. Sci. USA* **104**, 19005.
11. Dawson, R.J. and Locher, K.P. (2006) *Nature* **443**, 180.
12. Aller, S.G., Yu, J., Ward, A., Weng, Y., Chittaboina, S., Zhuo, R., Harrell, P.M., Trinh, Y.T., Zhang, Q., Urbatsch, I.L., and Chang, G. (2009) *Science* **323**, 1718.
13. Jardetzky, O. (1966) *Nature* **211**, 969.
14. Ambudkar, S.V., Kim, I.W., Xia, D., and Sauna, Z.E. (2006) *FEBS Lett.* **580**, 1049.
15. Zaitseva, J., Jenewein, S., Wiedenmann, A., Benabdelhak, H., Holland, I.B., and Schmitt, L. (2005) *Biochemistry* **44**, 9680.
16. Zaitseva, J., Jenewein, S., Jumpertz, T., Holland, I.B., and Schmitt, L. (2005) *EMBO. J.* **24**, 1901.
17. Zaitseva, J., Oswald, C., Jumpertz, T., Jenewein, S., Wiedenmann, A., Holland, I.B., and Schmitt, L. (2006) *EMBO. J.* **25**, 3432.
18. Orelle, C., Ayvaz, T., Everly, R.M., Klug, C.S., and Davidson, A.L. (2008) *Proc. Natl. Acad. Sci. USA* **105**, 12837.
19. Rodionov, D.A., Hebbeln, P., Eudes, A., ter Beek, J., Rodionova, I.A., Erkens, G.B., Slotboom, D.J., Gelfand, M.S., Osterman, A.L., Hanson, A.D., and Eitinger, T. (2009) *J. Bacteriol.* **191**, 42.
20. Loo, T.W. and Clarke, D.M. (1995) *J. Biol. Chem.* **270**, 843.

21. Chang, X.B., Hou, Y.X., Jensen, T.J., and Riordan, J.R. (1994) *J. Biol. Chem.* **269**, 18572.
22. Bakos, E., Hegedus, T., Hollo, Z., Welker, E., Tusnady, G.E., Zaman, G.J., Flens, M.J., Varadi, A., and Sarkadi, B. (1996) *J. Biol. Chem.* **271**, 12322.
23. Schrodt, S., Koch, J., and Tampe, R. (2006) *J. Biol. Chem.* **281**, 6455.
24. Zolnerciks, J.K., Wooding, C., and Linton, K.J. (2007) *Faseb. J.* **21**, 3937.
25. Kimura, Y., Kodan, A., Matsuo, M., and Ueda, K. (2007) *J. Bioenerg. Biomembr.* **39**, 447.
26. Mourez, M., Hofnung, M., and Dassa, E. (1997) *EMBO. J.* **16**, 3066.
27. Benabdelhak, H., Kiontke, S., Horn, C., Ernst, R., Blight, M.A., Holland, I.B., and Schmitt, L. (2003) *J. Mol. Biol.* **327**, 1169.
28. Rosenberg, M.F., Velarde, G., Ford, R.C., Martin, C., Berridge, G., Kerr, I.D., Callaghan, R., Schmidlin, A., Wooding, C., Linton, K.J., and Higgins, C.F. (2001) *EMBO. J.* **20**, 5615.
29. Rosenberg, M.F., Oleschuk, C.J., Wu, P., Mao, Q., Deeley, R.G., Cole, S.P., and Ford, R.C. (2010) *J. Struct. Biol.* **170**, 540.
30. Hellmich, U.A. and Glaubitz, C. (2009) *Biol. Chem.* **390**, 815.
31. Borbat, P.P., Surendhran, K., Bortolus, M., Zou, P., Freed, J.H., and McHaourab, H.S. (2007) *PLoS Biol.* **5**, e271.
32. Martin, C., Higgins, C.F., and Callaghan, R. (2001) *Biochemistry* **40**, 15733.
33. Higgins, C.F. and Linton, K.J. (2004) *Nat. Struct. Mol. Biol.* **11**, 918.
34. Janas, E., Hofacker, M., Chen, M., Gompf, S., van der Does, C., and Tampe, R. (2003) *J. Biol. Chem.* **278**, 26862.
35. Gao, M., Cui, H.R., Loe, D.W., Grant, C.E., Almquist, K.C., Cole, S.P., and Deeley, R.G. (2000) *J. Biol. Chem.* **275**, 13098.
36. Karttunen, J.T., Lehner, P.J., Gupta, S.S., Hewitt, E.W., and Cresswell, P. (2001) *Proc. Natl. Acad. Sci. USA* **98**, 7431.
37. Hrycyna, C.A., Ramachandra, M., Ambudkar, S.V., Ko, Y.H., Pedersen, P.L., Pastan, I., and Gottesman, M.M. (1998) *J. Biol. Chem.* **273**, 16631.
38. Tombline, G., Bartholomew, L.A., Urbatsch, I.L., and Senior, A.E. (2004) *J. Biol. Chem.* **279**, 31212.

Chapter 2

ABCA1-Mediated HDL Formation: Mechanism and Regulation

Kohjiro Nagao, Masako Hozoji-Inada, Yuya Azuma and Kazumitsu Ueda

1. ABCA1 and HDL Formation

Maintenance of cellular cholesterol homeostasis is important for normal human physiology, and its disruption can lead to a variety of pathological conditions, including cardiovascular disease.[1] ATP-binding cassette protein A1 (ABCA1) is involved in high-density lipoprotein (HDL) formation,[2,3] which is the only known pathway that can eliminate excess cholesterol from peripheral cells, thus is one of the key proteins for cholesterol homeostasis. It is clear that ABCA1 is important for HDL generation, because at least 70 mutations, including 44 missense, and 11 insertions or deletions, have been identified in the ABCA1 gene, leading to the allelic disorders Tangier disease (TD) and familial hypoalphalipoproteinemia (FHA),[4–6] in which patients have a near-absence or decrease of circulating HDL, prominent cholesterol-ester accumulation in tissue macrophages, and premature atherosclerotic vascular disease.[1,7] More than 15 mutations of them examined showed the high correlation between phospholipids (PL) and cholesterol efflux,[2,8,9] indicating that ABCA1 influences efflux of both PL and free cholesterol.

ABCA1 mediates secretion of both types of lipids to an extracellular acceptor in the plasma, apolipoprotein A-I (apoA-I), to form HDL.[2,3]

2. Structural Feature of ABCA1

ABCA1 is a member of the ABCA subclass of ABC transporters, which contain the basic architecture of "full-length" ABC transporters organized into two tandemly arranged halves. Each half contains six transmembrane α-helices (TMs) followed by a cytoplasmic nucleotide-binding domain (NBD). The ABCA subclass is distinguished from other ABC transporter subclasses by the presence of large extracellular domains (ECDs) (Fig. 1).[10,11] Although TD and FHA mutations occur throughout the gene, they apparently cluster in five regions (Fig. 1): four mutations in cluster #1, six in cluster #2, eight in cluster #3, five in cluster #4 and five in cluster #5. The clusters #1 and #2 locate at ECD1, the cluster #3 at NBD1, the cluster #4 at the

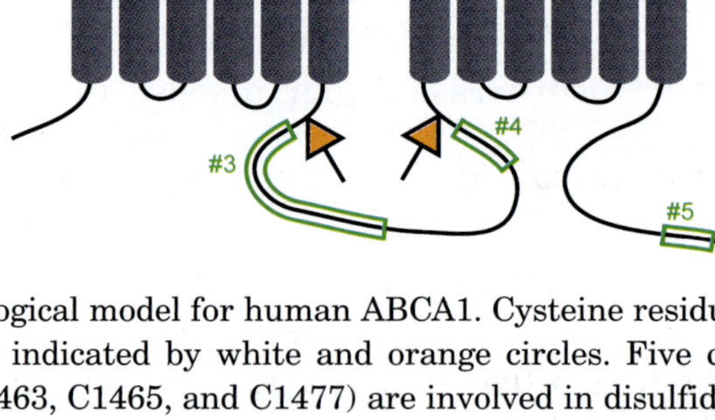

Fig. 1. Topological model for human ABCA1. Cysteine residues in extracellular regions are indicated by white and orange circles. Five cysteine residues (C75, C309, C1463, C1465, and C1477) are involved in disulfide bonds between ECD1 and ECD2. A TD mutation W590S is indicated by green circle. Five regions where TD and FHA mutation cluster are indicated by green boxes. Two cleaved sites by limited trypsin digestion are indicated by arrows.

linker region, and the cluster #5 at the C-terminal region, suggesting the importance of these regions for ABCA1 structure and functions.

When ABCA1 is partially digested by trypsin, it is cleaved at two sites, one just C-terminal to the 6th transmembrane α-helix (TM6) and the other just N-terminal to TM7 (indicated by arrows in Fig. 1).[12] When these fragments are analyzed by SDS-PAGE under non-reducing conditions, they co-migrate with undigested ABCA1. Thus, just as in ABCA4 (ABCR) the N and C halves of ABCA1 are connected by disulfide bond(s).[13] ECDs of ABCA1 contain 16 cysteine residues, and these cysteine residues are well conserved among ABCA1, ABCA4, and ABCA7 transporters. A Tangier disease mutation, C1477R located in ECD2 has been reported to abolish apoA-I binding and HDL formation.[8,9,14] Several missense mutations in cysteine residues within ECD1 (C54Y, C75G) and ECD2 (C1488R, C1490Y) of ABCA4 have also been linked to Stargardt disease, inherited juvenile macular degeneration.[15–18] However, it remains unclear whether disulfide bond formation *per se* is important for the function of ABCA1.

To address these questions, we recently constructed more than 30 ABCA1 mutants, in which ECD cysteine residues were replaced with serines, and examined disulfide bond formation, apoA-I binding, and HDL formation in these mutants.[19] From the single cysteine replacements, two cysteine residues (C75 and C309) in ECD1 were found to be essential for apoA-I binding. In contrast, in ECD2, only C1477 was found to be essential for HDL formation, and no single cysteine replacement impaired apoA-I binding. The concurrent replacement of two cysteine residues, C1463 and C1465, impaired apoA-I binding and HDL formation. This suggested that four out of five extracellular cysteine residues (C75, C309, C1463, C1465, and C1477) are involved in these functions of ABCA1. Trypsin digestion experiments showed that one disulfide bond is not sufficient; rather, two intramolecular disulfide bonds (between C75 and C309 in ECD1, and either C1463 or C1465 and C1477 in ECD2) are required for ABCA1 to be fully functional.[19]

3. Mechanism of apoA-I-Dependent Cholesterol Secretion by ABCA1

Cholesterol is synthesized on the endoplasmic reticulum (ER) membrane, but the content of free cholesterol in ER membrane is only 5% mol of total ER lipids.[20] Approximately 65–80% of free cholesterol in cells is suggested to be in the plasma membrane,[21] and free cholesterol makes up 20–30% of the plasma membrane lipids. Cholesterol is transferred from ER to the plasma membrane, and from the plasma membrane to ER, via both vesicular and non-vesicular pathways, but the detail mechanisms are largely unknown.

Because the major reservoir of cholesterol in cells is the plasma membrane, HDL formation is suggested to occur on the plasma membrane. Several models have been proposed for the mechanism of ABCA1-mediated HDL formation: (i) a two-step process model: ABCA1 first mediates PL efflux to apoA-I, and this apolipoprotein-PL complex accepts free cholesterol (FC) in an ABCA1-independent manner; (ii) a concurrent process model: FC and PL efflux by ABCA1 to apoA-I are tightly coupled to each other; and (iii) ABCA1 generates a specific apoA-I binding site on the plasma membrane and subsequent translocation of PL and cholesterol to apoA-I.

When it was proposed by Fielding, P. E. *et al.*[22] and Wang, N. *et al.*,[23] the two-step model looked most plausible, because photoactive PL could be cross-linked with ABCA1 whereas direct binding of photoactive cholesterol to ABCA1 could not be detected.[23] However, analysis of functions of ABCA7 raised questions about this model. Human ABCA7, which has the highest homology (69%) to ABCA1, mediates the apoA-I-dependent efflux of PL and cholesterol, just as ABCA1;[24] however, human ABCA7 mediates cholesterol release much less efficiently than ABCA1,[25] and PLs but not cholesterol are loaded onto apoA-I by mouse ABCA7.[26] These results cannot be explained by the two-step process model. Instead, they suggest that ABCA1 have a higher affinity for cholesterol to transport than ABCA7 does, and are consistent with the concurrent process model.[27]

The third model was originally proposed by Chimini *et al.*:[28] ABCA1 mediates the translocation of phosphatidylserine (PS) to the outer leaflet, and extracellular exposure of PS promotes apoA-I binding to the cell surface and subsequent translocation of phosphatidylcholine (PC) and cholesterol to apoA-I. Similar models have been proposed by several groups, as described in the next chapter.

4. Membrane Meso-Domains and ABCA1 Functions

Landry *et al.*[29] reported that ABCA1 re-organizes the plasma membrane and generates loosely packed domains, which facilitate apoA-I association with cells and, subsequently, PL and cholesterol acquisition by apoA-I to form nascent HDL particles. Furthermore, macrophages from ABCA1-deficient mice exhibit increased lipid rafts on the cell surface;[30] induction of increased synthesis of ABCA1 by 8Br-cAMP makes RAW264.7 cells more sensitive to methyl-β-cyclodextrin (MβCD) treatment;[27] and ABCA1 makes cholesterol available to exogenous cholesterol oxidase and MβCD.[29,31,32] Additionally, ABCA1 expression increases the Triton X-100 solubility of sphingomyelin (SM) but not that of PC.[29] The action of cholesterol oxidase on membrane cholesterol is highly dependent on the environment of the substrate and loosely packed cholesterol is likely more accessible to the oxidase;[33] MβCD is thought to extract cholesterol from non-raft domains. SM has high affinity for cholesterol, and tends to form raft domains in the plasma membrane; therefore, a change in SM content likely affects membrane dynamics. It has also been proposed that ABCA1 generates a high curvature phospholipid-rich apoA-I binding site on the plasma membrane, and that apoA-I binds this site and solubilizes membrane PL and cholesterol.[34] Indeed, treating rat fibroblasts with SMase increased apoA-I- or apo-E-dependent cholesterol efflux;[35] therefore, ABCA1 functions might include reducing or redistributing SM, which facilitates apoA-I-dependent HDL formation.

5. SM Content and Functions of ABCA1

To test the model in which ABCA1 redistributes SM in the plasma membrane, we utilized a CHO-K1 mutant cell line, LY-A,[36] which has a missense mutation in the ceramide transfer protein CERT, allowing the SM level in the plasma membrane to be altered.[31] When the SM content of LY-A cells was brought to ~65% by culture in a sphingolipid-deficient medium, apoA-I-dependent cholesterol efflux by ABCA1 became 1.65-fold higher. Since exogenously added SM reduced cholesterol efflux, a decrease in the SM content of the plasma membrane stimulates cholesterol efflux by ABCA1.[31] However, in the absence of ABCA1 expression, though the SM content was ~65%, no cholesterol was loaded onto apoA-I spontaneously. These results suggest that ABCA1 makes cholesterol available to apoA-I and MβCD extraction in non-raft domains, rather than reducing or redistributing SM in the plasma membrane as previously suggested.[27,29] Mendez, A.J. *et al.*[37] also proposed that ABCA1 preferentially functions in non-raft domains containing more PC and less SM.

As opposed to ABCA1-mediated cholesterol efflux, ABCG1-mediated cholesterol efflux from LY-A cells was reduced when the SM content of LY-A cells was ~65%.[38] ABCG1 has been suggested to mediate the efflux of cholesterol to preβ-HDL,[39–41] which ABCA1 generates by loading lipids on lipid-free apoA-I. Although no genetic diseases are known to be caused by mutations of ABCG1, chow-fed mice lacking ABCG1 showed an accumulation of PL and neutral lipids, including cholesterol and triglyceride, in liver and lung.[42,43] Furthermore, combined deficiency of ABCA1 and ABCG1 promotes foam cell accumulation and accelerates atherosclerosis in mice.[44] Therefore, ABCG1 and ABCA1 function cooperatively in the removal of excess cholesterol from peripheral tissues including macrophages. The difference in the functional consequences of SM content suggests that ABCG1-mediated cholesterol efflux occurs in raft domains, while ABCA1-mediated cholesterol efflux occurs in non-raft domains.

Alternatively, ABCG1 may co-transport cholesterol with SM as substrates for transport, while ABCA1 co-transports cholesterol with PC as the more favored substrates for transport. Indeed, the medium of ABCA1-expressing cells contains more PC species than SM species in the presence of apoA-I.[45]

6. Substrate Transport Specificity

As discussed in the previous chapter, the comparison among ABCA1, ABCA7, and ABCG1 suggested that they have different substrate specificities for transport. To look deeply at the substrate specificity of ABCA1, we purified human ABCA1 as a detergent-soluble form and examined ATPase activity.[12] Purified ABCA1 showed robust ATPase activity when reconstituted in liposomes made of synthetic PC or SM, suggesting that ABCA1 recognizes PLs with choline head groups. We could not determine whether ABCA1 prefers PC or SM in this experiment. Furthermore, ATPase activity was reduced by the addition of cholesterol.[12] ATPase activity of the ABCG5/ABCG8 complex, which excretes sitosterol and cholesterol into bile, is also suppressed by the addition of cholesterol.[46] These results suggest that cholesterol binds directly to ABCA1; however, the mechanism by which ABCA1 mediates lipid efflux to apoA-I, and the identity of the lipids ABCA1 recognizes, have yet to be clarified.

When we analyzed the ATPase activity of ABCB1 (MDR1) using purified human ABCB1 reconstituted in liposomes containing 0–20% (w/w) cholesterol,[47] cholesterol affected not only basal ATPase activity but also the drug-stimulated ATPase activity of ABCB1. ABCB1 is a plasma membrane-located glycoprotein that confers multidrug resistance to cancer cells by actively excreting structurally diverse chemotherapeutic compounds from cells.[48–51] ABCB1 is clinically important because it not only confers multidrug resistance but also affects the pharmacokinetics of various drugs. Our data suggested that cholesterol fills the empty space of the drug-binding site, aids in the recognition of small drugs, and facilitates the ability of ABCB1 to recognize compounds with various structures and molecular weights.[47] We thus proposed the "cholesterol fill-in model":[52] ABC proteins, either lipid or drug transporters, have similar substrate binding sites that recognize PL and cholesterol (or drugs and cholesterol), thereby playing an important role in substrate recognition.

7. ApoA-I and Sodium Taurocholate, Lipid Acceptors for ABCA1 and ABCB4

Cholesterol secretion by ABCA1 is strongly dependent on the presence of extracellular acceptors, such as lipid-free apoA-I. A direct interaction between ABCA1 and apoA-I has been shown by several groups via cross-linking experiments,[8,53–55] even with a cross-linker as small as 3 Å.[56] However, only a portion of the apoA-I associated with the cell surface binds directly to ABCA1.[57,58] In this section, we discuss the apoA-I-dependent and independent functions of ABCA1 by comparing them with the functions of ABCB4 [multidrug resistance (MDR)3 P-glycoprotein].

ABCB4 is expressed in the canalicular membrane of hepatocytes, and is required for PC secretion into bile, which is a prerequisite for bile formation.[59] ABCB4 mediates the secretion of PC and cholesterol in the presence of sodium taurocholate (NaTC) in the medium.[60] However, ABCB4 cannot mediate the efflux of PL and cholesterol in the presence of apoA-I. The related protein ABCB1 (MDR1) cannot mediate the efflux of PL or cholesterol in the presence of NaTC,[60] although it is 75.8% identical and 86.6% similar to ABCB4 in amino acid sequence. Thus, ABCA1 and ABCB4 are considered to mediate cholesterol and PC efflux to different extracellular acceptors, apoA-I and NaTC, respectively.

We examined directly whether NaTC serves as an acceptor for lipids secreted by ABCA1, and found that NaTC extracts both cholesterol and PC as efficiently as apoA-I from cells expressing ABCA1, but does not extract lipids from HEK293 host cells.[61] In other words, NaTC indeed functions as a lipid acceptor for

ABCA1, which mimics apoA-I, in the real sense of the term. ABCB1 and ABCB4 are predicted to contain very short extracellular loops, which are thought not to form specific binding sites for acceptors. It has been reported that synthetic amphiphilic helical peptides can serve as lipid acceptors, and act as efficiently as apoA-I; furthermore, there is no sequence or stereoselective interaction with ABCA1 in the lipid release process.[62,63] NaTC, as well as amphiphilic helical peptides, may therefore extract PC and cholesterol translocated by ABCA1 and ABCB4 without engaging in specific interactions with those proteins.

8. Four-Step Model for HDL Formation

ABCA1 mediates NaTC-dependent efflux of cholesterol and PL *in vitro*, but it is not physiologically relevant, because it is unlikely that the plasma concentration of NaTC reaches 1 mM. However, this system allowed us to analyze the role of apoA-I and dissect the steps in HDL formation.

The Tangier mutation W590S (Fig. 1), which impairs apoA-I-dependent lipid efflux, also greatly decreases NaTC-dependent cholesterol and PL efflux.[61] Since ABCA1-W590S interacts in a normal manner with ATP,[2] the W590S mutation may abolish the coupling of ATP-induced conformational changes of ABCA1 to lipid translocation. The kinetics of apoA-I binding to cells expressing ABCA1-W590S are similar to those for cells expressing wild-type ABCA1,[61] consistent with a previous report showing that the W590S mutation does not impair apoA-I binding.[8,9,14,53] However, the W590S mutation delayed the dissociation of apoA-I from ABCA1.[61] This mutation may impair the transition to the conformation required for apoA-I release. Alternatively, the conformational transition caused by lipid loading of ABCA1 facilitates the dissociation of apoA-I from ABCA1. ApoA-I undergoes a conformational transition in response to lipid,[64] and lipidated apoA-I does not interact with ABCA1.[55,65]

From these results, we propose that HDL formation mediated by ABCA1 consists of at least four separable steps (Fig. 2)

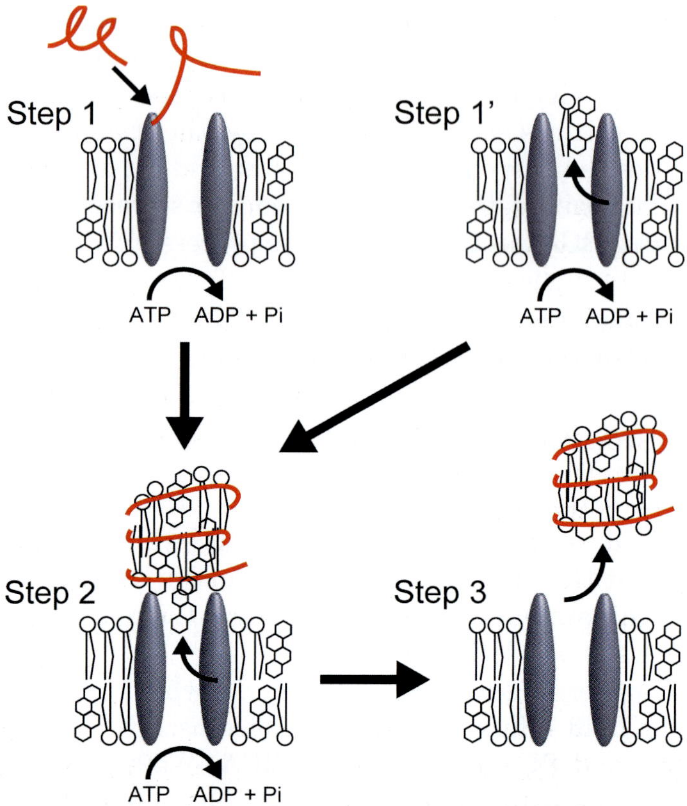

Fig. 2. Four-step models for HDL formation.

step 1: ATP binding and/or hydrolysis (presumably triggered by lipid binding) causes conformational changes within the extracellular domains of ABCA1, to which apoA-I directly binds; step 1: lipid translocation by ABCA1, which is apoA-I-independent; step 2: lipid loading of apoA-I bound to ABCA1; and step 3: dissociation of lipid-loaded apoA-I.

9. Endocytosis and Recycling of ABCA1

In section 5, we described how ABCA1 interacts with apoA-I in special micro-domains on the plasma membrane in order to generate HDL. In 1999, however, it had already been shown by

Takahashi *et al.*[66] that following internalization, apoA-I is recycled back to the cell surface to be re-secreted. Other studies also showed that apoA-I and ABCA1 are colocalized in endosomal compartments, and that ABCA1 rapidly shuttles between intracellular compartments and the plasma membrane.[67,68] Trapping ABCA1 on the plasma membrane by cyclosporine treatment reduces apoA-I-mediated cholesterol efflux.[69] In addition, deletion of a PEST sequence in the linker region of ABCA1 protein blocks its endocytosis and decreases apoA-I-mediated efflux of cholesterol derived from endocytosed lipoproteins.[70] Together, these results support the idea that ABCA1-bound apoA-I is delivered to late endosomal and/or lysosomal compartments, where it forms nascent lipoprotein particles that are subsequently secreted from the cell.

Several groups have recently examined internalization and recycling of apoA-I, and concluded that the retroendocytosis pathway does not contribute significantly to HDL formation. The majority of internalized apoA-I is directly transported to late endosomes and lysosomes for degradation; furthermore, blocking endocytosis does not decrease apoA-I-dependent cholesterol efflux from BHK cells expressing exogenous ABCA1 and RAW264.7 macrophages.[71] Although apoA-I is specifically taken up by macrophages, only a small fraction of apoA-I is re-secreted from these cells. These results suggest that the mass of retroendocytosed apoA-I is not sufficient to account for HDL produced by cholesterol efflux.[72] The direct observation of apoA-I binding structures protruding from the cell surface, which was proposed to be an intermediate step of HDL formation, also suggested that apoA-I removes excess cholesterol from the cell surface.[34,73]

The analysis of the endocytic and recycling routes of newly internalized ABCA1 molecules from the cell surface demonstrated that ABCA1 and apoA-I are endocytosed via a clathrin- and Rab5-mediated pathway and recycled rapidly back to the cell surface, at least in part via a Rab4-mediated route; approximately 30% of the endocytosed ABCA1 is recycled back to the cell surface.[74] When clathrin-dependent endocytosis is inhibited by monodansyl cadaverine (MDC) or hypertonic sucrose, the level of ABCA1 at the cell surface increases and apoA-I internalization is blocked. Under these conditions, apoA-I mediated cholesterol efflux from THP-1 macrophages that have accumulated lipoprotein-derived cholesterol is decreased, whereas efflux from cells without excess cholesterol is increased.[74] These results suggest that the retroendocytosis pathway of ABCA1/apoA-I does in fact contribute to HDL formation when excess lipoprotein-derived cholesterol has accumulated in cells.

These studies also suggest that ABCA1 is constantly internalized, irrespective of the presence of apoA-I.[75] ApoA-I binding may not be a trigger for internalization of ABCA1, although endocytosis of ABCA1 is increased in the presence of apoA-I.[76] ApoA-I binding may alter the conformation of ABCA1, or the environment of surrounding lipids, to prevent the proteolytic degradation of ABCA1 as reported previously.[77,78] Many membrane receptors, such as the LDL receptor and transferrin receptors, constitutively internalize and recycle to the plasma membrane irrespective of whether they have bound their ligands. It is likely that ABCA1 follows essentially the same constitutive recycling pathway as the LDL receptor and transferrin receptor. It is intriguing that ABCA1, involved in lipid secretion from cells, follows the same constitutive pathway as the LDL receptor, involved in cholesterol uptake. Thus, ABCA1 is a unique membrane protein that functions as both a receptor and a transporter.

10. Degradation of ABCA1

Since cholesterol is an essential component of cells, excessive elimination of cholesterol could result in cell death, and ABCA1-mediated cholesterol efflux is highly regulated at both the transcriptional and post-transcriptional levels. Thus, the degradation of ABCA1 is carried out via several pathways: (i) the cell surface ABCA1 is endocytosed and delivered to the lysosomes;[67] (ii) calpain protease degrades ABCA1 on the plasma membrane[77,78] and intracellularly, especially when apoA-I does not bind to ABCA1;[75] and (iii) ABCA1 is also degraded through the ubiquitin-proteasome pathway.[79,80]

In progressed atherosclerotic plaques, macrophages accumulate large quantities of free cholesterol as well as esterified cholesterol, likely due to a combination of perturbed cholesterol esterification and diminished cholesterol efflux. Proteasome inhibitors increase ABCA1 in free cholesterol-loaded macrophages;[79] therefore, the ubiquitin-proteasome pathway could be the main pathway of ABCA1 degradation under those conditions. The COP9 signalosome (CSN complex) is suggested to associate with ABCA1 and controls ubiquitinylation and deubiquitinylation of ABCA1, and consequently controls the proteasomal degradation of ABCA1.[81]

ApoA-I prevents calpain-mediated degradation of ABCA1 and increases the levels of cell surface ABCA1.[75,77,78] Synthetic amphiphilic helical peptides that promote ABCA1-dependent lipid efflux, as well as NaTC, also stabilize ABCA1.[62,63] In addition, ABCA1-W590S, which shows normal apoA-I binding but does not translocate lipids,[61] was reported not to be stabilized by apoA-I.[78] These results suggest that removal of cellular cholesterol and PL by apoA-I and NaTC stabilizes ABCA1, and that direct interaction between ABCA1 and apoA-I may not be essential for stabilization of ABCA1.

ABCA1 and ABCG1 play synergistic roles in mediating macrophage cholesterol efflux and protecting against macrophage foam cell formation, inflammatory cell infiltration, apoptosis, and atherosclerosis.[44] Inhibition of ABCA1 degradation represents one potential way to suppress plaque formation, by exporting free cholesterol from cells and thereby breaking the vicious circle that connects cholesterol accumulation, apoptosis, and inflammatory responses.

11. Regulation of ABCA1 by Interacting Proteins

The functions and the rate of degradation of ABCA1 are regulated by the action of various kinases.[82–85] Furthermore, several proteins, including apoA-I, α1-syntrophin and β1-syntrophin, interact with ABCA1.[77,78,86,87] Syntrophins interact directly with ABCA1 via the C-terminal three amino acids SYV (a PDZ protein-binding motif). These interactions slow the degradation of ABCA1; increase the surface expression of ABCA1, consequently increasing the apoA-I-mediated release of cholesterol; and play important roles in regulating the apoA-I-dependent cholesterol efflux (and thus in lipid homeostasis) in both brain[86] and liver.[87] The mechanisms of the delay in ABCA1 degradation have not yet been elucidated.

We performed a yeast two-hybrid screen to search for additional proteins associated with the C-terminal region of human ABCA1, one of the regions where TD and FHA mutations cluster (Fig. 1). Surprisingly, given that ABCA1 is a plasma membrane protein, the screen identified a nuclear receptor, LXRβ.[88] In WI-38 and THP-1 cells, endogenous LXRβ interacts with ABCA1 under conditions in which cholesterol does not accumulate, i.e., when cholesterol is not in excess. In contrast to other interacting proteins, LXRβ suppresses ABCA1-mediated cholesterol efflux, and thereby blocks HDL formation.[88]

When cholesterol accumulates in cells, intracellular concentrations of oxysterols increase, and LXRβ, activated via the binding of oxysterols, stimulates the gene expression of ABCA1, ABCG1, and other proteins that remove cholesterol from cells.[89–91] Newly synthesized ABCA1 protein is distributed in the plasma membrane as well as in intracellular membrane-bound compartments, and turns over rapidly, with a half-life of 1–2 h.[86,87,92] Therefore, after excess cholesterol is eliminated, ABCA1 would be degraded rapidly, and cells would have little ABCA1 in the plasma membrane. Due to the large size of the gene and protein (>2000 amino acid residues), transcription, splicing, translation and maturation of ABCA1 takes several hours to complete following transcriptional activation. Therefore, cells would not be able to cope with an acute accumulation of cholesterol for several hours prior to the vigorous transcriptional activation of ABCA1 via the LXRα auto-regulatory loop.

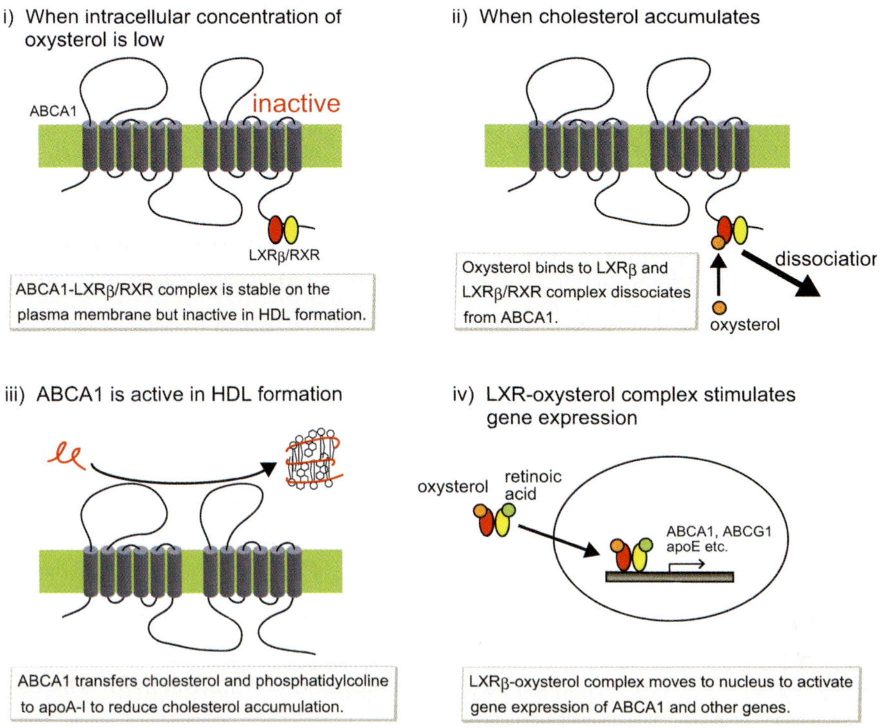

Fig. 3. Post-translational regulation mechanism for ABCA1 by LXRβ.

We proposed a novel regulatory mechanism of ABCA1 (Fig. 3): (i) When the intracellular concentration of oxysterols is low, the LXRβ/RXR complex binds to ABCA1. The ABCA1-LXRβ/RXR complex stays on the plasma membrane stable but is inert in terms of cholesterol efflux, which prevents the excessive elimination of cholesterol from cells. (ii) When cholesterol accumulates and the intracellular concentration of oxysterols increases, oxysterols bind to LXRβ, and the LXRβ/RXR complex dissociates from ABCA1. (iii) Once free from LXRβ/RXR, ABCA1 is now active in the formation of HDL, and immediately decreases the local cholesterol concentration. (iv) Upon binding oxysterols in the cytosol, LXRβ/RXR is translocated to the nucleus and activates the transcription of *ABCA1* and other genes. Therefore, cells exert two regulatory pathways of ABCA1-mediated cholesterol efflux: (i) the slow indirect pathway by degradion/resynthesis of ABCA1 and (ii) the immediate direct, ON or OFF, regulation of ABCA1 activity. Together, these two responses allow the cell to maintain cholesterol homeostasis. This novel mechanism would be important for macrophages, because they must cope with rapid increases in intracellular cholesterol when they phagocytose apoptotic cells.

12. Perspectives for the Future

Membrane lipids are essential for various cell functions, and their improper distribution or metabolism can have serious consequences for cells and organisms. Cholesterol is a structural component of eukaryote membranes that influences fluidity, permeability and formation of microdomains, which are important for various cellular functions and signal transduction, and mammals have developed sophisticated but complex mechanisms to maintain cellular cholesterol levels in membranes within a narrow range. The mechanisms include tight regulation of cholesterol synthesis, esterification, uptake via receptors, normal localization in the bilayer and elimination of excess through efflux transporters. Among these mechanisms, ABCA1 is one of the key molecules that regulate cholesterol concentration and distribution. Therefore, the activity of ABCA1 is regulated at multiple levels, including gene expression, protein modification, trafficking, degradation, and even direct modulation of the transport activity. We believe that the precise understanding of these mechanisms will allow us to develop therapeutic techniques for control of ABCA1 and stimulation of HDL formation. Such techniques might help prevent or even cure atherosclerosis and many other related diseases.

Acknowledgments

This work was supported by a Grant-in-aid for Scientific research (S) from the Ministry of Education, Culture, Sports, Science and Technology of Japan, and by the Program for Promotion of Basic and Applied Researches for Innovations in Bio-oriented Industry (BRAIN) of Japan. This work was supported by the World Premier International Research Center Initiative, MEXT, Japan.

References

1. Oram, J.F. and Vaughan, A.M. (2006) *Circ. Res.* **99**(10), 1031.
2. Tanaka, A.R., Abe-Dohmae, S., Ohnishi, T., Aoki, R., Morinaga, G., Okuhira, K.I., Ikeda, Y., Kano, F., Matsuo, M., Kioka, N., Amachi, T., Murata, M., Yokoyama, S., and Ueda, K. (2003) *J. Biol. Chem.* **278**, 8815.
3. Yokoyama, S. (2000) *Biochim. Biophys. Acta.* **1529**, 231.
4. Brooks-Wilson, A., Marcil, M., Clee, S., Zhang, L., Roomp, K., van Dam, M., Yu, L., Brewer, C., Collins, J., Molhuizen, H., Loubser, O., Ouelette, B., Fichter, K., Ashbourne-Excoffon, K., Sensen, C., Scherer, S., Mott, S., Denis, M., Martindale, D., Frohlich, J., Morgan, K., Koop, B., Pimstone, S., Kastelein, J., and Hayden, M. (1999) *Nat. Genet.* **22**, 336.
5. Bodzioch, M., Orso, E., Klucken, J., Langmann, T., Bottcher, A., Diederich, W., Drobnik, W., Barlage, S., Buchler, C., Porsch-Ozcurumez, M., Kaminski, W., Hahmann, H., Oette, K., Rothe, G., Aslanidis, C., Lackner, K., and Schmitz, G. (1999) *Nat. Genet.* **22**, 347.
6. Rust, S., Rosier, M., Funke, H., Real, J., Amoura, Z., Piette, J., Deleuze, J., Brewer, H., Duverger, N., Denefle, P., and Assmann, G. (1999) *Nat. Genet.* **22**, 352.
7. Singaraja, R.R., Brunham, L.R., Visscher, H., Kastelein, J.J., and Hayden, M.R. (2003) *Arterioscler Thromb. Vasc. Biol.* **23**(8), 1322.
8. Fitzgerald, M.L., Morris, A.L., Rhee, J.S., Andersson, L.P., Mendez, A.J., and Freeman, M.W. (2002) *J. Biol. Chem.* **277**, 33178.
9. Singaraja, R.R., Visscher, H., James, E.R., Chroni, A., Coutinho, J.M., Brunham, L.R., Kang, M.H., Zannis, V.I., Chimini, G., and Hayden, M.R. (2006) *Circ. Res.* **99**(4), 389.

10. Peelman, F., Labeur, C., Vanloo, B., Roosbeek, S., Devaud, C., Duverger, N., Denefle, P., Rosier, M., Vandekerckhove, J., and Rosseneu, M. (2003) *J. Biol. Chem.* **325**(2), 259.
11. Tanaka, A.R., Ikeda, Y., Abe-Dohmae, S., Arakawa, R., Sadanami, K., Kidera, A., Nakagawa, S., Nagase, T., Aoki, R., Kioka, N., Amachi, T., Yokoyama, S., and Ueda, K. (2001) *Biochem. Biophys. Res. Commun.* **283**(5), 1019.
12. Takahashi, K., Kimura, Y., Kioka, N., Matsuo, M., and Ueda, K. (2006) *J. Biol. Chem.* **281**(16), 10760.
13. Bungert, S., Molday, L.L., and Molday, R.S. (2001) *J. Biol. Chem.* **276**(26), 23539.
14. Rigot, V., Hamon, Y., Chambenoit, O., Alibert, M., Duverger, N., and Chimini, G. (2002) *J. Lipid. Res.* **43**, 2077.
15. Lewis, R.A., Shroyer, N.F., Singh, N., Allikmets, R., Hutchinson, A., Li, Y., Lupski, J.R., Leppert, M., and Dean, M. (1999) *Am. J. Hum. Genet.* **64**(2), 422.
16. Stone, E.M., Webster, A.R., Vandenburgh, K., Streb, L.M., Hockey, R.R., Lotery, A.J., and Sheffield, V.C. (1998) *Nat. Genet.* **20**(4), 328.
17. Webster, A.R., Heon, E., Lotery, A.J., Vandenburgh, K., Casavant, T.L., Oh, K.T., Beck, G., Fishman, G.A., Lam, B.L., Levin, A., Heckenlively, J.R., Jacobson, S.G., Weleber, R.G., Sheffield, V.C., and Stone, E.M. (2001) *Invest. Ophthalmol. Vis. Sci.* **42**(6), 1179.
18. Maugeri, A., van Driel, M.A., van de Pol, D.J., Klevering, B.J., van Haren, F.J., Tijmes, N., Bergen, A.A., Rohrschneider, K., Blankenagel, A., Pinckers, A.J., Dahl, N., Brunner, H.G., Deutman, A.F., Hoyng, C.B., and Cremers, F.P. (1999) *Am. J. Hum. Genet.* **64**(4), 1024.
19. Hozoji, M., Kimura, Y., Kioka, N., and Ueda, K. (2009) *J. Biol. Chem.* **284**(17), 11293.
20. Radhakrishnan, A., Goldstein, J.L., McDonald, J.G., and Brown, M.S. (2008) *Cell. Metab.* **8**(6), 512.
21. Lange, Y., Ye, J., Rigney, M., and Steck, T. (2000) *J. Biol. Chem.* **275**(23), 17468.
22. Fielding, P.E., Nagao, K., Hakamata, H., Chimini, G., and Fielding, C.J. (2000) *Biochemistry* **39**, 14113.
23. Wang, N., Silver, D.L., Thiele, C., and Tall, A.R. (2001) *J. Biol. Chem.* **276**(26), 23742.

24. Abe-Dohmae, S., Ikeda, Y., Matsuo, M., Hayashi, M., Okuhira, K., Ueda, K., and Yokoyama, S. (2004) *J. Biol. Chem.* **279**(1), 604.
25. Hayashi, M., Abe-Dohmae, S., Okazaki, M., Ueda, K., and Yokoyama, S. (2005) *J. Lipid. Res.* **46**(8), 1703.
26. Wang, N., Lan, D., Gerbod-Giannone, M., Linsel-Nitschke, P., Jehle, A.W., Chen, W., Martinez, L.O., and Tall, A.R. (2003) *J. Biol. Chem.* **278**(44), 42906.
27. Smith, J.D., Le Goff, W., Settle, M., Brubaker, G., Waelde, C., Horwitz, A., and Oda, M.N. (2004) *J. Lipid. Res.* **45**(4), 635.
28. Hamon, Y., Broccardo, C., Chambenoit, O., Luciani, M., Toti, F., Chaslin, S., Freyssinet, J., Devaux, P., McNeish, J., Marguet, D., and Chimini, G. (2000) *Nat. Cell. Biol.* **2**, 399.
29. Landry, Y.D., Denis, M., Nandi, S., Bell, S., Vaughan, A.M., and Zha, X. (2006) *J. Biol. Chem.* **281**(47), 36091.
30. Koseki, M., Hirano, K.I., Masuda, D., Ikegami, C., Tanaka, M., Ota, A., Sandoval, J.C., Nakagawa-Toyama, Y., Sato, S.B., Kobayashi, T., Shimada, Y., Ohno-Iwashita, Y., Matsuura, F., Shimomura, I., and Yamashita, S. (2007) *J. Lipid. Res.* **48**, 299.
31. Nagao, K., Takahashi, K., Hanada, K., Kioka, N., Matsuo, M., and Ueda, K. (2007) *J. Biol. Chem.* **282**(20), 14868.
32. Vaughan, A.M. and Oram, J.F. (2003) *J. Lipid. Res.* **44**(7), 1373.
33. Wang, M.M., Olsher, M., Sugar, I.P., and Chong, P.L. (2004) *Biochemistry* **43**(8), 2159.
34. Vedhachalam, C., Duong, P.T., Nickel, M., Nguyen, D., Dhanasekaran, P., Saito, H., Rothblat, G.H., Lund-Katz, S., and Phillips, M.C. (2007) *J. Biol. Chem.* **282**(34), 25123.
35. Ito, J., Nagayasu, Y., and Yokoyama, S. (2000) *J. Lipid. Res.* **41**(6), 894.
36. Hanada, K., Hara, T., Fukasawa, M., Yamaji, A., Umeda, M., and Nishijima, M. (1998) *J. Biol. Chem.* **273**(50), 33787.
37. Mendez, A.J., Lin, G., Wade, D.P., Lawn, R.M., and Oram, J.F. (2001) *J. Biol. Chem.* **276**(5), 3158.
38. Sano, O., Kobayashi, A., Nagao, K., Kumagai, K., Kioka, N., Hanada, K., Ueda, K., and Matsuo, M. (2007) *J. Lipid. Res.* **48**(11), 2377.
39. Wang, N., Lan, D., Chen, W., Matsuura, F., and Tall, A.R. (2004) *Proc. Natl. Acad. Sci. USA* **101**(26), 9774.
40. Vaughan, A.M. and Oram, J.F. (2006) *J. Lipid. Res.* **47**(11), 2433.

41. Gelissen, I.C., Harris, M., Rye, K.A., Quinn, C., Brown, A.J., Kockx, M., Cartland, S., Packianathan, M., Kritharides, L., and Jessup, W. (2006) *Arterioscler Thromb. Vasc. Biol.* **26**(3), 534.
42. Kennedy, M.A., Barrera, G.C., Nakamura, K., Baldan, A., Tarr, P., Fishbein, M.C., Frank, J., Francone, O.L., and Edwards, P.A. (2005) *Cell. Metab.* **1**(2), 121.
43. Baldan, A., Tarr, P., Vales, C.S., Frank, J., Shimotake, T.K., Hawgood, S., and Edwards, P.A. (2006) *J. Biol. Chem.* **281**(39), 29401.
44. Yvan-Charvet, L., Ranalletta, M., Wang, N., Han, S., Terasaka, N., Li, R., Welch, C., and Tall, A.R. (2007) *J. Clin. Invest.* **117**(12), 3900.
45. Kobayashi, A., Takanezawa, Y., Hirata, T., Shimizu, Y., Misasa, K., Kioka, N., Arai, H., Ueda, K., and Matsuo, M. (2006) *J. Lipid. Res.* **47**(8), 1791.
46. Wang, Z., Stalcup, L.D., Harvey, B.J., Weber, J., Chloupkova, M., Dumont, M.E., Dean, M., and Urbatsch, I.L. (2006) *Biochemistry* **45**(32), 9929.
47. Kimura, Y., Kioka, N., Kato, H., Matsuo, M., and Ueda, K. (2007) *Biochem. J.* **401**(2), 597.
48. Chen, C., Chin, J.E., Ueda, K., Clark, D.P., Pastan, I., Gottesman, M.M., and Roninson, I.B. (1986) *Cell* **47**, 381.
49. Gottesman, M.M. and Pastan, I. (1993) *Annu. Rev. Biochem.* **62**, 385.
50. Kimura, Y., Morita, S.Y., Matsuo, M., and Ueda, K. (2007) *Cancer Sci.* **98**(9), 1303.
51. Ueda, K., Cardarelli, C., Gottesman, M.M., and Pastan, I. (1987) *Proc. Natl. Acad. Sci. USA* **84**(9), 3004.
52. Kimura, Y., Kodan, A., Matsuo, M., and Ueda, K. (2007) *J. Bioenerg. Biomembr.* **39**(5–6), 447.
53. Fitzgerald, M.L., Morris, A.L., Chroni, A., Mendez, A.J., Zannis, V.I., and Freeman, M.W. (2004) *J. Lipid. Res.* **45**(2), 287.
54. Oram, J., Lawn, R., Garvin, M., and Wade, D. (2000) *J. Biol. Chem.* **275**, 34508.
55. Wang, N., Silver, D., Costet, P., and Tall, A. (2000) *J. Biol. Chem.* **275**, 33053.
56. Chroni, A., Liu, T., Fitzgerald, M.L., Freeman, M.W., and Zannis, V.I. (2004) *Biochemistry* **43**(7), 2126.

57. Vedhachalam, C., Ghering, A.B., Davidson, W.S., Lund-Katz, S., Rothblat, G.H., and Phillips, M.C. (2007) *Arterioscler Thromb. Vasc. Biol.* **27**(7), 1603.
58. Hassan, H.H., Denis, M., Lee, D.Y., Iatan, I., Nyholt, D., Ruel, I., Krimbou, L., and Genest, J. (2007) *J. Lipid. Res.* **48**(11), 2428.
59. Smit, J.J., Schinkel, A.H., Oude Elferink, R.P., Groen, A.K., Wagenaar, E., van Deemter, L., Mol, C.A., Ottenhoff, R., van der Lugt, N.M., van Roon, M.A., van der Valk, M.A., Offerhaus, G.J.A., Berns, A.J.M., and Borst, P. (1993) *Cell* **75**(3), 451.
60. Morita, S.Y., Kobayashi, A., Takanezawa, Y., Kioka, N., Handa, T., Arai, H., Matsuo, M., and Ueda, K. (2007) *Hepatology* **46**(1), 188.
61. Nagao, K., Zhao, Y., Takahashi, K., Kimura, Y., and Ueda, K. (2009) *J. Lipid. Res.* **50**(6), 1165.
62. Arakawa, R., Hayashi, M., Remaley, A.T., Brewer, B.H., Yamauchi, Y., and Yokoyama, S. (2004) *J. Biol. Chem.* **279**(8), 6217.
63. Remaley, A.T., Thomas, F., Stonik, J.A., Demosky, S.J., Bark, S.E., Neufeld, E.B., Bocharov, A.V., Vishnyakova, T.G., Patterson, A.P., Eggerman, T.L., Santamarina-Fojo, S., and Brewer, H.B. (2003) *J. Lipid. Res.* **44**(4), 828.
64. Davidson, W.S. and Thompson, T.B. (2007) *J. Biol. Chem.* **282**(31), 22249.
65. Mulya, A., Lee, J.Y., Gebre, A.K., Thomas, M.J., Colvin, P.L., and Parks, J.S. (2007) *Arterioscler Thromb. Vasc. Biol.* **27**(8), 1828.
66. Takahashi, Y. and Smith, J.D. (1999) *Proc. Natl. Acad. Sci. USA* **96**(20), 11358.
67. Neufeld, E.B., Remaley, A.T., Demosky, S.J., Stonik, J.A., Cooney, A.M., Comly, M., Dwyer, N.K., Zhang, M., Blanchette-Mackie, J., Santamarina-Fojo, S., and Brewer, H.B., Jr. (2001) *J. Biol. Chem.* **276**(29), 27584.
68. Neufeld, E.B., Stonik, J.A., Demosky, S.J., Jr., Knapper, C.L., Combs, C.A., Cooney, A., Comly, M., Dwyer, N., Blanchette-Mackie, J., Remaley, A.T., Santamarina-Fojo, S., and Brewer, H.B., Jr. (2004) *J. Biol. Chem.* **279**(15), 15571.
69. Le Goff, W., Peng, D.Q., Settle, M., Brubaker, G., Morton, R.E., and Smith, J.D. (2004) *Arterioscler Thromb. Vasc. Biol.* **24**(11), 2155.
70. Chen, W., Wang, N., and Tall, A.R. (2005) *J. Biol. Chem.* **280**(32), 29277.
71. Denis, M., Landry, Y.D., and Zha, X. (2008) *J. Biol. Chem.* **283**, 16178.

72. Faulkner, L.E., Panagotopulos, S.E., Johnson, J.D., Woollett, L.A., Hui, D.Y., Witting, S.R., Maiorano, J.N., and Davidson, W.S. (2008) *J. Lipid Res.* **49**(6), 1322.
73. Lin, G. and Oram, J.F. (2000) *Atherosclerosis* **149**(2), 359.
74. Azuma, Y., Takada, M., Shin, H.W., Kioka, N., Nakayama, K., and Ueda, K. (2009) *Genes to Cells* **14**, 191.
75. Lu, R., Arakawa, R., Ito-Osumi, C., Iwamoto, N., and Yokoyama, S. (2008) *Arterioscler Thromb. Vasc. Biol.* **28**(10), 1820.
76. Hassan, H.H., Bailey, D., Lee, D.Y., Iatan, I., Hafiane, A., Ruel, I., Krimbou, L., and Genest, J. (2008) *J. Biol. Chem.* **283**(17), 11164.
77. Arakawa, R., and Yokoyama, S. (2002) *J. Biol. Chem.* **277**(25), 22426.
78. Wang, N., Chen, W., Linsel-Nitschke, P., Martinez, L.O., Agerholm-Larsen, B., Silver, D.L., and Tall, A.R. (2003) *J. Clin. Invest.* **111**(1), 99.
79. Feng, B. and Tabas, I. (2002) *J. Biol. Chem.* **277**(45), 43271.
80. Tanaka, A.R., Kano, F., Yamamoto, A., Ueda, K., and Murata, M. (2008) *Genes Cells* **13**(8), 889.
81. Azuma, Y., Takada, M., Maeda, M., Kioka, N., and Ueda, K. (2009) *Biochem. Biophys. Res. Commun.* **382**(1), 145.
82. See, R.H., Caday-Malcolm, R.A., Singaraja, R.R., Zhou, S., Silverston, A., Huber, M.T., Moran, J., James, E.R., Janoo, R., Savill, J.M., Rigot, V., Zhang, L.H., Wang, M., Chimini, G., Wellington, C.L., Tafuri, S.R., and Hayden, M.R. (2002) *J. Biol. Chem.* **277**(44), 41835.
83. Yamauchi, Y., Hayashi, M., Abe-Dohmae, S., and Yokoyama, S. (2003) *J. Biol. Chem.* **278**(48), 47890.
84. Wang, Y. and Oram, J.F. (2007) *J. Lipid Res.* **48**(5), 1062.
85. Tang, C., Vaughan, A.M., and Oram, J.F. (2004) *J. Biol. Chem.* **279**(9), 7622.
86. Munehira, Y., Ohnishi, T., Kawamoto, S., Furuya, A., Shitara, K., Imamura, M., Yokota, T., Takeda, S., Amachi, T., Matsuo, M., Kioka, N., and Ueda, K. (2004) *J. Biol. Chem.* **279**(15), 15091.
87. Okuhira, K., Fitzgerald, M.L., Sarracino, D.A., Manning, J.J., Bell, S.A., Goss, J.L., and Freeman, M.W. (2005) *J. Biol. Chem.* **280**(47), 39653.
88. Hozoji, M., Munehira, Y., Ikeda, Y., Makishima, M., Matsuo, M., Kioka, N., and Ueda, K. (2008) *J. Biol. Chem.* **283**(44), 30057.

89. Venkateswaran, A., Laffitte, B.A., Joseph, S.B., Mak, P.A., Wilpitz, D.C., Edwards, P.A., and Tontonoz, P. (2000) *Proc. Natl. Acad. Sci. USA* **97**(22), 12097.
90. Repa, J.J., Turley, S.D., Lobaccaro, J.M.A., Medina, J., Li, L., Lustig, K., Shan, B., Heyman, R.A., Dietschy, J.M., and Mangelsdorf, D.J. (2000) *Science* **289**, 1524.
91. Costet, P., Luo, Y., Wang, N., and Tall, A.R. (2000) *J. Biol. Chem.* **275**, 28240.
92. Wang, Y. and Oram, J.F. (2002) *J. Biol. Chem.* **277**(7), 5692.

Chapter 3

Bile Formation and the Enterohepatic Circulation

Peter H. Dixon and Catherine Williamson

1. The Enterohepatic Circulation

Bile acid (BA) synthesis, circulation and reabsorption are tightly regulated homeostatic processes, essential for normal body function. In humans, multiple feedback systems operate through a variety of signalling mechanisms to control these pathways.[1,2] Disruptions (inherited or acquired) can result in serious, or even fatal disease, due to the inherently toxic nature of bile acids. Bile acids together with other biliary constituents are actively transported across cellular membranes by a number of specific transporters, which include a number of members of the ABC family. This efflux is driven by ATP hydrolysis and is central to bile formation. This chapter will consider the roles of the key players in bile formation, secretion and recycling known as the enterohepatic circulation. In particular the role of two transporters responsible for the formation of the majority of bile, namely the bile salt export pump (BSEP, *ABCB11*) and the multidrug resistance protein 3 (MDR3, *ABCB4*), will be explored.

Bile acids play a key role in facilitating digestion of dietary nutrients by solubilising dietary fats and vitamins. The synthesis of the primary bile acids cholic acid (CA) and chenodeoxycholic acid (CDCA) from cholesterol is also the primary route of cholesterol catabolism. Bile also acts as an important disposal route for lipophilic xenobiotics and endobiotics such as drug metabolites and dietary toxins.

In addition, bile acids act as signalling molecules that influence pathways of cholesterol, lipid and carbohydrate homeostasis and also influence the immune system. These actions are in part facilitated by the interaction of bile acids with a network of specific nuclear receptors. There is also a G-protein coupled receptor, TGR5, whose principal endogenous ligand is bile acids.[3] Furthermore, bile acids can interact and signal through muscarinic receptors[4] and a formyl-peptide receptor.[5] Thus it is clear that bile acid biology is much more complex than was previously thought.

Following secretion from the hepatocyte, bile drains into the gallbladder through the common bile duct. Upon digestion of a meal, the hormone cholecystokinin (CCK) causes gallbladder contraction and release of bile into the small intestine. Here bile acids act to emulsify dietary fats and fat-soluble vitamins. Bacterial modification in the gut results in deconjugation and conversion to secondary bile acids (e.g., deoxycholic acid (DCA) or lithocholic acid (LCA)). Bile acids are reabsorbed and are transported back to the liver where the majority are taken up by the sodium-dependant taurocholate co-transporter peptide (NTCP, *SLC10A1*) and the remainder by the organic anion co-transporting polypeptides (OATPs). These basolateral transporters are members of other membrane transporter families (the SLC and SLCO families respectively). Secondary bile acids are detoxified in the enterocytes and liver and re-enter the enterohepatic cycle, or targeted for excretion via alternative modification.

The enterohepatic circulation reabsorbs 95% of bile acids; the remaining 5% are lost in faeces. These are replaced by *de novo* synthesis. Any bile acids not uptaken during the first hepatic pass transit to the kidney. Filtration by the renal glomerulous is followed by reabsorption by the epithelial cells of the proximal convoluted tubule allowing for a return to the hepatic uptake sites via the systemic circulation. The efficiency of this system means that in normal healthy individuals BAs are almost undetectable in urine. Fig. 1 summarises the major transporter systems involved in the enterohepatic circulation.

In humans the primary bile acids CA and CDCA are synthesised in the liver using cholesterol as the starting material;

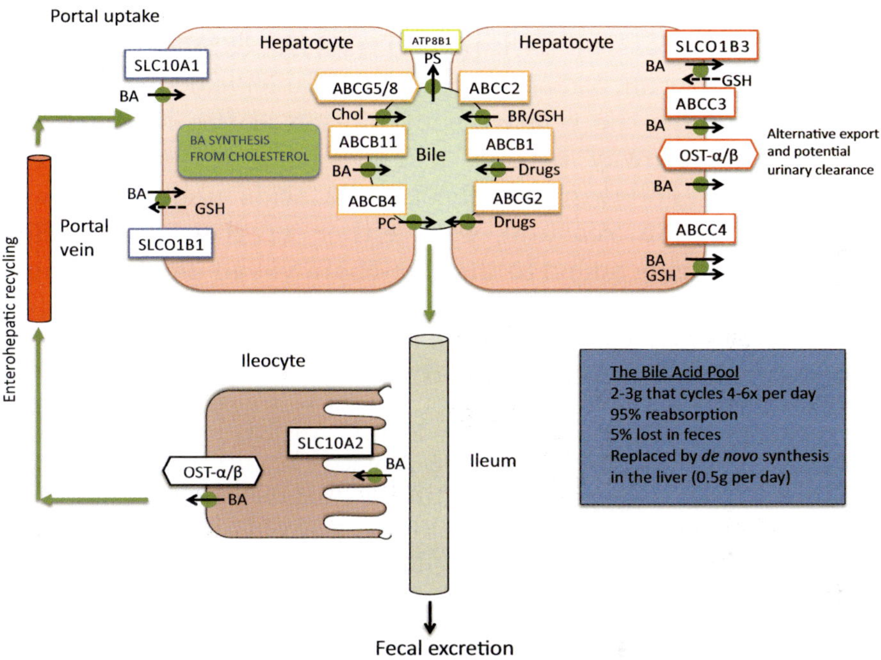

Fig. 1. The molecular basis of the enterohepatic circulation of bile acids. Schematic representation of the transporters that influence cycling of bile acids from a hepatocyte couplet through biliary excretion, reabsorption in the terminal ileum, recycling via the portal vein and uptake into the hepatocyte. Sinusoidal bile acid uptake (>80%), is facilitated by the sodium-dependent taurocholate uptake pump (NTCP, systematic gene name SLC10A1). In addition, SLCO1B1, a member of the organic anion transporter polypeptide family, imports <20% of the sinusoidal bile acids. Using cholesterol as the starting point, *de novo* synthesis of the primary bile acids cholic acid and chenodeoxycholic acid occurs in the hepatocyte. ABC transporters involved in the export of biliary components are outlined in orange. In addition to the bile salt export pump (BSEP, ABCB11), phosphatidyl choline is secreted into bile through the action of a specific floppase (MDR3, ABCB4).

indeed this is the major body route of cholesterol catabolism.[6] All bile acids have a similar structure with a steroid ring backbone and various side chain modifications together with hydroxylation of the different ring positions. [Rodents have a different bile acid pool as a result of different hydroxylation pathways resulting in muricholic (α, β and γ) acids.] The enzyme CYP7A1, a member of the cytochrome P450 super-family, is the first and rate-limiting step in this pathway, known as the classical pathway.[7,8] CYP27A1 catalyses the first step in the alternative, acidic pathway of bile acid synthesis. In healthy humans the relative contribution of the classical and alternative pathway to bile acid synthesis is 94% and 6% respectively.[9] Following synthesis, bile acids undergo a two-step conjugation reaction with either of the amino acids taurine or glycine added to the side-chain via an amide linkage.

Fig. 1. (*Continued*) Other ABC transporters involved in biliary formation and secretion are indicated including the drug transporters multidrug resistance protein 1 (P-glycoprotein, MDR1, ABCB1) and breast cancer resistance protein (BCRP, ABCG2) together with the heterodimeric cholesterol and sterol transporter complex ABCG5/8 and the multidrug resistance associated protein 2 (MRP2, ABCC2), which exports bilirubin into bile. Also indicated is the phosphatidyl serine flippase, ATP8B1 (the familial intrahepatic cholestasis 1 gene/protein), a type 4 P-type ATPase, which is also crucial for bile formation and canalicular membrane integrity.

Reabsorption of bile acids (following bacterial modification in the intestine) across the ileal enterocyte brush border membrane is facilitated by the apical sodium-dependant bile acid transporter ASBT, systematic gene name SLC10A2. Following activation of enterocyte FXR and induction of FGF19 expression, export from the basolateral membrane of the ileocyte back into the portal circulation is via the heterodimeric organic solute transporter OSTα/β.

Under cholestatic conditions hepato-protective mechanisms are induced to protect the hepatocyte from raised levels of bile acids. In this circumstance alternative export mechanisms are activated (transporters outlined in red) which return bile acids into the systemic circulation. In addition to OST-α/β and the organic anion transporter SLCO1B3, two ABC transporters, namely the multidrug resistance related protein 3 and 4 (ABCC3 and ABCC4) act to clear bile acids. BA — bile acid; Chol — cholesterol; PC — phosphatidyl choline; PS — phosphatidyl serine; BR — bilirubin; GSH — reduced glutathione.

This reaction greatly increases the hydrophilicty of bile acids, and in healthy humans the ratio of glycine to taurine conjugates is approximately 3:1.[6] In addition to glycine and taurine, bile acids can be metabolised via other pathways to generate alternative metabolites. Under conditions of reduced bile flow (cholestasis) these pathways act to reduce levels of bile acids by targeting them for excretion. This commonly occurs by sulphation, in particular for the highly toxic secondary BA lithocholic acid (LCA), through the activity of the hepatic sulphotransferase enzyme, SULT2A1.[10] Although glucuronidation pathways also exist, in humans their importance has not been conclusively demonstrated.[11]

Bile is composed of two principal components, namely the conjugated bile acids described above together with phosphatidyl choline (PC). When actively transported into the lumen, they form mixed micelles or vesicles. Other components of bile include cholesterol, haem breakdown products (primarily bilirubin) and drugs, together with small quantities of other, rarer bile acids with non-standard modifications (sulphation, glucuronidation, amidation and others). Each of these components is exported from the hepatocyte to the canalicular lumen via specific ABC transporters.

Active ATP-driven export into the lumen takes place against a strong concentration gradient of up to several thousand-fold. Bile acids drain into the gallbladder and are subsequently released as a result of endocrine signalling by CCK after eating. Release of bile into the duodenum is thus concurrent with the transit of food.

1.1 *Hepatic ABC transporters involved in bile formation*

The two key ATP-driven transporters involved in bile formation are the bile salt export pump, BSEP (*ABCB11*) and the phosphatidyl choline floppase or multidrug resistance protein 3, MDR3 (*ABCB4*). Together they generate the majority of bile. In order to maintain normal bile flow, bile salt export by ABCB11 must be balanced by the presence of PC, flopped into the canaliculus by ABCB4. The identification and characterisation of these two transporters is discussed in other sections. In addition, a number of other members of the ABC family are expressed in hepatocytes and play a role in biliary transport and excretion of endo and xeno-biotics. It is worth noting that in other cell types these other ABC transporters play different roles and may be transporting other substrates. Some ABC transporters, which are normally expressed at very low levels in hepatocytes, are up-regulated under cholestatic conditions. These transporters are expressed on the basolateral membrane and provide an alternative export route for toxic bile acids back into the blood. These other hepatic ABC transporters are:

1.1.1 *ABCG5/8 — Transports plant sterols and cholesterol*

Export of cholesterol and plant sterols at the canalicular membrane occurs through the heterodimeric ABCG5/G8 complex. These are half transporters consisting of one transmembrane (TM) domain and one nucleotide-binding domain (NBD). The structure and function of these two half transporters, together with genetic aspects, is described in Chapter 13.

1.1.2 *ABCC2 — Transports bilirubin and other organic ion conjugates*

Bile also serves as an excretory route for bilirubin, drug conjugates, and other organic ion conjugates. These organic anions are exported from the hepatocyte by ABCC2 (formerly known as MRP2 — the multidrug resistance-associated protein 2). This protein is responsible for bile-salt independent bile flow. In addition to some steroid hormone reproductive metabolites (e.g., estrogen 17β-glucoronide), ABCC2 exports divalent bile acids that have additional modifications (e.g., sulphation) to the standard conjugation. The structure and function of ABCC2 is described in further detail in Chapter 8.

1.1.3 *ABCG2 — Transports organic anions*

ABCG2 is a half transporter that forms homodimers and its expression confers a multidrug resistance phenotype on cells. Canalicular expression of ABCG2 (formerly known as the breast cancer resistance protein BCRP) at low levels has been reported. ABCG2 is capable of transporting a wide range of organic anions including some bile acid species. It may be of considerable importance with respect to drug transport into the biliary excretion route but this needs to be clarified. A role for the export of porphyrin conjugates into bile has also been proposed. Further details concerning ABCG2 can be found in Chapter 4.

1.1.4 *ABCC3/4 — Transports bile acids, reduced glutathione*

Under cholestatic conditions (i.e., reduced bile flow) the hepatocyte responds to reduce intracellular build-up of toxic bile acids. One of the key mechanisms involved is the activation of alternative export pathways, involving (to a lesser or greater degree depending on the species) the ABC transporters ABCC3 and ABCC4. These transporters export bile acids back into the sinusoidal space (i.e., the blood) to protect the hepatocyte. See Chapter 8 for further details.

1.2 *Feedback, signalling and regulation of expression through nuclear receptors*

As outlined above, bile acids are extremely cytotoxic, and hence their synthesis and transport is under tight regulation via negative feedback. The nuclear hormone receptor FXR, NR1H4 (farnesoid-X receptor) is the principal bile acid receptor in hepatocytes (and in enterocytes) and acts to tightly regulate intracellular bile acid levels within a narrow range (human serum $<10\,\mu M$). Bile acids bind FXR and following heterodimerisation with RXR (retinoid-X receptor, ligand 9-cis retinoic acid) the receptor complex translocates to the nucleus and binds to response elements in the promoters of target genes.[12–14] Detailed studies of all the known natural ligands of FXR have demonstrated that gene expression is controlled in a ligand and promoter selective fashion, with CDCA being the most potent natural ligand. Other natural bile acids range from agonist to partial agonist to antagonist.[15]

Hence FXR acts as a master regulator of bile acid levels by sensing intracellular bile acid concentrations and regulating bile acid metabolism and transport. Direct targets of FXR activation are ABCB11(BSEP) and ABCB4(MDR3).[16,17] Thus, transactivation of these promoters by activated FXR results in increased export of bile acids and phosphatidylcholine from the hepatocyte.

In addition to promoting export, FXR acts to reduce import by activating the promoter of the short heterodimeric partner gene (SHP, NR0B2). This is a negative repressor, which acts to reduce transcription of CYP7A1 (to reduce bile acid synthesis) via HNF4a and LRH-1 interactions, and SLC10A1 to reduce uptake.[18]

Other nuclear receptors have been identified as being responsive to bile acids, including the pregnane-X receptor (PXR),[19,20] the constitutive andrane receptor (CAR)[20] and the vitamin D receptor (VDR).[6,21] Although these receptors respond either very weakly or not at all to primary bile acids, secondary bile acids can activate them strongly. The high toxicity of secondary bile acids makes their rapid detoxification essential and this is achieved by activation of these nuclear receptors (PXR, CAR, VDR). The essential nature of these pathways is highlighted by the multiple overlapping ligands and target genes of these xeno and endobiotic sensing nuclear receptors. Exploitation of these regulators of bile acid levels as therapeutic targets is considered further in Chapter 14.

1.3 *Intestinal reabsorption and gut feedback via FGF19 action*

Following transit through the gut, bile acids (some of which will have been modified by bacterial action) are reabsorbed. Although some passive diffusion takes place, the majority (especially conjugated forms) are actively taken up by the enterocytes via the apical sodium-dependant bile acid transporter, ASBT (*SLC10A2*).[22] Trans-enterocyte passage (and cellular protection from bile acid detergent action) is facilitated by the ileal bile-acid binding protein IBABP (*FAPB6*),[23] but this process is poorly understood. The identity of the export system for release back into the systemic circulation (and hence back to the liver for re-uptake) was unknown until recently with the identification of the heterodimeric transporter[24] organic solute transporter (OST) α/β.[24] During passage through the enterocytes, BAs bind to FXR causing nuclear translocation and target gene activation. In these cells a different set of target genes are expressed and, in particular, activated FXR/RXR complexes transactivate the *FGF19* promoter. FGF19 (fgf15 in mice) is a peptide hormone member of the fibroblast growth factor (FGF) family and is released into the bloodstream from the enterocytes. It acts via the FGFR4 receptor on the hepatocyte.[25,26] Stimulation of FGFR4 in combination with the β-klotho accessory protein[27] leads to the down-regulation of CYP7A1 (via SHP independent mechanisms and the JNK phosphorylation cascade)[28] and hence reduced BA synthesis. In contrast the mouse orthologue, fgf15, is believed to act through SHP-dependent mechanisms.[26] FGF19 thus represents a feedback signal from the gut to the hepatocyte and is another layer of regulation of bile acid homeostasis. In addition, FGF19 also acts upon the gallbladder in an opposite way to CCK to cause the gallbladder muscle to relax and hence fill with bile.[29]

2. ABC Transporters and Bile Formation — The Central Role of *ABCB4* and *ABCB11*

2.1 *Identification of ABCB4/MDR3 and characterization of transport substrate*

Detailed analysis of multidrug resistant cellular phenotypes suggested the presence of more than one cell membrane protein (called P-glycoproteins), and in 1987 a second "multidrug resistance" gene was identified (in humans) and given the name *MDR3* (due to a higher homology with the third *mdr* gene in the hamster).[30] (Of note there is no MDR2 gene in humans leading to some nomenclature confusion.) Subsequent characterization of the genomic organization of the locus identified a 28 exon gene spanning over 70 kilobases on chromosome 7q21, downstream of the *MDR1/ABCB1* locus and transcribed in the same orientation.[31] The *MDR3* locus has recently been reclassified as *ABCB4*. It encodes a four domain polypeptide with two transmembrane domains[6] and two nucleotide–binding domains (NBDs) joined by a linker region (Chapter 1 describes current theories of the function of these domains in the transport cycle.)

Despite the high level of identity (77% at the amino acid level), ABCB4 does not exhibit the broad transport-substrate specificity exhibited by ABCB1. High expression in hepatocytes together with localisation on the canalicular membrane indicated a more specific role in bile formation. Generation of an $mdr2^{-/-}$ mouse (the equivalent murine locus) identified phosphatidyl choline (PC) as the primary transport substrate of this transporter,[32] and demonstrated its role in PC flopping into the biliary canaliculus; work confirmed by analysis of transgenic mouse fibroblasts.[33] Transport of a small number of other substrates (several cytotoxic drugs) has been demonstrated but the physiological relevance of these findings in normal physiology is not clearly understood.[34] ABCB4 expression has been detected at low levels, and in some cases localized in other tissues (for example human placenta) but again the physiological relevance of these findings is unclear.[35]

A detailed characterisation of the molecular mechanisms of lipid flopping by ABCB4 has not yet been achieved. This is in part due to the problems in establishing a reliable assay; efficient culture and expression systems for analysis of ABCB4 function remain to be established. Expression studies to study the consequences of mutations have been limited although some studies have been performed using homologous mutants generated in ABCB1 vectors, for which assays of efflux activity are available.[36,37]

2.2 *Identification of ABCB11/BSEP and characterisation of transport substrate*

Homology screening of pig liver cDNA libraries with P-glycoprotein (ABCB1) probes at low stringency identified a gene initially called sister of P-glycoprotein, *spgp*.[38] The high and specific hepatic expression suggested a key role in liver function. Subsequently this gene was cloned in the rat and demonstrated to function as an ATP-dependant bile salt transporter through functional expression in insect (Sf9) cells.[39] It exhibits the same domain structure as ABCB1 and ABCB4 with two TMDs and two NBDs. Following linkage studies of familial cholestasis syndromes, *spgp* mutations were identified in progressive familial intrahepatic cholestasis type 2 (PFIC2), confirming its identity as the human bile salt export pump.[40]

ABCB11 (BSEP) is a highly specific transporter. While mono-anionic conjugated bile salts are the primary transport substrate, additional substrates are transported to a limited extent, e.g., pravastatin; and some species differences have been identified. Transport studies performed in insect cells utilising bacalovirus systems established the relative affinities each different BAs and conjugates, together with the identification of inhibitors.[41,42] In addition, and of probable relevance to disease susceptibility, two studies have compared expression levels of several ABC-transporters in healthy liver tissue, revealing considerable inter-individual variability in ABCB11 levels in the normal human population.[43,44]

2.3 *Murine studies of transporter function — ABCB4/MDR3*

The generation of transgenic mice has given insights into the function of ABCB4. Initial generation of a total knockout confirmed the identity of phosphatidyl choline as the major endogenous transporter substrate, with the null mice rapidly developing liver disease as a consequence of their complete inability to secrete phospholipid into bile.[32] Transgenic humanized mice cells expressing ABCB4 under a vimentin promoter confirmed these findings.[33] Subsequent to the discovery of human mutations of ABCB4 as a cause of PFIC type 3,[45] further studies have investigated the potential of the mdr2$^{-/-}$ mouse as a model for human liver diseases. These mice develop hepatic damage similar to the human disease primary sclerosing cholangitis (PSC) and have proved a valuable model in investigating the mechanism of this disease.[46,47] Detailed study of the gallbladder bile of these mice revealed the consistent finding of spontaneous gallstone formation, also making them an appropriate model for low-phospholipid associated cholethiasis (LPAC).[48] By 16 months of age 100% of the null mice develop liver tumors and hence they have also been used to study mechanisms of hepatocelluar carcinoma (HCC) formation.[49]

2.4 *Murine studies of transporter function — ABCB11/BSEP*

While humans with two mutant alleles causing loss of, or extremely low level, expression of ABCB11 protein inevitably develop severe cholestasis with extremely low levels of bile acid secretion (less than 1% of that in normal individuals),[40] the abcb11$^{-/-}$ mouse exhibits a phenotype of mild cholestasis which is non-progressive, together with reduced but substantial bile flow (30% of wild type).[50] This can be, in part, explained by compensatory mechanisms in rodents not present in humans. These mice generate considerable quantities of the much less toxic tetrahydroxy bile acid 3α,6β,7β,12α-tetrahydroxy-5β-cholan-24-oic acid.[51] Considerable up-regulation of mdr1a and mdr1b (the mouse orthologues of ABCB1) also occurs, suggesting that compensatory metabolic and transport adaptations result in the much milder phenotype observed in these animals.[52] This conclusion is supported by a triple knock-out mouse (abcb11$^{-/-}$, mdr1a$^{-/-}$, mdr1b$^{-/-}$) which has a significantly more severe phenotype, even when maintained on a standard chow diet.[53]

Dietary supplementation is an effective approach to generate a closer model of the human disease in the abcb11$^{-/-}$ mice. When they are fed a diet supplemented with cholic acid (typically 0.5%), they develop a more severe cholestatic phenotype that is much closer to the human PFIC2 disease.[54] In addition to these knockout animals, colonies of transgenic mouse that over-expresses hepatic abcb11 have been generated (transthyretin(TTR)-abcb11 mice and abcb11-BAC mice). Interestingly, these mice are resistant to hepatic lipid accumulation, suggesting that targeting ABCB11 therapeutically may be beneficial.[55]

2.5 *Progressive familial intrahepatic cholestasis and benign recurrent intrahepatic cholestasis — clinical aspects*

Cholestasis describes impaired or absent bile flow into the small intestine. It can occur as a secondary phenomenon due to obstruction or malignancy, or as a primary syndrome. The PFIC syndromes are a group of autosomal recessive disorders characterised by severe progressive cholestatic liver disease resulting in end-stage liver failure. The diseases are inevitably fatal unless a transplant is performed. Three types are recognised clinically (PFIC1-3), together with two less severe forms, benign recurrent intrahepatic cholestasis 1 and 2 (BRIC1 and 2).[56] Patients often present in the first year of life with cholestasis characterised by very high serum bile acids and recurrent jaundice. Each of these syndromes is caused by mutations of genes that encode hepatic transporters involved in bile formation. The genes mutated in PFIC1, 2 and 3 are *ATP8B1* (FIC1), *ABCB11* (BSEP) and *ABCB4* (MDR3) respectively.[57] It is becoming increasingly apparent that mutations in these genes cause a spectrum of diseases ranging from PFIC, to the episodic cholestatic disorder benign recurrent intrahepatic cholestasis (BRIC). Furthermore, evidence is accumulating for variation in these loci playing a role in acquired forms of cholestasis such as intrahepatic cholestasis of pregnancy and drug-induced cholestasis.

PFIC1 disease, although not caused by mutations of an ABC transporter, is of relevance to the enterohepatic circulation and is considered briefly here. The *ATP8B1* gene on chromosome 18 encodes a P-type ATPase, previously called FIC1 that acts to flip phosphatidyl serine from the outer to the inner leaflet of the canalicular membrane. The resulting asymmetric membrane is believed to be important for protection against the detergent action of the high bile acid concentrations in the lumen. Exactly how *ATP8B1* mutations cause cholestasis is not yet clear. It has been proposed that they influence FXR function either as a direct or indirect effect, with subsequent ABCB11 down-regulation. The ATP8B1 protein is also expressed in a number of extra-hepatic tissues, which explains the presence of additional extrahepatic symptoms in PFIC1 patients compared to PFIC2 and PFIC3 cases. In particular the protein has recently been shown to be essential for maintaining normal hearing.[58]

2.6 *Inherited cholestasis ABCB11 — PFIC2/BRIC2*

The mapping of a large number of pedigrees with PFIC to chromosome 2q24[59] was followed by the identification of mutations in what was initially called "sister of p-glycoprotein" (SPGP) and subsequently renamed BSEP, and more recently ABCB11.[40] In the first study reporting mutations in 25 families with PFIC, 16 families had homozygous mutations, one compound heterozygote was identified and in seven families only one of the mutant alleles was identified.[59] Subsequent intensive mutation screening of the *ABCB11* gene in affected families has led to the identification of a large number of different mutations. These include missense, nonsense, truncations, deletions, insertions and splice mutations. Homozygous and compound heterozygous mutations have been detected, and a number of recurrent mutations observed. In particular two mutant alleles, E297G and

D482G, have been found frequently in European families, and one or both of these alleles present in 58% of affected individuals.[60–65] Clinical follow-up of these patients has led to the realisation that homozygous individuals, especially those carrying two null alleles, are at much greater risk of hepatobiliary malignancies.[66,67] Following the identification of the PFIC1/BRIC1 locus (*ATP8B1*, see above) it was observed that a number of BRIC patients had normal *ATP8B1* alleles and so were investigated for *ABCB11* mutations. This led to the identification of BRIC2 as a "BSEP-deficiency syndrome".[68,69] Figure 2 summarises the spectrum of reported disease-associated mutations in *ABCB11*.[66,67] A number of studies have investigated the functional consequences of *ABCB11* mutations on bile acid transport. Detailed characterization of the substrate specificity and transport properties of the wild-type protein was followed by analysis of the effect of mutations on these characteristics in rat, mouse and human *in vitro* systems,[70–75] together with a large study of mini-gene constructs to determine splicing effects prior to translation.[76] Specifically with respect to splicing, twenty mutations/SNPs of *ABCB11* from a study of 84 showed reduced wild-type splicing, with eight of these showing mRNA levels reduced to 5% or less.

Taken together these functional studies have given important information into the structure/function relationship in ABCB11. It has also been proposed that the severity of disease in the "BSEP deficiency" continuum is directly related to the levels of expression of the protein at the canalicus and some of these studies have supported this hypothesis. This work to functionally correlate patient genotypes with molecular phenotype and disease severity will potentially lead to individual therapies for specific mutation carriers. For example, if a mutant retains wild-type transport properties but is mis-trafficked, this could be overcome by the use of chemicals that can restore expression of cell surface targeted proteins. 4-Phenylbutyrate is one such compound, which has been demonstrated to enhance both wild-type and mutant ABCB11 proteins.[77]

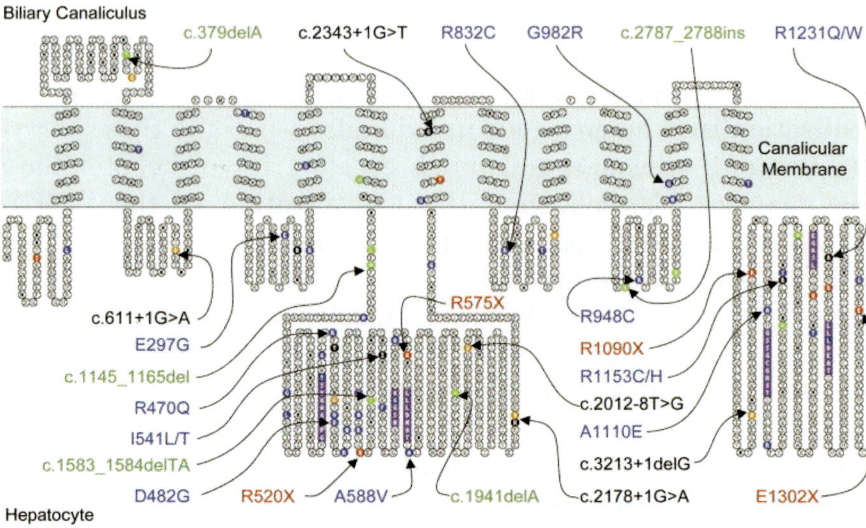

Fig. 2. The spectrum of mutations of *ABCB11*, exemplified by the results of the most comprehensive study of PFIC/BRIC 2 to date. This diagrammatic representation of the protein shows each amino acid residue as a circle, and the predicted protein topology organised (similar to other family members) with twelve hydrophobic transmembrane helices and two cytoplasmic nucleotide-binding domains. The arrows indicate mutations observed in more than one individual whereas mutations identified only once are colored. A color code has been used to identify the type of mutation; the residue at the site of indels is colored, in the case of splicing mutations the adjacent amino acid is colored. Color code: Blue — missense, Red — Stop, Green — indel, Orange — splice site, black — multiple alternate mutations identified at this site. The location of the Walker A and B motifs and the ABC signature is indicated by purple coloring. (*Reproduced with permission from Gastroenterology 2008, 134(4):1203–14, license 2357740208129.*)

2.7 *Inherited cholestasis ABCB4 — PFIC3*

Following the identification of the second multidrug resistance gene in humans (MDR3)[30] it was established to be a specific phosphatidyl choline transporter, by disruption of the murine homologue (*mdr2*). Study of these null mice revealed the absence of phospholipid secretion into bile, leading to a lack of protection against the detergent toxicity of bile acids.[32] A subgroup of patients with PFIC and raised γ-glutamyltransferase (γ-GT) levels were noted to have similar biochemical and histochemical patterns to these mice, leading to mutational screening and identification of mutations of *ABCB4* in patients with PFIC3. Initially two patients were identified with homozygous mutations in small pedigrees.[45] As with the genetic studies of *ABCB11*, a large number of families have now been analysed and a wide range of mutations described.[78–80] This includes the recurrent mutation S320F, which has been associated with several different cholestatic diseases. The elucidation of the three-dimensional structure of a bacterial homologue of *ABCB4* has facilitated the mapping of the location of mutated residues onto the protein structure.[81,82] For a summary of *ABCB4* mutations identified in a number of cholestatic disorders see Fig. 3.

It remains methodologically challenging to perform functional studies of ABCB4 *in vitro*. Two groups have used the closest homologue of ABCB4; namely ABCB1, to express and characterize the effect of mutations identified in patients.[36,37] Although this work has given some insight into the effect of mutations on protein function, this approach is limited somewhat by the dependence on the mutated residue performing the same function in each of the transporters. In another approach, a stable cell line expressing ABCB4 derived from HEK (human embryonic kidney) 293 cells has been established.[83] Studies of this cell line have demonstrated a marked increase in the flux of phospholipid into the culture medium in the presence of sodium taurocholate. This approach however is not practical for studying all of the missense mutations identified in *ABCB4*. Recently, a transient expression system has been developed in which the PC floppase ABCB4 is co-expressed with the PS flippase complex ATP8B1/CDC50. This appears to allow the recipient cells to tolerate the function of ABCB4 and may be more suitable for the high through-put analysis of ABCB4 mutants (ref. Groen *et al.* as submitted).

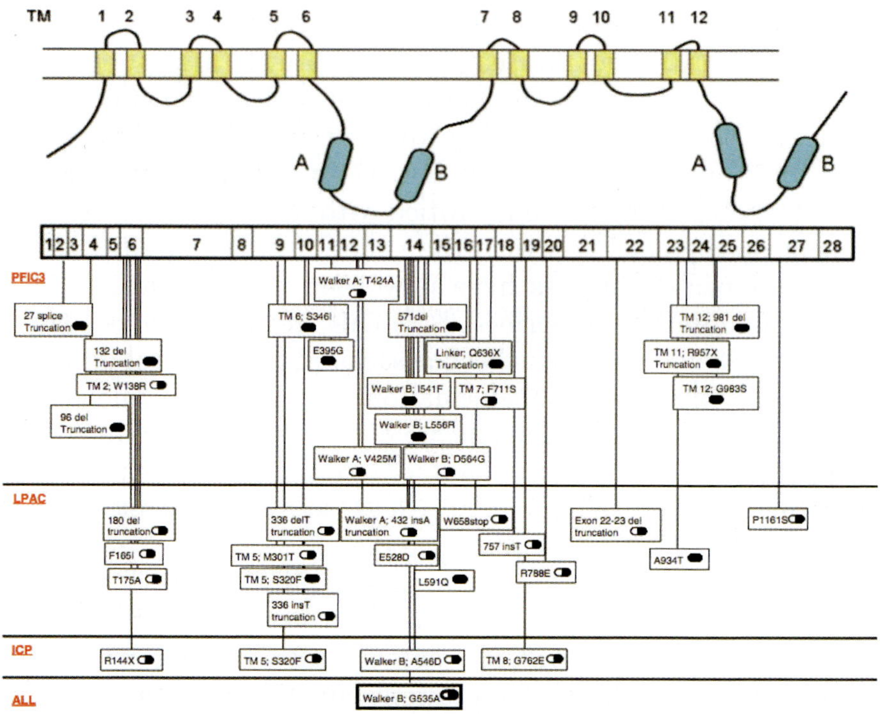

Fig. 3. A diagram of identified mutations of ABCB4 and their protein localization. The schematic diagram at the top part of the figure indicates the twelve transmembrane helices (in yellow) and the two nucleotide-binding domains with the consensus motifs (Walker A and B) indicated in blue. These mutations have been described in individuals with progressive intrahepatic cholestasis type 3 (PFIC3), low-phospholipid associated choleolithiasis (LPAC) and intrahepatic cholestasis of pregnancy (ICP). Careful phenotyping of these patients has identified individuals in whom the clinical features of these three diseases overlap; in one patient all three syndromes have been described (indicated by "ALL"). The use of half-filled ovals next to the mutation description indicates heterozygous mutations; the use of filled ovals, homozygous. Predicted protein domains essential for normal function are also indicated if this is where the location of the mutations occurs (TM; transmembrane helix, A and B; Walker A and B motifs). In addition to the mutations summarised here further mutations have been identified in a number of subsequent studies. (*Reproduced with permission from Pfulgers Archiv, European Journal of Physiology 2007, 453(5):601–10, license number 2357731157733.*)

2.8 *The spectrum of cholestatic phenotypes for ABCB11 and ABCB4*

It is becoming apparent from the large amount of patient phenotype and genotype data now available that *ABCB4* and *ABCB11* mutation carriers can have a complex range of phenotypes. Several investigators have coined the phrases "BSEP (ABCB11) deficiency syndrome" and "MDR3 (ABCB4) deficiency syndrome", which better describes the spread of phenotypes than the classical terminology "PFIC" and "BRIC". Aside from autosomal recessive homozygous and compound heterozygous individuals, disease associations have been described for heterozygous mutations. This is particularly true for muta-tions of *ABCB4*: which appears to have a role in a number of inherited forms of cholestasis, and/or cholestatic-related phenotypes.

The identification of a mixture of heterozygous and homozygous mutations in a subgroup of patients with gallstones led to the description of a new syndrome, "low-phospholipid-associated cholethiasis", LPAC. This group of patients is defined by a specific combination of symptoms and is estimated to account for 5% of all gallstone disease. Several studies have identified a spectrum of *ABCB4* mutations in LPAC.[84–86] Investigations continue as to the precise contribution of ABCB4-associated cholethiasis to the population.[87] Differences in the presentation of this form of cholethiasis have been noted which suggest the biliary symptoms are more complicated than those caused simply by the presence of detectable gallstones. The most recent review of these series of patients reported a mutation rate of 56% with the current clinical definition. As a result of these studies, an alternative name for the *ABCB4* locus, gall bladder disease 1 (GBD1), has been proposed.

Together with this well-defined group of patients, *ABCB4* mutations have been identified in small numbers of other individuals with hepatolithiasis,[88] recurrent biliary pancreatitis,[89] unexplained cholestasis in association with fibrosis[90] and in a large family with cholestasis, ductopenia and cirrhosis.[91]

Thus, heterozygous (and in a few cases, homozygous) ABCB4 mutations can result in a very wide spectrum of cholestatic liver disease that can present across a wide age range.

2.9 *Intrahepatic cholestasis of pregnancy (ICP)*

Intrahepatic cholestasis of pregnancy (ICP), also known as obstetric cholestasis, presents with maternal pruritus and abnormal liver function. It affects 0.7% of pregnancies in the UK and has a higher prevalence in women of Chilean, Scandinavian and South Asian origin. Women commonly present in the third trimester of pregnancy, although severe cases may present in the first trimester.[92] ICP has a complex aetio-logy with a significant component.[93] The primary phenotype of cholestasis indicated possible involvement of genetic variation at the *ABCB4* and *ABCB11* loci, together with clinical observations of ICP more commonly occurring in heterozygous mothers of children with PFIC disease. ICP-associated mutations in *ABCB4* were initially identified in a familial case[94] and then a sporadic case.[36] Since these initial findings a range of other mutations have been described and the role of heterozygous *ABCB4* mutations in ICP firmly established, albeit in a limited number of cases (in some populations, up to 10%).[95–102] In addition to the identification of a number of mutations that are inherited in an apparently Mendelian fashion in ICP, a haplotype study in 52 Swedish cases has identified a further role for *ABCB4* variation by defining haplotypes across this locus which are overrepresented or underrepresented in cases compared to controls.[103]

The function of ABCB11 and the role of *ABCB11* mutations in PFIC and BRIC type 2 indicate that variation in this gene could also be involved in the aetiology of ICP. This was initially proposed in the Finnish population following analysis of a two SNP haplotype in 57 cases.[104] However, a subsequent study in a larger cohort of 142 ICP cases failed to replicate the initial findings, leaving the role of ABCB11 open to question.[105]

A study of the entire coding region of the *ABCB11* gene identified a single potential mutation (N591S) in a cohort of

21 women with ICP.[98] In addition, a polymorphism (c.1331C>T, p. V444A, rs2287622) that may affect hepatic ABCB11 expression has been suggested to play a role in ICP.[43,98] A recent large study of 491 Caucasian cases has confirmed and expanded the role of *ABCB11* variation in ICP. Rare occurrences of the common ABCB11 PFIC/BRIC 2 mutations (E297G and D482G) together with the previously identified N591S mutation were identified in a small number of cases. In the same study the 444A variant allele of ABCB11 was demonstrated to be a significant risk factor for ICP in this cohort.[106]

Research in the field of cholestasis of pregnancy has led to the consideration of other factors that may affect ABCB11 expression and be involved in the disease, namely reproductive hormone metabolites. During pregnancy there are enormous rises in the levels of the hormones progesterone and estrogen, and consequently also their metabolites. Evidence is accumulating that these metabolites can affect the expression of canalicular transporters. The cholestatic estrogen metabolite, estradiol-17β-D-glucuronide, has been shown to cause endocytic internalization of ABCB11 protein,[107] and various estrogen and progesterone metabolites cause a reduction in ABCB11 mediated bile acid transport from *Xenopus oocytes*.[108] Interestingly, the estradiol-17β-D-glucuronide metabolite appears to interact with ABCB11 from the exofacial surface and inhibition *in vivo* may be dependent upon prior excretion into the canaliculus by ABCC2.[135]

Hence evidence is accumulating that reproductive hormones have a cholestatic effect by acting directly upon ABCB11, causing both a reduction in transporter density and activity leading to reduced bile flow. Clinical reports of the cholestatic effects of the contraceptive pill in some women with a history of ICP support these observations.[92]

2.10 *Drug-induced cholestasis (DIC)*

Drug-induced cholestasis is a potentially serious, and occasionally fatal, complication of drug treatment and is a common medical problem. However, the origins of this condition are poorly understood. A minority of cases are caused by specific hepatotoxic effects of drugs (for example methotrexate), which cause dose-dependant liver damage. Given that biliary excretion is a key excretory route for drugs/drug metabolites, it is plausible that susceptibility to drug-induced cholestasis could in part be determined by genetic variation at *ABCB11* and *ABCB4*. In addition, various studies of transporter substrates have identified a growing list of drugs and xenobiotics that have been demonstrated in model systems to inhibit ABCB11.[109] A recent study identified the 444A polymorphism, the same as is implicated in ICP susceptibility, as being associated with the condition in a small cohort.[110] This finding is not surprising given the aetiological overlap between DIC and ICP and indicates that this predisposing factor may be implicated in a much wider disease spectrum. In the same study two missense mutations of *ABCB11* were identified in the drug-induced cholestasis cohort. The involvement of mutations of *ABCB4* in drug-induced cholestasis was also investigated in the same study. Two missense mutations were identified, I764L and L1082Q (the latter being associated with drug-induced hepatocellular injury).

2.11 *Common variation around the ABCB4 and ABCB11 loci and other diseases*

The murine *abcb4* and *abcb11* loci have been identified as candidate lithogenic loci in inbred studies.[111] However as yet evidence has not emerged to suggest a role for common variation at these loci and gallstone risk.[112] However, ABCB4 is established as a mongeneic risk factor in humans for gallstones (LPAC, see above).[87]

Clinical progression of primary biliary cirrhosis has been associated with SNP-based haplotypes of *ABCB4*.[113] However, genetic variation in ABCB4 has not been found to cause susceptibility to the disease itself, nor does it predispose to primary sclerosing cholangitis.[114]

2.12 *Regulation of expression through FXR*

The key role played by FXR-mediated control of *ABCB11* transcription outlined earlier is considered in more detail in Chapter 14. In addition to FXR, other transcription factors regulate the activity of the *ABCB11* promoter. Specifically with regard to *ABCB11*, Ku proteins have been identified as functional co-repressors,[118] which suppress the ability of FXR to transactivate the *ABCB11* gene and are associated with FXR on the ABCB11 promoter.

Observations in the liver receptor homologue-1 (Lrh-1) hepatocyte-specific knock-out mice indicate that this receptor can regulate *ABCB11* expression.[115] This observation was confirmed by in silico and molecular studies of the *ABCB11* promoter. A functional LRH-1 response element (LRHRE) was identified which acts to modulate bile acid/FXR mediated regulation of *ABCB11*.[116] Additionally, there is some residual *ABCB11* expression in $fxr^{-/-}$ and $lrh1^{-/-}$ mice, indicating the involvement of other regulatory transcription factors. Subsequent studies have identified nuclear factor erythroid 2-related factor 2 (Nrf2) as a positive regulator of *ABCB11* expression.[117] An active MARE (musculo-aponeurotic fibrosarcoma recognition element) was identified in the *ABCB11* promoter at position –175 to –195 relative to the translational start site, together with a second inactive site. Nrf2 was demonstrated to be bound directly to this site and have a positive effect on *ABCB11* transcription using cell lines, siRNA CHiP assys and EMSAs.

2.13 *Regulation of activity by membrane cholesterol content*

Local membrane composition at the canaliculus can have an effect on ABCB11 activity. Analysis of the effects of mutations of another canalicular-expressed membrane transporter (*ATP8B1)*, have led to the discovery that ABCB11 activity can be altered significantly by varying cholesterol content of the canalicular membrane.[119] Homozygous *ATP8B1* mutations cause a similar phenotype to ABCB11 deficiency but the mechanism is poorly understood given the phosphatidylserine flipping function of this transporter. Analysis of the canalicular membrane of atp8b1 deficient mice[120] showed a marked reduction of cholesterol content compared to phospholipids. *In vitro* studies using targeted cholesterol depletion supported these findings by showing a clear relationship between cholesterol content and ATP-dependant BA transport. Another study focused upon the effect of cholesterol loading of the baculovirus-infected Sf9 system, using human, rat and mouse clones, and demonstrated similar effects upon ABCB11 activity.[121]

2.14 *Transporter interacting proteins and intracellular reservoirs*

Correct protein folding and chaperoning to the canalicular membrane is clearly necessary for normal functioning of ABCB11 and ABCB4. However the understanding of these processes is currently limited. Both proteins must be correctly targeted to the canalicular membrane of the polarised hepatocyte; this process involves protein/protein interactions and trafficking mechanisms. Analysis of protein-protein interactions using yeast two-hybrid screens identified HAX-1[122] as a binding partner of rat ABCB11, ABCB1 and ABCB4 which were confirmed by GST pull-down assays from rat liver homogenates, co-localization studies and co-immunoprecipitation studies.[123] The interaction was mapped to the region linking the two homologous halves of ABCB11. Furthermore, RNAi studies suggested that HAX-1 depletion retarded ABCB11 internalisation. Similar studies using the same yeast two-hybrid technology with rat cDNA libraries identified the myosin II regulatory light chain (MLC2) as an interacting partner for the same three ABC transporters.[124]

As with ABCB11, the mechanisms controlling the intracellular distribution and trafficking of ABCB4 have not been extensively characterised. Using similar approaches as those described above, receptor for activated C-kinase 1 (RACK1) was identified as an interacting partner of ABCB4.[125] The linker region between the two half-transporter domains was used to screen a human cDNA library from adult liver for protein interactions again using the yeast two-hybrid system. From this screen the RACK1 protein was identified and the nature of the interaction investigated further using GST pull-down and siRNA knock-down of RACK1. RACK1-depeleted cells showed reduced ABCB4 localization to the plasma membrane and reduced capacity for PC efflux. Another ABC transporter, ABCC7 (the cystic fibrosis conductance regulator), has been shown to have its apical regulation regulated by RACK1 although no direct interaction between the two proteins has been observed.[126]

The speed at which canalicular transporters in rats responded to treatment with taurocholate suggested the presence of intracellular reservoirs of transporters, from which extra transporters could be recruited to cope with sudden increased demand for canalicular excretion[127] to allow the cell to respond to rapidly changing physiological conditions. This observation has led to studies to determine the nature of these reservoirs and the mechanism of their cycling.[128] Of particular note, Rab11-positive recycling endosomes have been identified as a ABCB11 reservoir in polarized WIF-B9 cells.[129]

In addition to these interacting proteins, the role of N-linked glycosylation in maintaining ABCB11 activity has been explored and it was determined that at least two N-linked glycans are required for protein stability, trafficking and function.[130]

2.15 *Epigenetics and enterohepatic circulation*

Relatively little is known about the contribution of epigenetic regulatory mechanisms to the function of *ABCB4* and *ABCB11*. A recent study[131] performed a differential analysis of tissue specific transporters with respect to DNA methylation and histone marks near the transcriptional start sites of a comparative analysis of three tissues including liver. *ABCB11* (together with *ABCG5* and *ABCG8*) exhibited liver hypomethylation and hyperacetylated histone H3, suggesting involvement of these mechanisms in tissue-specific expression.

3. Perspectives for the Future

The identification of biliary transporters together with the mechanisms that regulate their expression has led to an understanding of the enterohepatic circulation at a molecular level. The identification of the nuclear receptor-mediated pathways has opened new areas for research and new potential for therapeutic targets. Genetic studies have identified Mendelian diseases of the enterohepatic circulation and the role of mutations of ABC transporters in this.

However, much remains to be uncovered regarding the mode of action of these transporters. Many studies have analysed these proteins individually, yet *in vivo* a number of ABC transporters are expressed at the canalicular membrane. It is therefore crucial for future studies to analyze combinations of transporters rather than to study them in isolation. This is particularly relevant for ABCB4 where direct studies of expressed clinically-relevant mutations has not yet proved possible. Initial studies focussed on unravelling of the protein-protein interactions of ABCB11 have led to investigations of therapeutic applications of chaperone-like molecules. In the case of mutations where transport activity is maintained but the protein is incorrectly trafficked, the opportunity exists to override these effects and ensure functional ABCB11 reaches the membrane. Studies *in vitro* and *in vivo* have demonstrated this to be possible with two mutant forms of *ABCB11*.[77]

The advent of genome-wide association studies and whole-genome resequencing technologies will undoubtedly reveal novel involvement of ABCB4 and ABCB11 in some unexpected pathologies, as well as confirming their role in cholestatic disease. Future work will reveal combinations of mutations at different transporter loci (together with the nuclear receptors that regulate them) as having a cumulative effect on disease as has already started with studies of intrahepatic cholestasis of pregnancy[132,133] and of combined LPAC and PFIC3.[134]

References

1. Houten, S.M. and Auwerx, J. (2004) *Ann. Med.* **36**(7), 482.
2. Hoffman, A.F. (2009) *Front Biosci.* **14**, 2584.
3. Maruyama, T., *et al.* (2002) *Biochem. Biophys. Res. Commun.* **298**(5), 714.
4. Raufman, J.P., Cheng, K., and Zimniak, P. (2003) *Dig. Dis. Sci.* **48**(8), 1431.
5. Chen, X., *et al.* (2000) *Inflamm. Res.* **49**(12), 744.
6. Russell, D.W. (2003) *Annu. Rev. Biochem.* **72**, 137.
7. Schwarz, M., *et al.* (1996) *J. Biol. Chem.* **271**(30), 18024.
8. Ishibashi, S., *et al.* (1996) *J. Biol. Chem.* **271**(30), 18017.
9. Pullinger, C.R., *et al.* (2002) *J. Clin. Invest.* **110**(1), 109.
10. Chatterjee, B., Echchgadda, I., and Song, C.S. (2005) *Methods Enzymol.* **400**, 165.
11. Hofmann, A.F. (2007) *Hepatology* **45**(4), 1083.
12. Makishima, M., *et al.* (1999) *Science* **284**(5418), 1362.
13. Parks, D.J., *et al.* (1999) *Science* **284**(5418), 1365.
14. Wang, H., *et al.* (1999) *Mol. Cell* **3**(5), 543.
15. Lew, J.L., *et al.* (2004) *J. Biol. Chem.* **279**(10), 8856.
16. Ananthanarayanan, M., *et al.* (2001) *J. Biol. Chem.* **276**(31), 28857.
17. Huang, L., *et al.* (2003) *J. Biol. Chem.* **278**(51), 51085.
18. Goodwin, B., *et al.* (2000) *Mol. Cell* **6**(3), 517.
19. Staudinger, J.L., *et al.* (2001) *Proc. Natl. Acad. Sci. U.S.A.* **98**(6), 3369.
20. Zhang, J., *et al.* (2004) *J. Biol. Chem.* **279**(47), 49517.
21. Makishima, M., *et al.* (2002) *Science* **296**(5571), 1313.
22. Dawson, P.A., *et al.* (2003) *J. Biol. Chem.* **278**(36), 33920.
23. Gong, Y.Z., *et al.* (1994) *J. Proc. Natl. Acad. Sci. U.S.A.* **91**(11), 4741.
24. Dawson, P.A., *et al.* (2005) *J. Biol. Chem.* **280**(8), 6960.
25. Lundasen, T., *et al.* (2006) *J. Intern. Med.* **260**(6), 530.

26. Inagaki, T., *et al.* (2005) *Cell Metab.* **2**(4), 217.
27. Kurosu, H. and Kuro, O.M. (2009) *Mol. Cell. Endocrinol.* **299**(1), 72.
28. Holt, J.A., *et al.* (2003) *Genes Dev.* **17**(13), 1581.
29. Jones, S. (2008) *Mol. Pharm.* **5**(1), 42.
30. Van der Bliek, A.M., *et al.* (1987) *EMBO J.* **6**(11), 3325.
31. Lincke, C.R., *et al.* (1991) *J. Biol. Chem.* **266**(8), 5303.
32. Smit, J.J., *et al.* (1993) *Cell* **75**(3), 451.
33. Smith, A.J., *et al.* (1994) *FEBS Lett.* **354**(3), 263.
34. Smith, A.J., *et al.* (2000) *J. Biol. Chem.* **275**(31), 23530.
35. Patel, P., *et al.* (2003) *Placenta* **24**(1), 39.
36. Dixon, P.H., *et al.* (2000) *Hum. Mol. Genet.* **9**(8), 1209.
37. Delaunay, J.L., *et al.* (2009) *Hepatology* **49**(4), 1218.
38. Childs, S., *et al.* (1995) *Cancer Res.* **55**(10), 2029.
39. Gerloff, T., *et al.* (1998) *J. Biol. Chem.* **273**(16), 10046.
40. Strautnieks, S.S., *et al.* (1998) *Nat. Genet.* **20**(3), 233.
41. Noe, J., Stieger, B., and Meier, P.J. (2002) *Gastroenterology* **123**(5), 1659.
42. Byrne, J.A., *et al.* (2002) *Gastroenterology* **123**(5), 1649.
43. Meier, Y., *et al.* (2006) *Hepatology* **44**(1), 62.
44. Ho, R.H., *et al.* (2010) *Pharmacogenet. Genomics* **20**(1), 45.
45. de Vree, J.M., *et al.* (1998) *Proc. Natl. Acad. Sci. U.S.A.* **95**(1), 282.
46. Fickert, P., *et al.* (2004) *Gastroenterology* **127**(1), 261.
47. Popov, Y., *et al.* (2005) *J. Hepatol.* **43**(6), 1045.
48. Lammert, F., *et al.* (2004) *Hepatology* **39**(1), 117.
49. Katzenellenbogen, M., *et al.* (2007) *Mol. Cancer Res.* **5**(11), 1159.
50. Wang, R., *et al.* (2001) *Proc. Natl. Acad. Sci. U.S.A.* **98**(4), 2011.
51. Perwaiz, S., *et al.* (2003) *J. Lipid Res.* **44**(3), 494.
52. Lam, P., Wang, R., and Ling, V. (2005) *Biochemistry* **44**(37), 12598.
53. Wang, R., *et al.* (2009) *Hepatology* **50**(3), 948.
54. Wang, R., *et al.* (2003) *Hepatology* **38**(6), 1489.
55. Figge, A., *et al.* (2004) *J. Biol. Chem.* **279**(4), 2790.
56. Davit-Spraul, A., *et al.* (2009) *Orphanet J. Rare Dis.* **4**, 1.
57. van Mil, S.W., Houwen, R.H., and Klomp, L.W. (2005) *J. Med. Genet.* **42**(6), 449.
58. Stapelbroek, J.M., *et al.* (2009) *Proc. Natl. Acad. Sci. U.S.A.* **106**(24), 9709.
59. Strautnieks, S.S., *et al.* (1997) *Am. J. Hum. Genet.* **61**(3), 630.
60. Strautnieks, S.S., *et al.* (2008) *Gastroenterology* **134**(4), 1203.
61. Takahashi, A., *et al.* (2007) *Eur. J. Gastroenterol. Hepatol.* **19**(11), 942.

62. Treepongkaruna, S., *et al.* (2009) *J. Gastroenterol.* **15**(34), 4339.
63. Chen, H.L., *et al.* (2002) *J. Pediatr.* **140**(1), 119.
64. Goto, K., *et al.* (2003) *J. Pediatr. Gastroenterol. Nutr.* **36**(5), 647.
65. Chen, H.L., *et al.* (2008) *J. Pediatr.* **153**(6), 825.
66. Scheimann, A.O., *et al.* (2007) *J. Pediatr.* **150**(5), 556.
67. Knisely, A.S., *et al.* (2006) *Hepatology* **44**(2), 478.
68. van Mil, S.W., *et al.* (2004) *Gastroenterology* **127**(2), 379.
69. Kubitz, R., *et al.* (2006) *J. Clin. Gastroenterol.* **40**(2), 171.
70. Plass, J.R., *et al.* (2004) *J. Hepatol.* **40**(1), 24.
71. Kagawa, T., *et al.* (2008) *Am. J. Physiol. Gastrointest. Liver Physiol.* **294**(1), G58.
72. Wang, L., Soroka, C.J., and Boyer, J.L. (2002) *J. Clin. Invest.* **110**(7), 965.
73. Lam, P., *et al.* (2007) *Am. J. Physiol. Cell Physiol.* **293**(5), 1709.
74. Hayashi, H., *et al.* (2005) *Hepatology* **41**(4), 916.
75. Noe, J., *et al.* (2005) *J. Hepatol.* **43**(3), 536.
76. Byrne, J.A., *et al.* (2009) *Hepatology* **49**(2), 553.
77. Hayashi, H. and Sugiyama, Y. (2007) *Hepatology* **45**(6), 1506.
78. Jacquemin, E., *et al.* (2001) *Gastroenterology* **120**(6), 1448.
79. Chen, H.L., *et al.* (2001) *Pediatr. Res.* **50**(1), 50.
80. Degiorgio, D., *et al.* (2007) *Eur. J. Hum. Genet.* **15**(12), 1230.
81. Dawson, R.J. and Locher, K.P. (2006) *Nature* **443**(7108), 180.
82. Dawson, R.J. and Locher, K.P. (2007) *FEBS Lett.* **581**(5), 935.
83. Morita, S.Y., *et al.* (2007) *Hepatology* **46**(1), 188.
84. Rosmorduc, O., Hermelin, B., and Poupon, R. (2001) *Gastroenterology* **120**(6), 1459.
85. Rosmorduc, O., *et al.* (2003) *Gastroenterology* **125**(2), 452.
86. Nakken, K.E., *et al.* (2009) *Liver Int.* **29**(5), 743.
87. Rosmorduc, O. and Poupon, R. (2007) *Orphanet J. Rare Dis.* **2**, 29.
88. Kano, M., *et al.* (2004) *Hepatol. Res.* **29**(3), 160.
89. Fein, F., *et al.* (2007) *Gastroenterol. Clin. Biol.* **31**(1), 106.
90. Ziol, M., *et al.* (2008) *Gastroenterology* **135**(1), 131.
91. Gotthardt, D., *et al.* (2008) *Hepatology* **48**(4), 1157.
92. Geenes, V. and Williamson, C. (2009) *World J. Gastroenterol.* **15**(17), 2049.
93. Dixon, P.H. and Williamson, C. (2008) *Obstetric Medicine* **1**(2), 65.
94. Jacquemin, E., *et al.* (1999) *Lancet* **353**(9148), 210.
95. Gendrot, C., *et al.* (2003) *J. Med. Genet.* **40**(3), 32.
96. Mullenbach, R., *et al.* (2003) *J. Med. Genet.* **40**(5), e70.
97. Lucena, J.F., *et al.* (2003) *Gastroenterology* **124**(4), 1037.

98. Pauli-Magnus, C., *et al.* (2004) *Pharmacogenetics* **14**(2), 91.
99. Floreani, A., *et al.* (2006) *Aliment. Pharmacol. Ther.* **23**(11), 1649.
100. Floreani, A., *et al.* (2008) *Dig. Liver Dis.* **40**(5), 366.
101. Schneider, G., *et al.* (2007) *Hepatology* **45**(1), 150.
102. Bacq, Y., *et al.* (2009) *J. Med. Genet.* **46**(10), 711.
103. Wasmuth, H.E., *et al.* (2007) *Gut* **56**(2), 265.
104. Eloranta, M.L., *et al.* (2003) *Scand. J. Gastroenterol.* **38**(6), 648.
105. Painter, J.N., *et al.* (2004) *Scand. J. Gastroenterol.* **39**(7), 694.
106. Dixon, P.H., *et al.* (2009) *Gut* **58**(4), 537.
107. Crocenzi, F.A., *et al.* (2003) *Am. J. Physiol. Gastrointest. Liver Physiol.* **285**(2), 449.
108. Vallejo, M., *et al.* (2006) *J. Hepatol.* **44**(6), 1150.
109. Stieger, B. (2009) *Drug Metab. Rev.*
110. Lang, C., *et al.* (2007) *Pharmacogenet. Genomics* **17**(1), 47.
111. Wang, D.Q., Cohen, D.E., and Carey, M.C. (2009) *J. Lipid Res.* **50 Suppl**, S406.
112. Acalovschi, M., *et al.* (2009) *Lipids* **44**(6), 521.
113. Ohishi, Y., *et al.* (2008) *Hepatology* **48**(3), 853.
114. Pauli-Magnus, C., *et al.* (2004) *Hepatology* **39**(3), 779.
115. Mataki, C., *et al.* (2007) *Mol. Cell. Biol.* **27**(23), 8330.
116. Song, X., *et al.* (2008) *J. Lipid Res.* **49**(5), 973.
117. Weerachayaphorn, J., *et al.* (2009) *Hepatology* **50**(5), 1588.
118. Ohno, M., *et al.* (2009) *Biochem. Biophys. Res. Commun.* **390**(3), 738.
119. Paulusma, C.C., *et al.* (2009) *J. Biol. Chem.* **284**(15), 9947.
120. Pawlikowska, L., *et al.* (2004) *Hum. Mol. Genet.* **13**(8), 881.
121. Kis, E., *et al.* (2009) *Drug Metab. Dispos.* **37**(9), 1878.
122. Fadeel, B. and Grzybowska, E. (2009) *Biochim. Biophys. Acta* **1790**(10), 1139.
123. Ortiz, D.F., *et al.* (2004) *J. Biol. Chem.* **279**(31), 32761.
124. Chan, W., *et al.* (2005) *J. Biol. Chem.* **280**(25), 23741.
125. Ikebuchi, Y., *et al.* (2009) *Hepatol. Res.* **39**(11), 1091.
126. Auerbach, M. and Liedtke, C.M. (2007) *Am. J. Physiol. Cell Physiol.* **293**(1), C294.
127. Gatmaitan, Z.C., Nies, A.T., and Arias, I.M. (1997) *Am. J. Physiol.* **272**(5 Pt 1), G1041.
128. Wakabayashi, Y., Kipp, H., and Arias, I.M. (2006) *J. Biol. Chem.* **281**(38), 27669.
129. Wakabauyashi, Y., Lippincott-Schwartz, J., and Arias, I.M. (2004) *Mol. Biol. Cell* **15**(7), 3485.

130. Mochizuki, K., *et al.* (2007) *Am. J. Physio. Gastrointest. Liver Physiol.* **292**(3), G818.
131. Imai, S., *et al.* (2009) *Mol Pharmacol.* **75**(3), 568.
132. Keitel, V., *et al.* (2006) *Gastroenterology* **131**(2), 624.
133. Zimmer, V., *et al.* (2009) *Liver Int.* **29**(8), 1286.
134. Poupon, R., *et al.* (2009) *Liver Int.*
135. Stieger, B., *et al.* (2000) *Gastroenterology* **118**(2), 422.

Chapter 4

ABCB1 and ABCG2: Deciphering the Role of Human Efflux Proteins in Cellular and Tissue Permeability

Marcos M. Pires, Dana M. Emmert, Jean A. Chmielewski and Christine A. Hrycyna

1. Introduction

In 1976, Juliano and Ling observed that a glycoprotein expressed at the cell surface was responsible for the drug resistant profile of Chinese hamster ovary cells. Because this protein altered drug permeability, they named it P-glycoprotein (ABCB1).[1] Not only was this protein the first eukaryotic multidrug transporter discovered, in 1986 it became the first gene encoding a eukaryotic ATP Binding Cassette (ABC) protein to be cloned and characterized.[2] Human ABCB1, also known as MDR1 (for multidrug resistance 1), has been renamed ABCB1 by the Human Gene Nomenclature Committee. Later, in 1998, another ABC protein associated with multidrug resistance in cells in culture was identified and cloned almost simultaneously by three different laboratories. The protein was thus given several names; BCRP (breast cancer resistant protein), ABCP (ABC transporter located in the placenta) and MXR (mitoxantrone resistance-associated protein).[3–5] The Human Gene Nomenclature Committee renamed the protein ABCG2 because it was the second protein discovered in the ABC superfamily subgroup G.[6] Furthermore, although not covered in this review, another group of ABC transporters that includes the MRP family of ABC transporters, classified as the ABCC subgroup, have also been implicated in multidrug resistance.

ABCB1 is a 140 kDa protein that migrates at 170 kDa on SDS-PAGE due to secondary modifications, including N-linked glycosylation. It consists of 1,280 amino acids organized into two halves, each with a transmembrane domain (TMD) comprised of six α-helices and a cytoplasmic nucleotide binding domain (NBD) connected via the sixth TM helix of each half (Fig. 1).[7] The second half shares the same overall architecture and 43% primary sequence similarity to the first half.[2] Both NBDs of ABCB1 possess canonical sequence motifs that are conserved in all ABC transporters, such as the Walker A, Walker B, Q-loop, ABC Signature, stacking aromatic, D-loop, and the H-loop as described in Chapter 1. The motifs contribute to the recognition of the ATP molecules and Mg^{2+},[8] the close interaction of the NBDs (dimerization),[9] and the catalysis of the γ–phosphate release from ATP.[10] The dimerization event is believed to be mediated by the binding of two ATP molecules and the two NBDs function cooperatively to bind and hydrolyze ATP, releasing ADP, and inorganic phosphate. An indication of this co-operativity was illustrated by Urbatsch *et al.*, among others, who showed that single mutations in the individual NBDs in mouse MDR3 generated non-functional proteins (somewhat confusingly, rodents have two orthologues of ABCB1, encoded by the *mdr1* (*mdr1a*) and *mdr3* (*mdr1b*) genes).[11] Furthermore, they demonstrated that these mutant MDR3 isoforms had little measurable basal or drug-stimulated ATPase activities but were capable of hydrolyzing one ATP as shown by vanadate trapping experiments. The binding and hydrolysis of ATP is believed to energize the transport activity of ABCB1, although the exact mechanism by which ATP binding and hydrolysis are coupled with respect to the binding and efflux of substrates remains elusive.

It is believed that substrates most likely initially interact with the transmembrane helices, and therefore the entry into

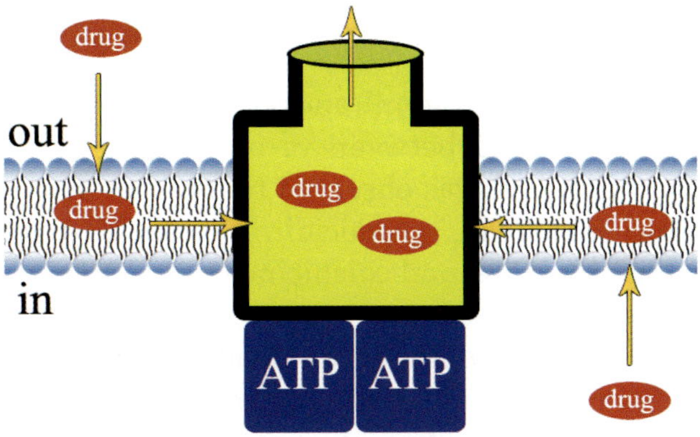

Fig. 1. A model demonstrating how ABCB1 in the cell membrane is able to intercept molecules from the plasma membrane of cells, thus maintaining low levels of intracellular drug. Adapted from Ref. 16.

the binding pockets of ABCB1 occurs in the membrane.[12–14] In fact, it has been shown that the TMDs of ABCB1 alone are sufficient to bind transport substrates although this mutated ABCB1 lacking NBDs is unable to confer the MDR phenotype on transfected cells.[15]

The drug binding pocket of ABCB1 has been roughly mapped out by several groups utilizing photo-affinity analogs of transport substrates and labeling of cysteine mutants with thiol-reactive substrates.[13,14] Through these studies, it has been postulated that drug binding involves an induced fit mechanism at the interface between the two TMDs.[17–21]

A striking property of ABCB1 is its ability to bind a wide-range of structurally diverse transport substrates and inhibitors. These data suggest the existence of multiple binding sites, and this property of ABCB1 has been studied in several labs using different approaches. Early studies showed that a photo-reactive derivative of azidopine was able to modify both the N and the C termini of the protein.[22] Further support for the theory of drug binding multiplicity came in the form of drug competition experiments. In 1997, Stein and co-workers, using the drug-stimulated ATPase activity of ABCB1 to report drug interactions, discovered that cooperative, competitive, and non-competitive interactions occurred between different "sites".[23] Non-competitive interactions were found between verapamil and vanadate while competitive inhibition was observed between verapamil and cyclosporin A. Additionally, verapamil was observed to interact in a cooperative manner with epirubicin, gramicidin S and D, daunorubicin, amiodarone, vinblastine, and colchicines.

In a cleverly designed experiment using crude membranes from CH^RB30 cells, Shapiro *et al.* loaded vesicles with either Hoescht 33342 or rhodamine 123 and monitored efflux of these fluorophores upon the addition of various other ABCB1 transport substrates and inhibitors.[24] They found that while colchicine and quercetin stimulated rhodamine 123 efflux and inhibited Hoechst 33342 efflux, both daunorubicin and doxorubicin induced Hoechst 33342 efflux and caused a reduction in rhodamine 123 efflux. Meanwhile, vinblastine, actinomycin D, and etoposide all inhibited the transport of both Hoescht 33342 and rhodamine 123. Overall, they concluded that two distinct binding sites (H-site for Hoescht 33342 and R-site for rhodamine 123) exist.

These findings were extended using radioligand-binding techniques to characterize how the multiple binding sites of ABCB1 interact.[25] A minimum of four drug binding sites were assigned to ABCB1, each related to transport, regulation, or both. Additionally, drug binding at one site changed other sites between high- and low-affinity conformations. Unfortunately, whether ABCB1 has one large binding pocket with distinct areas of specificity, or whether it contains several distinct binding sites is a question that remains unanswered.

Loo *et al.*[26] pioneered the use of cysteine-scanning mutagenesis of ABCB1 to map areas of the protein that could be covalently modified by the thiol-reactive rhodamine derivative, methanethiosulfonate-rhodamine (MTS-rhodamine), thus defining the substrate binding regions. Competition experiments determined that the MTS-rhodamine site overlapped with those of other substrates such as colchicine and calcein-AM, but not with the binding sites of verapamil and Hoechst 33342.

2. Structural Studies of ABCB1 and ABCG2

It is still debated how ABCB1 is able to recognize and transport such a wide array of different substrates. A high resolution crystal structure would help to shed light on this conundrum. Although extensive biochemical studies have provided insight into the ABCB1 structure, a high resolution crystal structure of ABCB1 remains elusive. Starting in 2001, three structures of the bacterial ABCB1 homolog MsbA were published revealing a structure contradictory to previously published biochemical data.[27–29] Scientists started to model ABCB1 based on these structures. However, the publication in 2006 of a high resolution (3 Å) structure of Sav1866, another ABCB1 bacterial homolog, suggested inaccuracies with the MsbA structures,[30] which were subsequently retracted owing the incorrect structures to data processing errors.[31]

These unfortunate circumstances confirmed the need for structural data of ABCB1 itself and tremendous progress has been made in the past few years, culminating most recently with the publication of a 3.8 Å crystal structure of mouse MDR1a.[32] This structure revealed valuable insights into the mechanism by which ABCB1 can simultaneously bind multiple molecules. Whereas a single inhibitor molecule, QZ59-RRR, was found to co-crystallize with ABCB1, two molecules of its stereoisomer, QZ59-SSS, were found to co-crystallize with ABCB1 to sites which overlapped with the binding "pocket" of QZ59-RRR but which were not identical. The authors proposed that the drug binding site formed at the interface between transmembrane helices was a large and fluid internal cavity capable of accommodating the binding of multiple molecules. The resolution of the structure was insufficient to allow for exact localization of amino acid residues in the drug binding site, therefore limiting the possibility of structure-based drug design. However, the study provides the first structural evidence that ABCB1 can bind multiple drugs simultaneously and can bind stereoisomers.

These data fit very well with the hypothesis proposed by Kimura *et al.* that cholesterol may fill the empty space of a singular large binding site to aid in the recognition and transport of small molecular weight substrates.[33] The binding affinity of low molecular weight (350–500 Da) drugs is increased in the presence of 20% cholesterol while binding of larger drugs (800–900 Da) remains unchanged.[34] The authors suggest that the ABCB1 drug binding site may best accommodate molecules with 800–900 Da mass, while cholesterol supports the recognition and transport of smaller drugs by adjusting the size of the site. This hypothesis was termed the "cholesterol fill-in model".[33] The emerging picture of ABCB1 drug binding is one consisting of a large drug binding cavity that has overlapping specificity for different drugs, rather than numerous distinct pockets as initially proposed.

Consistent with this hypothesis, there appears to be a domain swap resulting in an intertwined interface between the two halves of ABCB1, as revealed by the crystal structure.[32] Transmembrane helices 4 and 5 (and 10 and 11) crossover and contact the opposite α-helical bundle, enabling the burial of ~6900 Å^2 which stabilizes the dimer interface and creates a central cavity presumably where drugs can bind. This domain swap is a conserved motif also seen in the high resolution structures of Sav1866 and MsbA.[30,35]

As pointed out by Gottesman *et al.* however, while 3.8 Å may be the highest resolution structure of a mammalian multidrug transporter to date, there are potential pitfalls inherent in these static structures.[36] While the structural data support the "hydrophobic vacuum cleaner" model[37] and the biochemical data regarding the interplay of transmembrane helices 4, 5, 10 and 11,[22,38,39] all three structures (apo, one and two inhibitors bound) are nearly the same. One would have expected a conformational change with inhibitor binding. Additionally, in order to obtain crystals, ABCB1 was expressed as a glycosylation-deficient mutant in the yeast *Pichia pastoris* and crystallized in the absence of ATP. The authors argue that ATP concentrations in the cell are sufficient enough to doubt whether ABCB1 would ever exist in a nucleotide-free state.

While the crystal structures of ABCB1 have led to great insight, as exemplified above, these need to be rectified with biochemical data. Through the combination of these efforts, a more complete image of ABCB1 structure should be discerned in the future.

In contrast, much less is known about the three-dimensional structure of ABCG2. The *ABCG2* gene is highly conserved and is found in the genomes of nearly all vertebrates sequenced to date.[40–42] In humans, the *ABCG2* gene encodes a 655 amino acid plasma membrane glycoprotein with a molecular mass of 72 kDa.[40] ABCG2 is considered to be a half-transporter (comprised of a single N-terminal cytosolic NBD followed by a C-terminal TMD containing six transmembrane α-helices), that must dimerize in order to function.[43–47] Current knowledge of the structure of ABCG2 is limited to electron microscopy data with structural resolution at 18 Å.[46] The authors suggest that the protein may purify as a tetramer (i.e., a dimer of dimers), but higher resolution structures are needed.

3. Physiological Roles of ABCB1 and ABCG2

The discovery of ABCB1 in cancer cells was initially thought to be solely a mechanism by which cancerous cells evaded cytotoxic agents. Upon further examination, it was found that ABCB1 expression was not limited to chemotherapy-resistant cells, thus opening the door to the exploration of ABC transporters in homeostatic and disease states throughout the human body.[48]

ABCB1 mRNA was found in the liver, kidney, gastrointestinal tract, adrenal gland, lung, rectum, brain, prostate, heart, muscle, skin and ovary.[48] Notably, ABCB1 expression was especially prominent in blood-tissue barriers such as the blood-brain barrier, the blood-testis barrier, the blood-placenta barrier, and the blood-ocular barrier.

ABCG2 is similarly localized to the plasma membrane of cells in a variety of tissues throughout the body. Due to varied analysis methods, the reported expression levels and locations of

ABCG2 differ greatly in literature, however it is agreed upon that the highest levels of protein expression exist in the placenta, liver, gut and brain, with lesser expression in the breast, kidney, and hematopoietic stem cells.[40,49]

3.1 *Knock-out mouse models to study ABCB1 and ABCG2*

Knock-out mouse models of ABCB1 and ABCG2 have provided excellent tools for elucidating the physiological roles of these transporters. In mice, the situation is complicated as there are two isoforms of the multidrug resistance P-glycoproteins. Mdr1a is found primarily in the intestinal brush border epithelium and in the endothelial cells in the brain, while mdr1b is expressed primarily in the adrenal glands, in the uterus during pregnancy and in the placenta.[50]

The first knockout models revealed that *mdr1a (–/–)* or *mdr1b (–/–)* mice had a relatively benign phenotype.[50] By serendipity, the researchers who created the mdr1a knockout mice found that they were extremely susceptible to toxicity from treatment with ivermectin, a drug often used to treat mite infestations in mouse populations.[51] Analysis revealed that these mice were 50 to 100-fold more susceptible to ivermectin toxicity than the wild-type mice. The finding identified ivermectin as an ABCB1 transport substrate and also provided an explanation for the extreme sensitivity that had been observed in the collie dog breed almost ten years earlier, when it was determined that these dogs were homozygous for a four base pair deletion in the canine *MDR*1 gene.[52]

There are now single knockout mice for each mouse isoform and the double knockout. Through the use of these mice, it has been possible to study the pharmacokinetics of numerous ABCB1 transport substrates and inhibitors *in vivo* using a variety of methods. This work is vital to drug development, and can be used in combination with other protein knockouts, for example ABCG2, to determine the role played by multiple proteins in the regulation of xenotoxin or drug availability in the body.

A recent study highlighted the importance of the use of multiple mouse models to determine the physiological roles of ABCB1 and ABCG2.[53] The effect of these proteins on the bioavailability of topotecan was studied in wild — type, *mdr1a/mdr1b* knockout (henceforth defined as *ABCB1* knockout for convenience), *ABCG2* knockout and *ABCB1/ABCG2* knockout mice. While the area under the plasma and tissue concentration curve (AUC) in brains for the *ABCB1* and *ABCG2* knockout mice was 1.5-fold higher than wild-type, the AUC for the *ABCB1/ABCG2* knockout mice was 12-fold higher. These results suggest co-operativity between ABCB1 and ABCG2 in the elimination of substrates from the brain. Moreover, due to decreased secretion elsewhere in the body, there may be higher topotecan levels in the circulation leading to increased brain penetration. Finally, some substrates appear to be preferred by one protein or the other, but loss of both proteins may result in drastically altered tissue concentrations.

Due to their expression in areas of the body that are important in absorption, metabolism and elimination such as the liver, gut and kidney,[40,54,55] it is accepted that the major roles of ABCB1 and ABCG2 are to restrict penetration of their transport substrates. Although much more is known about ABCB1 and ABCG2-mediated transport of exogenous substrates, including many pharmaceuticals such as anti-cancer, anti-epileptic and immunosuppressive drugs, steroids and HIV protease inhibitors, information regarding endogenous substrates is surprisingly limited. The physiological substrates of ABCB1 may include steroid hormones, lipids, peptides and small cytokines,[56–64] while those of ABCG2 may include porphyrins and vitamins.[65,66] Many of the suggested ABCB1 endogenous substrates remain controversial, and more studies are needed to determine the true physiological roles of both ABCB1 and ABCG2.

4. Drug Bioavailability

4.1 *Roles of ABCB1 and ABCG2 in the gut*

Initial pharmacological experiments indicated that ABCB1 played a role in the elimination of compounds from the bloodstream due to its presence in the gastrointestinal (GI) tract.[67–69] Several ABC transporters expressed in the GI tract, such as ABCB1, ABCC2, ABCC4 and ABCG2, are expressed on the luminal membrane of villous epithelial cells at varying levels.[48,70–74]

Critical evidence that ABCB1 plays a major role in determining the oral bioavailability of certain drugs was provided through the use of knockout mouse models. For example, Sparreboom *et al.* demonstrated that lack of ABCB1 expression altered the bioavailability of Taxol, an ABCB1 substrate, in mice lacking the *mdr1a* gene.[75] Specifically, they found that the area under the plasma concentration curve was two- and six- times higher in *mdr1a (–/–)* mice than in wild-type mice after intravenous (i.v.) and oral administration, respectively. Many other studies have utilized *ABCB1*-null mice to delineate the role ABCB1 plays in the oral bioavailability of certain drugs.[67,76,77]

One method to circumvent the efflux of drugs by ABCB1 and thus allow greater levels of the compound to accumulate from the digestive tract would be to co-administer a known ABCB1 inhibitor. Westphal *et al.* demonstrated that when healthy volunteers were given digoxin, an ABCB1 transport substrate, in the presence of the ABCB1 inhibitor talinolol, a significant increase in the bioavailabilty of digoxin was observed.[78] More specifically, they found that when patients were given 100 mg talinolol in the presence of digoxin, the area under the concentration-time curve (AUC) from 0 to 6 hours and the AUC from 0 to 72 hours of digoxin increased by 18% and 23%, respectively and the concentration of digoxin in the serum was elevated by 45% compared to the administration of digoxin alone. Many other studies have found similar results, indicating ABCB1 inhibition is a viable method to increase the oral bioavailability of drugs that are also ABCB1 transport substrates.[75,79–85]

ABCG2 is also expressed in the GI tract. Expression analysis in humans specifically revealed that ABCG2 mRNA was highest in the duodenum and decreased only slightly throughout the rest of the GI tract, suggesting that the transporter may limit the uptake of xenotoxins.[71]

Jonker and colleagues showed that topotecan concentration in the plasma was six-fold higher in ABCB1 knockout mice co-administered with the inhibitor GF120918 when compared to mice without inhibitor treatment,[80] suggesting another transporter (such as ABCG2) may be playing a role. Subsequently, numerous studies have confirmed the limitation of oral drug absorption through ABCG2 function in the GI tract.[83,86–94] The bioavailability of the anticancer drug topotecan when co-administered with GF120918 has also been confirmed in humans.[95]

4.2 *Roles of ABCB1 and ABCG2 in the liver*

The liver is the primary site of xenobiotic metabolism and excretion in the body.[55] ABCB1 and ABCG2, along with several other transporters including members of the ABCC family, are vital for this due to their expression at the canalicular membranes of the liver (Fig. 2).[3,48,73,74,96] ABCG2 is also expressed in the bile ducts and blood vessel endothelium of the human liver.[97] Studies of the available knockout mice have shown that both ABCB1 and ABCG2 efflux toxins into the bile, and have a dramatic influence on drug pharmacokinetics.[90,94,98–103]

In 1998, the role of ABCB1 in the hepatobiliary elimination of therapeutic agents was investigated using ABCB1 knockout mice.[105] Tritiated Tri-butylmethyl ammonium (TBuMA), azidoprocainamide methoiodide (APM) and vecuronium were i.v.-administered to both wild-type and ABCB1 knockout mice and the secretion of the drugs into the bile was monitored. It was found that the knockout mice had significantly reduced hepatic elimination relative to wild-type mice, thus indicating that ABCB1 plays a role in the elimination of drugs not only in the GI tract, but also in the liver.

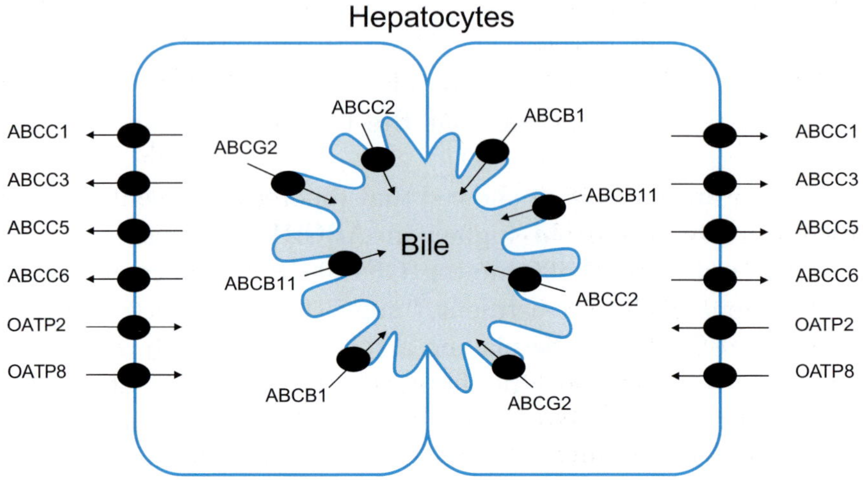

Fig. 2. Representation of liver hepatocytes (adapted from Kusuhara *et al.*).[104] ABCB1 and ABCG2 are expressed at the canalicular membranes of the liver, along with a number of other ABC transporters.

Many other reports have shown that ABCB1 can play a significant role in the elimination of its substrates in the liver. However, in 1997, Schinkel and colleagues observed that the loss of ABCB1 expression alone could not account for the elimination of digoxin and taxol by the liver[106] and ABCG2, along with other transporters, such as ABCC2 (see Chapter 8) and BSEP (ABCB11) (see Chapter 3), contributes to this elimination process.[104] Making use of the dual *ABCB1* and ABCG2 inhibitor GF120918 in the *ABCB1*-deficient mice, Jonker *et al.* showed that ABCG2 was partly responsible for the hepatic clearance of topotecan.[80]

4.3 *Roles of ABCB1 and ABCG2 in the kidney*

While it is agreed that ABCB1 is expressed at high levels in the human kidney,[48,107] reports of ABCG2 expression vary between no expression,[4] low expression[73] and moderate expression,[108,109] while expression in the mouse kidney is high.[110] Specifically in humans, ABCB1 and ABCG2 are expressed at the proximal tubule brush border membrane, thus indicating that they could be important in renal drug excretion.[108–110] The low to moderate expression of ABCG2 in the kidney suggests a minor efflux role, but it still may function to transport specific substrates.

The first methods employed to determine the role of ABCB1 in the renal excretion of its substrates utilized specialized cell lines. Porcine kidney (LLC-PK_1) epithelial cells transfected with human ABCB1 exhibited a concentration dependent decrease in digoxin transport when incubated with increasing concentrations of the inhibitor cyclosporin A.[111] In order to confirm the results *ex vivo*, the authors tested the ability of cyclosporin A to modulate ABCB1 activity in isolated perfused rat kidney. It was observed that treatment with 1 μM cyclosporin A was enough to lower digoxin efflux by half.

Surprisingly, some studies revealed that mice deficient in ABCB1 did not lose the ability to efflux several drugs, in fact, an increase in efflux was observed in some cases.[105,112] It was suggested that the loss of ABCB1 expression in these knockout mice resulted in an increase in the expression of other transport systems.[111] It is now known that *ABCB1* knockout mice express ~3 times the normal levels of ABCG2, most likely due to compensation for the loss of ABCB1.[113]

While ABCG2 expression in the human kidney is much lower than that of ABCB1, the possibility remains that it may play a minor role in the efflux of toxins and drugs into the urine. Mizuno and colleagues tested this hypothesis and demonstrated that urinary clearance of benzothiazole sulfates was reduced 2.4-fold in ABCG2 knockout versus wild-type mice.[114] ABCG2 has more recently been shown to affect renal availability of edaravone sulfate, ciprofloxacin and grepafloxacin in mice,[98,115] but ABCG2-mediated renal excretion of xenotoxins has yet to be shown in humans.

4.4 *Role of ABCG2 in the mammary gland*

ABCB1 is not expressed in healthy mammary tissue, however, because ABCG2 was originally cloned from a human breast cancer cell line it was believed that there must be some link to drug resistant breast cancer. To date, no link has been made in clinical studies, however, the protein was found to be up-regulated strongly during pregnancy and lactation,[116] but is not expressed in the mammary epithelial tissue of virgin mice.[117] ABCG2 is localized on the apical side of mammary alveolar epithelial cells which is the primary site of milk production.[116] Perhaps surprisingly, ABCG2 has been shown to transport substrates and toxins into breast milk.[40,117]

Given that the generally accepted role of ABCG2 is to protect the body from exposure to xenotoxins, it remains unclear why ABCG2 is up-regulated during pregnancy and lactation, which results in the accumulation of toxins in breast milk. Five hypotheses were considered to explain this observation and a recent review addresses this topic in detail.[118] The first hypothesis is that ABCG2 expression in the mammary gland is coincidental. This hypothesis suggests that expression in the mammary tissue is biologically irrelevant and merely a consequence of transcriptional elements present at specific stages of development. However, this was considered unlikely due to the fact that this up-regulation is conserved in at least four divergent species.

The second hypothesis suggests that ABCG2 is present to protect the mother from xenobiotics.[118] The evolutionary pressure for this seems improbable because newborns would be more susceptible to xenotoxins than the nursing mother and the mother expresses ABCG2 elsewhere in the body to secrete xenotoxins thus the expression of ABCG2 in mammary tissue should be redundant. However, this hypothesis has been given credence by the observation that ABCG2 in mammary tissue increased the milk-to-plasma ratio of nitrofurantoin over 80-fold in lactating mice, making this one of the primary mechanisms of elimination for this drug.[90]

The third hypothesis suggests that xenobiotic exposure to the newborn prepares it for solid food. Exposure to xenotoxins would induce detoxification systems in the newborn in a preemptive fashion. The only caveat with this hypothesis is that to date there are no known ABCG2 substrates that induce the xenobiotic receptors PXR (pregnane X receptor) and CAR (constitutive androstane receptor) that regulate detoxification pathways.

The fourth hypothesis is that exposure to xenotoxins through breast milk reduces the chance of developing allergies later in life.[118] Suckling wild-type BALB/c mouse pups have recently been shown to receive airborne antigens from the mother through breast milk, and this resulted in an induced tolerance that ultimately led to protection from those allergens.[119] However ABCG2 transport substrates are small molecules unlikely to elicit an allergenic effect (although some small molecule allergens have been described).[119]

The final hypothesis considered the possibility that ABCG2 is necessary to transfer nutrients to the newborn through the breast milk. In line with this, it has been shown that ABCG2 is able to transport riboflavin and other vitamins; however, there has been no evidence of vitamin deficiencies in the pups of ABCG2 knockout mice.[66,117,120–122] The evolutionary role of ABCG2 in mammary tissue remains unclear, and it is evident that more work needs to be done to explain this expression pattern.

4.5 *Roles of ABCB1 and ABCG2 in the maternal-fetal barrier (Placenta)*

The placenta links the maternal and fetal circulations while keeping the two separate. It allows nutrients and waste to pass through, while simultaneously protecting the fetus by blocking the entrance of toxins from the maternal blood supply.[55,123] The outer layer, which contains the syncytiotrophoblasts, serves the main protective function through the expression of efflux proteins including ABCB1, ABCC2 and ABCG2 in the apical membrane (Fig. 3).[55,123,124]

An early example of ABCB1 action in the placenta was provided through analysis of an *mdr1a* deficient subpopulation of the CF-1 mouse strain.[125] This subpopulation has an insertion of a solo long terminal repeat of the murine leukemia virus in the *mdr1a* gene, causing abnormal splicing and disruption of the

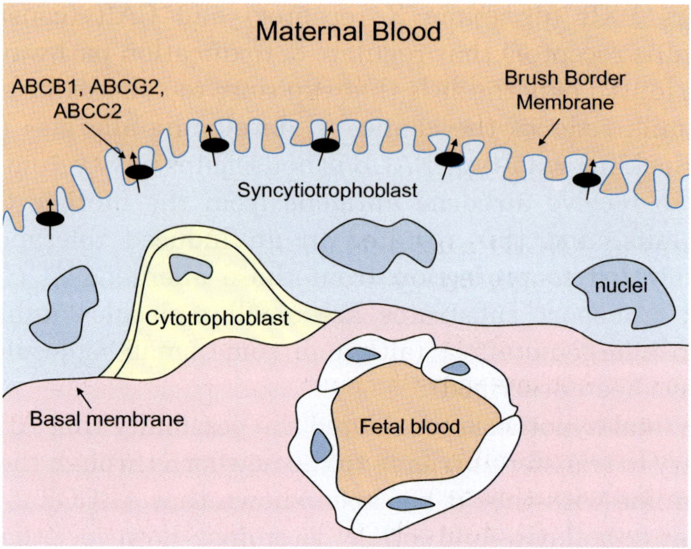

Fig. 3. Representation of ABCB1, ABCG2 and ABCC2 (MRP2) expression at the syncytiotrophoblast (adapted from Ganapathy *et al.*).[123] These proteins efflux their sub-strates into the maternal blood supply in order to protect the developing fetus.

gene at exon 23.[126] Heterozygous *mdr1a (+/–)* mice were mated, and the pregnant mothers were exposed to the ABCB1 substrate L-652,280, which produces a cleft palate in mice. It was found that all of the *mdr1a (–/–)* fetuses developed a cleft palate, while the heterozygous littermates were less susceptible and the homozygous *mdr1a (+/+)* fetuses exhibited no effect.

ABCB1 transport of cytotoxins in the human placenta was indicated shortly after, when directional transport of the ABCB1 substrates [^{3}H]-vinblastine, [^{3}H]-vincristine and [^{3}H]-digoxin was measured in human placental choriocarcinoma epithelial (BeWo) cells.[127] The basolateral-to-apical transport of all three drugs was dramatically reduced in the presence of the ABCB1 inhibitor cyclosporin A, thus suggesting that ABCB1 plays a major role in fetal protection at the placenta.

ABCG2 also serves to protect the fetus from xenotoxins. Jonker and colleagues were the first to report that ABCG2 lowered the bioavailability of topotecan by showing that co-administration with the inhibitor GF120918 in the ABCB1 knockout mice resulted in a 2-fold increase in fetal penetration of topotecan.[80] Several other studies have now shown that ABCG2 is an important factor in fetal xenobiotic protection.[110,128–130]

4.6 *Roles of ABCB1 and ABCG2 in the blood-brain barrier*

As the importance of ABCB1 as a general mechanism of defense and clearance became more understood, research naturally moved to other tissues that are prone to the harmful effects of xenobiotics. The selective permeability of the blood-brain barrier (BBB) protects the central nervous system (CNS) but is also a major impediment to the delivery of potentially therapeutic drugs.[69] The effective treatment of schizophrenia, epilepsy, HIV-related dementia, brain cancer, depression, Alzheimer's and Parkinson's disease are fundamentally linked to the ability of their respective therapeutic agents to reach their target destination and the limiting step for many of these drugs lies in their effectiveness at crossing the BBB.

Although the surface area separating the blood and the brain is vast, the passage of compounds is tightly regulated and non-endogenous molecules encounter resistance in the form of a physical and biochemical barrier.[131] The physical barrier is primarily formed by the tight junctions between the endothelial cells that line the capillaries of the BBB. These limit the passive diffusion of molecules between cells. Only hydrophobic molecules of <500 Da (usually) can passively diffuse from the blood into the CNS.

The biochemical barrier refers to the proteins (for example, ABCB1, ABCC1, ABCC2 and ABCG2) expressed in the luminal membrane of these endothelial cells that modify and transport compounds back into the blood stream (Fig. 4).[40,55] The overlap in function of these proteins has made it difficult to separate their individual contributions.

Resistance to treatment of epilepsy and chronic pain at first appeared to be unrelated to ABCB1. However, the role of ABCB1

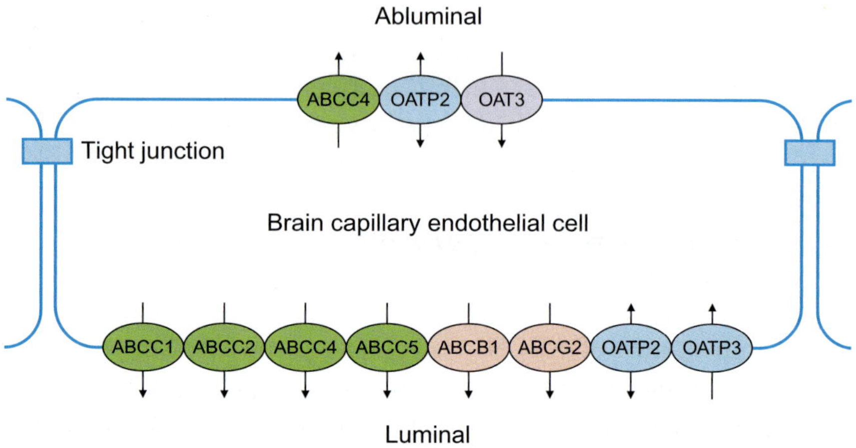

Fig. 4. Transporters located at the BBB.[132] ABCB1 and other transporters on the luminal face of the endothelial cells extrude molecules as they diffuse from the blood. The transporters shown are: ABCG2 (BCRP), the ABCC (MRP) family, OAT3 (organic anion transporter 3), and OATP2/3 (organic anion transporting polypeptide 2/3). Adapted from Ref. 132.

at the BBB became apparent with the generation of the *mdr1a (–/–)* mouse, which resulted in the striking increase in the brain penetration of both vinblastine and ivermectin.[51]

ABCB1 and ABCG2 have overlapping transport substrate specificities, therefore drug efflux by ABCG2 is often inferred when ABCB1 is absent. Studies with wild-type, ABCB1 knockout and ABCG2 knockout mice revealed a combined effect of both ABCG2 and ABCB1 in limiting brain penetration of imatinib. Imatinib, which is transported by both ABCB1 and ABCG2 *in vitro*, was administered to the mice and a 2.5 and 3.6-fold increase in brain uptake was observed for ABCG2 and ABCB1 knockout mice, respectively.[133] Co-administration of imatinib with GF120918, a dual ABCG2 and ABCB1 inhibitor, revealed that brain penetration was increased 4.2-fold in wild-type mice, while co-administration of imatinib with pantroprazole, an ABCG2 specific inhibitor, increased brain uptake slightly in wild-type and ABCB1 knockout mice, yet showed no effect in the ABCG2 knockout mice. Later studies confirmed these findings.[134,135] A recent study provided a further dramatic example of the combined effect of ABCG2 and ABCB1 at the BBB. Topotecan levels in the brain were measured for wild-type, ABCB1 knockout, ABCG2 knockout and ABCB1/ABCG2 knockout mice individually, and while the ABCB1 and ABCG2 single knockout mice exhibited ~1.5-fold increase in topotecan brain concentration, the combined knockout mouse exhibited a 12-fold increase.[53]

Taken together, these data demonstrate that ABCB1 and ABCG2 play a significant role at the BBB. Although ABCB1 appears to dominate in a "wild-type" situation, loss of ABCB1 expression can result in increased in ABCG2 activity to compensate, whereas loss of both transporters dramatically decreases barrier function.

4.7 *Roles of ABCB1 and ABCG2 in the blood-testis barrier*

The testis is a sanctuary site protected by the blood-testis barrier (BTB). The BTB protects germ cells from xenotoxins both passively, through tight junctions joining epithelial, myoid and sertoli cells, and actively, through the expression of efflux transporters.[118,136] Bart and colleagues revealed that both ABCG2 and ABCB1 are expressed on the apical side of myoid cells and on the luminal side of capillary epithelial cells, while ABCC1 is localized to the basal side of Sertoli and Leydig cells.[137] It appears that these proteins pump xenotoxins away from the seminiferous tubule, therefore protecting the developing germ cells (Fig. 5).[136,137]

The action of ABCB1 at the BTB in the transport of vinblastine and ivermectin was demonstrated using *mdr1a (–/–)* mice.[51] While the results were not as striking as the concentrations observed in the brain, it was clear that ABCB1 affects the availability of these drugs at the BTB. Since that time, several other studies have produced similar results with different ABCB1 transport substrates.

Shortly after this discovery, Enokizono and colleagues demonstrated that ABCG2 is able to efflux compounds from the testis.[130,136] Using the ABCG2 knockout mouse they found that loss of ABCG2 expression resulted in an increase in testis

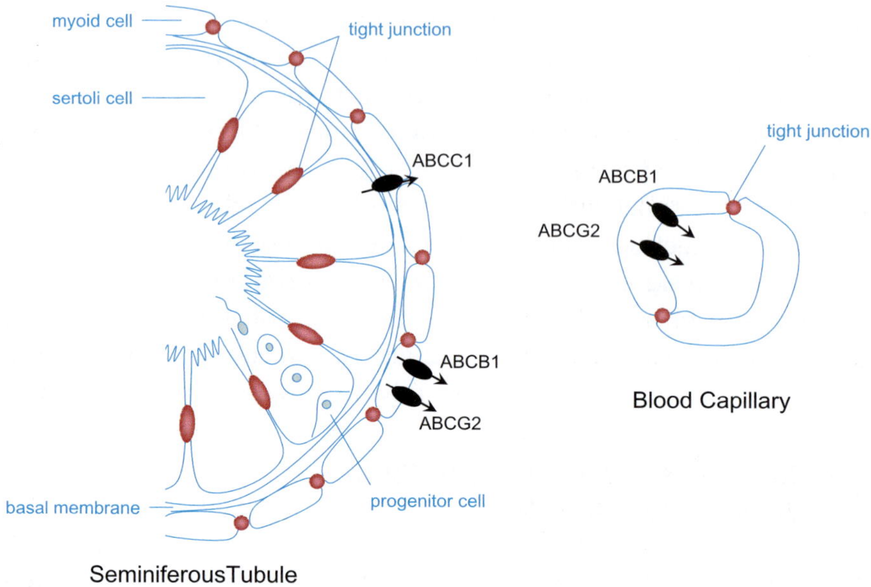

Fig. 5. Representation of the BTB (adapted from Vlaming *et al.*).[118] ABCG2 and ABCB1 are in the luminal membrane of blood capillaries and in the apical membrane of the myoid cells surrounding the seminiferous tubules. ABCC1 is expressed in the basal membrane of sertoli cells. The blood capillary cells are connected by tight junctions, while the myoid cells have inconsistent tight junctions.

concentrations of several phytoestrogens and xenobiotic compounds. Furthermore, as phytoestrogens are known to influence reproductive function (i.e., reducing testicular weight and sperm count), the role of ABCG2 at the BTB may turn out to be particularly important.[118]

5. Disease States Related to ABCB1 and ABCG2

5.1 *Roles of ABCB1 and ABCG2 with single nucleotide polymorphisms*

Over 20 single nucleotide polymorphisms (SNPs) have been identified in the *ABCB1* gene.[138] Many of these mutations could potentially alter both the function and expression of ABCB1 leading to variable drug pharmacokinetics and pre-disposition for certain diseases. For example, the *ABCB1* SNP 3435C > T is found in 36% of humans with early-onset Parkinson's Disease as compared to the control group (19%).[139] The higher susceptibility to Parkinson's Disease in this study may reflect the lower protective potential of ABCB1 at the BBB within these patients and indeed when individuals homozygous for 3435T were administered digoxin, the plasma level of the drug was 38% higher than those with the wild-type 3455CC genotype.[140] The C3435T SNP is silent with respect to the protein sequence but is proposed to exert an effect on the rate of translation of the protein.[141] Although few *ABCB1* SNPs have been characterized in detail, these data indicate that *ABCB1* variation is important, especially in individualized medicine.

Mutations in the *ABCG2* gene can also negatively affect the activity, transport specificity and expression of this protein.[142–154] The nonsynonymous SNP 421C > A which results in the amino acid substitution Q141K,[150] has an incidence rate of ~30% in Asian populations and ~10% in Caucasians.[144,148,155] Some investigators have reported decreased activity when compared with wild-type ABCG2 while others have reported reduced surface expression.[150,152,155] Elevated gefitinib and diflomotecan plasma levels and increased oral bioavailability of topotecan have also been reported in individuals heterozygous for Q141K.[156–158] The Q141K variant ABCG2 is also correlated with gefitinib-induced diarrhea,[159] suggesting that knowledge of the *ABCG2* genotype could help prevent dosing-induced side effects. These and other results indicate that patients could benefit greatly from dosing based on genotype.[160]

The *ABCG2* SNP 34G > A, which results in a V12M amino acid substitution, has also been studied extensively.[150] This SNP is fairly prevalent in Japanese, Caucasian and African populations (17.6%, 5.6% and 6.3%, respectively),[161] but reports about protein expression and activity are inconsistent.[150–152,155,161]

One interesting mutation observed in some drug-selected cell lines changed the substrate specificity of ABCG2.[162] Drug resistant cells expressing wild-type ABCG2 (R482) can efflux mitoxantrone, topotecan and SN-38, but not rhodamine or doxorubicin. However, drug resistant cells expressing ABCG2 T482 (MCF-7 AdVp cells) or G482 (S1-M1-80 cells) can transport all of the drugs above. These results were confirmed in transiently transfected cells. This altered substrate specificity has not been observed in humans to date.

6. Neurological Disease

6.1 *Alzheimer's disease*

The chemotherapeutic intervention of neurodegenerative diseases necessitates that sufficient quantities of the drugs reach their target sites. This may be one reason why drug design for such diseases has lagged behind other ailments. For example, few drugs have been shown to reverse or even slow the progression of Alzheimer's disease (AD).[163] Although the exact mechanism that causes the symptoms related to AD remains a controversial issue, it is currently believed that the accumulation of β-amyloid peptides (Aβ) in the brain tissue leads to decline in cognitive function and ultimately to patient death. α- and β-secretases are known to cleave amyloid precursor proteins (APPs), thus releasing Aβ which in turn aggregates into fibrils.[164] Molecules that inhibit α- and β-secretase hold the promise of inhibiting the formation of fibrils in the brain, and therefore alleviating the problems frequently associated with such aggregates.[165]

While α- and β-secretase inhibitors are potent disruptors of Aβ aggregation and fibril formation *in vitro*, little or no therapeutic effect is observed when these compounds are administered to laboratory animals or humans. One possible reason for the lack of *in vivo* activity of these inhibitors may be due to ABCB1 at the BBB. When the β-secretase inhibitor GSK188909 was administered to a mouse model of AD, it was observed that the brain function of the mice was similar to that of the untreated group.[166] However, upon co-administration of the ABCB1 inhibitor GF120918, the accumulation of GSK188909 in the brain was much greater. More significantly, the increased brain penetration of GSK188909 led to considerable reduction in the levels of Aβ observed in the brain. While it remains to be seen whether ABCB1 inhibition is a viable strategy for enhancing the accumulation of β-secretase inhibitors in humans, studies using small animal models highlight the possibility of utilizing drugs that would otherwise be discontinued due to poor brain penetration.

A recent development in AD research is the possibility that ABCB1 transports Aβ itself.[61] Some studies suggest that ABCB1 modulates the concentration of Aβ in the brain by moving these peptides from the brain to the blood. Initial proof-of-principle studies were provided by Lam *et al.*, who demonstrated that ABCB1 is capable of transporting Aβ in cultured cell lines.[61] Further evidence was provided by Cirrito *et al.*, who found that lower levels of Aβ were present in the blood of *ABCB1*-knockout mice.[167] This complicates the proposed use of ABCB1 inhibitors to improve delivery of α- and β-secretase inhibitors as loss of ABCB1 function may in itself increase β–amyloid concentration in the brain.

To date, no link between ABCG2 and AD has been made, however, because of the significant substrate overlap between ABCB1 and ABCG2, it is possible that ABCG2 plays a minor role in AD development and drug treatment.

6.2 *Psychotic disorders*

A number of anti-psychotic agents have been shown to interact with ABCB1, including reserpine, amisulpride, trifluoperazine, chlorpromazine, fluphenazine, pimozide, haloperidol, promethazine, quetiapine, triflupromazine, olanzapine, and risperidone.[168] In *mdr1a* knockout mice, Ejsing and co-workers[169] showed that the brain-to-serum ratio of risperidone was 10 times higher than in wild-type mice. It is therefore suspected that the poor response of patients to several of the FDA-approved anti-psychotic agents may be due to poor brain penetration caused by interaction with ABCB1, and that it may be possible to restore the efficacy of these agents by inhibiting the function of ABCB1 at the BBB.

To date, no clinical or *in vivo* study has linked ABCG2 and psychotic drug resistance, however, one research group has characterized several antipsychotic drugs that appear to interact with ABCG2, as determined by their effects on mitoxantrone efflux in ABCG2 over-expressing cells.[170]

6.3 *Epilepsy*

Epilepsy, a brain-related disorder characterized by recurrent seizures, affects millions of people worldwide. A number of antiepileptic drugs (AEDs) have been described that alleviate some of the symptoms,[171] however, approximately 30% of patients do not respond to AEDs, and many patients that respond initially become resistant over time.

With AEDs needing to cross the BBB to reach their site of action, possible interactions with ABCB1 may be an impediment to efficacy particularly where patients have acquired resistance to chemotherapeutic treatment.[132] The first example of a connection between AED resistance and ABCB1 showed that in 11 of 19 brain specimens of patients suffering from intractable epilepsy ABCB1 expression in the capillary endothelium was elevated more than 10-fold.[172] Furthermore, the concentration of the AED phenytoin was 4 times lower in those same specimens when compared to normal brain tissue. Several additional studies have since added substantial support for the hypothesis that ABCB1 serves a formidable barrier to permeation of AEDs into the brain.[173]

Direct evidence highlighting the role of ABCB1 in the reduction of AED penetration to the brain has been provided by various studies using ABCB1-null mice. It was found that brain tissue in these mice had elevated levels of phenytoin, lamotrigine, carbamazepine, phenobarbital, felbamate, and oxcarbazepine as compared to wild-type mice.[174–177] Specifically, Potschka and coworkers determined that ABCB1 was responsible for the limited penetration of phenobarbital, lamotrigine and felbamate in the cerebral cortex of rats.[174–177] Lombardo *et al.* have also demonstrated a mechanism by which resistance to AEDs may arise.[178] Immortalized rat brain microvascular endothelial cells were incubated with phenobarbital, topiramate, levetiracetam, carbamazepine, phenytoin, or tiagabine. It was found that carbamazepine, phenobarbital and phenytoin elevated the expression levels of PXR and CAR, which consequently led to an up-regulation of ABCB1 expression.

Although there is some disagreement among clinicians and researchers on the importance of ABCB1 as it relates of resistance to AEDs,[179] evidence linking ABCB1 to AED resistance merits the study of the use of ABCB1 inhibitors or regulators of ABCB1 expression to potentiate anti-epileptic treatment (this is discussed further in Chapter 14).

Some clinical trials have started to explore the possibility of using ABCB1 inhibitors to enhance AED penetration to the brain. For example, the seizure control of a 24-year-old woman with intractable epilepsy was greatly improved when the ABCB1 inhibitor verapamil was co-administered with carbamazepine (an AED and a known ABCB1 transport substrate).[180] Additional promising results were observed in a larger study utilizing epileptic rats[181] in which the ABCB1 inhibitor Tariquidar was co-administered with the AED phenytoin. Not only was phenytoin concentration in the brain increased, but seizure control was achieved in 94% of the rats.

It is interesting to note that most of the drugs designed to counteract the effects of epilepsy are small and hydrophobic in nature. This may be related to their target protein, but a more probable explanation may be that during the drug discovery phase of AEDs, these types of molecules were the only ones that could effectively cross the BBB. However, smaller drugs tend to be less potent and less specific, likely resulting in more side-effects. In other words, there may be a large number of much more potent and specific AEDs that simply cannot cross the BBB. Inhibition or down regulation of ABCB1 at the BBB may therefore open the door for the use of new classes of compounds in humans.

Researchers are still attempting to distinguish whether ABCG2 plays a role in epilepsy. One study revealed that ABCG2 is up-regulated in dysembryoplastic neuroepithelial tumors (DNT), which are relatively benign brain tumors that often cause medically intractable epilepsy,[182] while another report indicated that ABCG2 expression remained unchanged in cases of hippocampal sclerosis, (neuronal cell loss caused by temporal lobe epilepsy).[183] Another study tested the effects of several AEDs on ABCG2 activity and found no effect.[184] Overall, this evidence indicates that ABCG2 most likely has little to no effect on epilepsy treatment, except in the case of DNT.

7. Viral Disease

7.1 *HIV-1*

Although medication for patients infected with HIV-1 has improved greatly in the last decade, especially through the use of highly active antiretroviral therapy (HAART), there are reservoirs of the HIV virus located throughout the CNS that remain relatively unaffected by medication.[185] The virus persists due to the low levels of anti-HIV-1 drug penetration of this compartment. Although other transporters have been implicated in the efflux of anti-HIV-1 drugs from the brain, ABCB1 appears to play the most significant role, with ritonavir, amprenavir, abacavir, saquinavir, indinavir, and nelfinavir reported to be substrates of the transporter.[186,187] The evidence is indirect but nevertheless convincing with uptake of rhodamine 123 into endothelial cells from bovine brain microvessels modulated by saquinavir, nelfinavir, ritonavir, amprenavir, and indinavir.[186] Moreover, in a recent study, Kaddoumi and co-workers observed that the level of nelfinavir in macaque brains increased 146-fold following the co-administration of the ABCB1 inhibitor zosuquidar — adding strong evidence to the possible use of

ABCB1 inhibitors to increase brain penetration of HIV-1 drugs.[188]

The available data on the relevance of ABCG2 in relation to HIV-1 infection is more complicated. ABCG2 expression in a $CD4^+$ T-cell line resulted in resistance to nucleoside reverse transcriptase inhibitors (NRTIs) including zidovudine, which was reversed in the presence of the ABCG2 specific inhibitor fumitremorgin C.[189] In contrast, the HIV protease inhibitors (HIV PIs) ritonavir, saquinavir and nelfinavir have been shown to inhibit ABCG2, while indinavir and amprenavir had no effect on activity.[190] In 2007, a further study revealed that the ability of ABCG2 to efflux drugs was inhibited by several additional HIV PIs.[191]

Overall, current data indicate that NRTIs may be ABCG2 transport substrates, while HIV PIs are more likely to inhibit ABCG2 activity. However, further study including the direct measurement of transport of these compounds by ABCG2 is needed to clarify the specific role of ABCG2 in anti-HIV treatment.

8. Multidrug Resistance in Cancer

In 2003, it was determined that only 67% of people with cancer survived five years after their diagnosis.[192] Resistance to chemotherapy is most likely a main cause of these deaths. Acquired resistance of cancer cells to multiple structurally unrelated compounds is the hallmark of multidrug resistance and often results from the overexpression of ABC transporters such as ABCB1, ABCC1 and ABCG2. Due to high expression in barrier regions like the GI tract, BBB and placenta, ABCB1 and ABCG2 are poised to be vital players in the development and treatment of cancer.

Although several ABC proteins have been implicated in the multidrug resistance phenotype of cancer cells, ABCB1 is the most well-studied of all multidrug ABC transporters in this regard. Its possible significance to cancer survival rates emerged in a small study of 54 newly diagnosed patients suffering from acute myeloid leukemia (AML), which revealed that ABCB1 was responsible for the associated intrinsic resistance.[193] JSB-1, a monoclonal antibody that recognizes the extracellular surface of ABCB1, was used to analyze expression levels in bone marrow samples. Complete remission rates for patients showing elevated levels of ABCB1 expression were much less (60%) than those negative for ABCB1 (92%). Furthermore, the overall survival rate for ABCB1-positive patients was much improved for patients with low ABCB1 expression levels prior to treatment, with an average survival of 534 days *vs* 329 days. Similar trends have also been observed for patients suffering from Ewing's sarcoma,[194] breast cancer,[195] and neuroblastoma.[196]

However, there are contradictory reports that the level of ABCB1 expression in cancer tissue at the outset of treatment may not affect overall survival rates.[197] When the tissues of 57 patients with lymphoma were analyzed, it was observed that the presence of ABCB1 did not correlate with responsiveness to the chemotherapy. It should be noted, however, that the expression of *ABCB1* mRNA determined did not compare well with the immunoanalysis. Together, these results highlight the important issue concerning how the level of ABCB1 expression is evaluated.[198] Methods of detection tend to vary on whether the quantification is performed at the activity level, the protein level, or the mRNA level.

ABCB1-mediated acquired resistance to cancer drugs may arise following initial chemotherapy-based treatment. It is the more commonly known form of MDR since it usually, but not always, emerges as a consequence of selective pressure. In one particular study, it was determined that prior to treatment of patients with ovarian carcinoma, ABCB1 was expressed in the tumors of 15% of the patients analyzed.[199] However, following treatment with platinum and doxorubicin, 48% were positive for ABCB1.

In a separate study, the ABCB1 expression levels of patients with sarcoma pulmonary metastases were monitored shortly after treatment.[200] In four out five patients, a rapid and time dependent increase (3 to 15-fold) in *ABCB1* mRNA levels was observed in tumors 50 minutes after patients were administered doxorubicin. This study highlights the rapid response of cells to cytotoxic agents via the up-regulation of ABCB1. Low levels of *ABCB1* mRNA were found in normal lung tissues and these levels remained unchanged after doxorubicin treatment.

In a recent study, the relevance of ABCB1 expression in patients with renal clear cell carcinoma was evaluated. Two groups of patients were classified as either displaying low or high levels of ABCB1 expression. Their survival rates were assessed five years after the initial analysis. High levels of ABCB1 expression corresponded to lower survival rates, adding evidence to the prognostic ability of ABCB1 expression in patient survival.[201] The same observation was made in a study of 52 patients with ovarian carcinoma, 30 were positive for the presence of ABCB1 while the rest were not.[202] Following first line chemotherapeutic treatment, 11 of those who were ABCB1-positive had complete responses, four had partial responses, and the tumors of 15 of those patients had progressed. For those who were ABCB1-negative, a much better response profile was observed. Twenty of the patients had complete responses, while only one patient failed to respond to treatment.

Contradictory reports to the value of ABCB1 as a prognostic biomarker in cancer survival and responsiveness also exist. Schrader and co-workers observed no trend between ABCB1 expression and survival rates of patients suffering from metastatic non-seminomas.[203] Lu and co-workers arrived at a similar conclusion when assessing the impact of ABCB1 expression of ovarian cancer patients and their overall survival.[204] At present, it is difficult to conclude that ABCB1 can be a useful biomarker in all types of cancer and further study is required to define this relationship, yet the possibility of using ABCB1 as one of a larger set of biomarkers may yield more consistent results.

There is also controversy over the possible role that ABCG2 plays in the treatment of cancer and in the development of drug resistance. ABCG2 was discovered as a result of drug selection of cancer cell lines, and its expression in areas of protection and elimination make this protein a drug resistance candidate. The first transport substrates discovered for ABCG2 were the anticancer therapeutics mitoxantrone, topotecan, irinotecan, SN-38, anthracyclines and flavopiridol.[3,205–208] Later studies revealed that ABCG2 also transports carcinogens such as PhIP, MeIQx (2-Amino-3,8-dimethylimidazo[4,5-f]quinoxaline), IQ (2-amino-3-methylimidazo [4,5-f]quinoline), Trp-P-1 (3-amino-1,4-dimethyl-5H-pyrido[4,3-b]indole) and aflatoxin B1.[94,136,209] This ability to protect cells from carcinogens under normal circumstances, but to render abnormal cells resistant to chemotherapy marks ABCG2 as a double-edged sword and underlines the importance of determining the role that ABCG2 plays in cancer and chemotherapy.

More than ten years after the discovery of ABCG2, the role of this protein in cancer drug resistance is still unclear. It is believed that chromosomal rearrangements, gene amplification and translocations are the most common cause of ABCG2 overexpression in cell lines,[210,211] but unfortunately, studies analyzing ABCG2 expression in cancer have resulted in numerous conflicting reports. One of the first studies of ABCG2 expression in leukemia revealed that *ABCG2* mRNA was relatively high in 33% of the samples.[212] Furthermore, this high *ABCG2* mRNA expression did not correlate with *ABCB1* mRNA, which suggested that ABCG2 may be the source of resistance when ABCB1 expression is lacking.[212] However, other conflicting data have since been presented. Relatively high ABCG2 expression in leukemia was reported by some,[213,214] and low expression by others.[215–217] Additionally, some reports suggest that ABCG2 expression is related to therapeutic response,[213,218–221] while others find no correlation.[215,216,222]

Less is known regarding ABCG2 expression in solid tumors; however the most extensive study to date analyzed expression in 150 untreated solid tumors from 21 different tumor types.[223] The authors found ABCG2 in all tumor types, but most frequently in adenocarcinomas of the GI tract, endometrium, lung and melanoma. Other reports have been published, and as in the leukemia studies, ABCG2 testing methods and expression varied dramatically.[137,224–236] Small sample sizes and differing methods of analysis are the most likely explanations for the inconsistent data, therefore it is necessary to establish standardized testing methods and to increase sample numbers in order to determine levels accurately.

The effect of *ABCG2* SNPs should also be taken into account when attempting to establish the role of ABCG2 in chemoresistance in cancer. Variants of this protein could be very important in carcinogenesis, the development of chemotherapeutic resistance, and interindividual differences in dosage response. We already know that the Q141K SNP results in altered responses to chemotherapeutic agents such as topotecan, diflomotecan and 9-aminocamptothecin.[157,158,237] In fact, a small study of patients with prostate cancer revealed a significant association of the Q141K SNP with survival.[238] This SNP could result in higher sensitivity but also unexpected side effects due to its lower expression/activity.[40,118,159] Due to the prevalence of this SNP in Asian and Caucasian populations, this variant could have a direct effect on drug dosing and cancer treatment and genetic determination may be useful on an individual basis when deciding drug dosage.

Overall, a great deal of information is known about the role of ABCB1 and ABCC1, but there is much controversy over the possible role of ABCG2 in the development and treatment of cancer. If ABCG2 is in fact important in cancer chemotherapy, the prevalence of SNPs, mainly Q141K, should be taken into account. The link between ABCG2 and chemoresistance *in vitro* is undeniable and research is already focused on bypassing a likely problem *in vivo* with the development of chemotherapeutics that inhibit ABCG2 function and limit expression.

9. Inhibition of ABCB1 and ABCG2 Transport Activity in the Cancer Clinic

After the initial discovery that ABCB1 expression allowed cancer cell lines to become resistant to multiple unrelated cancer drugs, the search for inhibitors that could inactivate the transport activity of ABCB1 was a major goal in both academic and pharmaceutical laboratories. Lack of ABCB1 transport activity should, in theory, re-establish the accumulation of chemotherapeutic agents in resistant cancer cells. Direct evidence from knockout mouse models demonstrated the capacity of ABCB1 to limit the accumulation of certain drugs;[68,239] however another important observation was that the mice had a lower threshold of toxicity to most drugs. On the other hand, that the ABCB1 knockout mice were viable and showed minimal physiological dissimilarity to the wild-type mice showed the transporter was not essential, and implied that the inhibition of ABCB1 transport function may be a suitable strategy to enhance the permeation of various small synthetic molecules.[239]

In 1981, Tsuruo and co-workers were the first to show that a small molecule drug that could interfere with the transport function of ABCB1.[240] Verapamil, a calcium channel blocker used for the treatment of cardiac arrhythmia and hypertension, was found to restore the cytotoxic properties of vinblastine and vincristine in resistant P388 leukemia cells. Following these encouraging early results, the search for new and more potent inhibitors of ABCB1 intensified since there was a general belief that tumor resistance to chemotherapeutics relied heavily on the function of ABCB1.

A large number of novel synthetic inhibitors of ABCB1 have been discovered and, much like the substrates of ABCB1, these molecules are diverse in chemical structures and include lipids, cyclic peptides, transmembrane peptides, and small organic molecules. In fact, many transport substrates can be converted into inhibitors through modifications to the original molecules.[241] The number of ABCB1 inhibitors described in the literature thus far is large, yet it has proven to be difficult to construct a structure-activity relationship of these inhibitors due to their diversity and the lack of a high resolution structure of ABCB1.

A few promising ABCB1 inhibitors have been studied intensively and their potency and selectivity have been verified by separate laboratories. These inhibitors have also been selected for pre-clinical and clinical development. While most other inhibitors are also transport substrates of ABCB1, these are weak or non-substrates and include SCH66336, LY335979, GF120918, XR9576, and OC144–093.[242–246] To date, none has been approved for clinical use, and at this point, it remains unclear whether this approach will be a viable therapeutic strategy.

A possible reason that ABCB1 inhibitors have not fared well in clinical trials may be due to unexpected drug-drug interactions that affect the extremely narrow therapeutic window of most anti-cancer agents. A large majority of anti-cancer agents are extremely toxic and can only be safely used at specific concentrations (that have been determined in patients with normal ABCB1 transport activity). The addition of an ABCB1 inhibitor may allow greater accumulation of drugs in the cancerous cells, however, the impairment of ABCB1 in other tissues such as the liver will increase the circulating concentration and the retention time in the body. Impairment of ABCB1 barrier function in, for example, the BBB will allow anti-cancer agents to accumulate in sensitive tissues where they may cause systemic toxicity. The dosage of the drug must therefore be tailored to the new pharmacokinetic situation.[247]

A promising new approach to the disruption of ABCB1 function is through the knock-down of its expression via RNA-based technologies. Small interfering RNA's (siRNAs) are capable of down-regulating particular genes selectively, and therefore have sparked an interest in the medical community as a possible method of therapeutic intervention. Recent studies using siRNA demonstrated complete suppression of ABCB1 expression at the mRNA and protein levels, and complete reversal of the multidrug resistance phenotype intrinsic to a carcinoma cell line.[248,249] These seminal results show that siRNA may potentially provide a completely orthogonal approach at reducing the ability of ABCB1 to affect the distribution of certain small molecules in the body (this topic is discussed further in Chapter 14).

Since its discovery in 1998, the list of ABCG2 substrates and inhibitors has grown steadily. The first substrates reported were chemotherapy agents, however this has expanded to include antibiotics, antivirals, carcinogens and numerous other compounds. Inhibitors run the gamut as well. Two recent review articles address this topic in detail.[6,40,98,116,130,136,154,159,170,250–257]

Undoubtedly the list of substrates and inhibitors of ABCG2 will continue to grow. As of yet, there is no clear structure-activity relationship that defines a transport substrate or inhibitor of this protein. To make matters worse, some compounds are substrates at low concentrations and inhibitors at higher concentrations.

Given that many anti-cancer drugs are ABCG2 transport substrates, there is interest to develop anti-cancer therapies that bypass ABCG2-mediated efflux, or knock down protein expression. Similar strategies (to those proposed for control of ABCB1-mediated MDR) have been suggested for ABCG2 including co-administration of anti-cancer drugs with ABCG2 inhibitors, the development of bifunctional chemotherapies that inhibit ABCG2 function and have anti-proliferative properties, and the development of hammerhead ribozymes targeting the ABCG2 mRNA.[258] Several ABCG2 inhibitors have been identified recently, but none have yet been tested in clinical trial,[6] so the effects of these drugs when co-administered with anti-cancer drugs remains to be seen.

References

1. Juliano, R.L. and Ling, V. (1976) *Biochim. Biophys. Acta* **455**(1), 152.
2. Chen, C.J., *et al.* (1986) *Cell* **47**(3), 381.
3. Doyle, L.A., *et al.* (1998) *Proc. Natl. Acad. Sci. U.S.A.* **95**(26), 15665.
4. Allikmets, R., *et al.* (1998) *Cancer Res.* **58**(23), 5337.
5. Miyaki, K., *et al.* (1999) *Cancer Res.* **59**(1), 8.
6. Robey, R.W., *et al.* (2007) *Cancer Metastasis Rev.* **26**(1), 39.
7. Hamada, H. and Tsuruo, T. (1988) *J. Biol. Chem.* **263**(3), 1454.
8. Kerr, I.D. (2002) *Biochim. Biophys. Acta* **1561**(1), 47.

9. Smith, P.C., *et al.* (2002) *Mol. Cell* **10**(1), 139.
10. Moody, J.E., *et al.* (2002) *J. Biol. Chem.* **277**(24), 21111.
11. Urbatsch, I.L., *et al.* (2000) *Biochemistry* **39**(46), 14138.
12. Gros, P., *et al.* (1991) *Proc. Natl. Acad. Sci. U.S.A.* **88**(16), 7289.
13. Bruggeman, E.P., *et al.* (1992) *J. Biol. Chem.* **267**(29), 21020.
14. Greenberger, L.M., (1993) *J. Biol. Chem.* **268**(15), 11417.
15. Loo, T.W. and Clarke, D.M. (1999) *J. Biol. Chem.* **274**(35), 24759.
16. Gottesman, M.M. and Pastan, I. (1993) *Annu. Rev. Biochem.* **62**, 385.
17. Loo, T.W. and Clarke, D.M. (2001) *J. Biol. Chem.* **276**(40), 36877.
18. Loo, T.W. and Clarke, D.M. (2001) *J. Biol. Chem.* **276**(18), 14972.
19. Loo, T.W. and Clarke, D. M. (1997) *J. Biol. Chem.* **272**(51), 31945.
20. Loo, T.W. and Clarke, D.M. (2000) *J. Biol. Chem.* **275**(50), 39272.
21. Loo, T.W. and Clarke, D.M. (1999) *J. Biol. Chem.* **274**(50), 35388.
22. Bruggemann, E.P., *et al.* (1989) *J. Biol. Chem.* **264**(26), 15483.
23. Litman, T., *et al.* (1997) *Biochim. Biophys. Acta* **1361**(2), 169.
24. Shapiro, A.B. and Ling, V. (1997) *Eur. J. Biochem.* **250**(1), 130.
25. Martin, C., *et al.* (2000) *Mol. Pharmacol.* **58**(3), 624.
26. Loo, T.W., Bartlett, M.C., and Clarke, D.M. (2003) *J. Biol. Chem.* **278**(50), 50136.
27. Chang, G. and Roth, C.B. (2001) *Science* **293**(5536), 1793.
28. Pornillos, O., *et al.* (2005) *Science* **310**(5756), 1950.
29. Reyes, C.L. and Chang, G. (2005) *Science* **308**(5724), 1028.
30. Dawson, R.J. and Locher, K.P. (2006) *Nature* **443**(7108), 180.
31. Chang, G., *et al.* (2006) *Science* **314**(5807), 1875.
32. Aller, S.G., *et al.* (2009) *Science* **323**(5922), 1718.
33. Kimura, Y., *et al.* (2007) *J. Bioenerg. Biomembr.* **39**(5–6), 447.
34. Kimura, Y., *et al.* (2007) *Biochem. J.* **401**(2), 597.
35. Ward, A., *et al.* (2007) *Proc. Natl. Acad. Sci. U.S.A.* **104**(48), 19005.
36. Gottesman, M.M., Ambudkar, S.V., and Xia, D. (2009) *Nat. Biotechnol.* **27**(6), 546.
37. Raviv, Y., *et al.* (1990) *J. Biol. Chem.* **265**(7), 3975.
38. Dey, S., *et al.* (1997) *Proc. Natl. Acad. Sci. U.S.A.* **94**(20), 10594.
39. Loo, T.W. and Clarke, D.M. (2008) *Arch. Biochem. Biophys.* **476**(1), 51.
40. Robey, R.W., *et al.* (2009) *Adv. Drug Deliv. Rev.* **61**(1), 3.
41. Annilo, T., *et al.* (2006) *Genomics* **88**(1), 1.
42. Mickley, L., *et al.* (2001) *Mamm. Genome* **12**(1), 86.

43. Velamakanni, S., *et al.* (2007) *J. Bioenerg. Biomembr.* **39**(5–6), 465.
44. Sarkadi, B., *et al.* (2004) *FEBS Lett.* **567**(1), 116.
45. Bhatia, A., Schafer, H.J. and Hrycyna, C.A. (2005) *Biochemistry* **44**(32), 10893.
46. McDevitt, C.A., *et al.* (2006) *Structure* **14**(11), 1623.
47. Xu, J., *et al.* (2004) *J. Biol. Chem.* **279**(19), 19781.
48. Fojo, A.T., *et al.* (1987) *Proc. Natl. Acad. Sci. U.S.A.* **84**(1), 265.
49. Doyle, L.A. and Ross, D.D. (2003) *Oncegene* **22**(47), 7340.
50. Borst, P., *et al.* (1993) *Pharmacol.* **60**(2), 289.
51. Schinkel, A.H., *et al.* (1994) *Cell* **77**(4), 491.
52. Mealey, K.L., *et al.* (2001) *Pharmacogenetics* **11**(8), 727.
53. de Vries, N.A., *et al.* (2007) *Clin. Cancer Res.* **13**(21), 6440.
54. Haimeur, A., *et al.* (2004) *Curr. Drug Metab.* **5**(1), 21.
55. Leslie, E.M., Deeley, R.G., and Cole, S.P. (2005) *Toxicol. Appl. Pharmacol.* **204**(3), 216.
56. Bello-Reuss, E., *et al.* (2000) *Am. J. Physiol. Cell Physiol.* **278**(6), C1256.
57. Bosch, I., *et al.* (1997) *Biochemistry* **36**(19), 5685.
58. Eckford, P.D. and Sharom, F.J. (2005) *Biochem. J.* **389**(Pt 2), 517.
59. Eckford, P.D. and Sharom, F.J. (2006) *Biochem. Cell. Biol.* **84**(6), 1022.
60. Ernest, S. and Bello-Reuss, E. (1999) *J. Am. Soc. Nephrol.* **10**(11), 2306.
61. Lam, F.C., *et al.* (2001) *J. Neurochem.* **76**(4), 1121.
62. Liu, Y., *et al.* (1996) *Cancer Res.* **56**(21), 4992.
63. Pawlik, A., *et al.* (2005) *J. Pharm. Pharmacol.* **57**(11), 1421.
64. Romsicki, Y. and Sharom, F.J. (2001) *Biochemistry* **40**(23), 6937.
65. Krishnamurthy, P., *et al.* (2004) *J. Biol. Chem.* **279**(23), 24218.
66. van Herwaarden, A.E., *et al.* (2007) *Mol. Cell. Biol.* **27**(4), 1247.
67. Mayer, U., *et al.* (1996) *Br. J. Pharmacol.* **119**(5), 1038.
68. Schinkel, A.H., *et al.* (1995) *J. Clin. Invest.* **96**(4), 1698.
69. van Asperen, J., *et al.* (1996) *J. Natl. Cancer Inst.* **88**(14), 994.
70. Glavinas, H., *et al.* (2004) *Curr. Drug Deliv.* **1**(1), 27.
71. Gutmann, H., *et al.* (2005) *Biochem. Pharmacol.* **70**(5), 695.
72. Kruijtzer, C.M., Beijinen, J.H., and Schellens, J.H. (2002) *Oncologist* **7**(6), 516.
73. Maliepaard, M., *et al.* (2001) *Cancer Res.* **61**(8), 3458.
74. Thiebaut, F., *et al.* (1987) *Proc. Natl. Acad. Sci. U.S.A.* **84**(21), 7735.

75. Sparreboom, A., *et al.* (1997) *Proc. Natl. Acad. Sci. U.S.A.* **94**(5), 2031.
76. Kim, R.B., *et al.* (1998) *J. Clin. Invest.* **101**(2), 289.
77. Tahara, H., *et al.* (2005) *Drug Metab. Dispos.* **33**(7), 963.
78. Westphal, K., *et al.* (2000) *Clin. Pharmacol. Ther.* **68**(1), 6.
79. Bardelmeijer, H.A., *et al.* (2002) *Cancer Res.* **62**(21), 6158.
80. Jonker, J.W., *et al.* (2000) *J. Natl. Cancer Inst.* **92**(20), 1651.
81. Malingre, M.M., *et al.* (2001) *J. Clin. Oncol.* **19**(4), 1160.
82. Meerum Terwogt, J.M., *et al.* (1999) *Clin. Cancer. Res.* **5**(11), 3379.
83. Stewart, C.F., *et al.* (2004) *Cancer Res.* **64**(20), 7491.
84. van Asperen, J., *et al.* (1997) *Br. J. Cancer* **76**(9), 1181.
85. van Asperen, J., *et al.* (1998) *Clin. Cancer Res.* **4**(10), 2293.
86. Breedveld, P., Beijnen, J.H., and Schellens, J.H. (2006) *Trends Pharmacol. Sci.* **27**(1), 17.
87. Kondo, C., *et al.* (2005) *Pharm. Res.* **22**(4), 613.
88. Marchetti, S., *et al.* (2008) *Mol. Cancer Ther.* **7**(8), 2280.
89. Merino, G., *et al.* (2006) *Drug Metab. Dispos.* **34**(4), 690.
90. Merino, G., *et al.* (2005) *Mol. Pharmacol.* **67**(5), 1758.
91. Pulido, M.M., *et al.* (2006) *J. Vet. Pharmacol. Ther.* **29**(4), 279.
92. Seamon, J.A., *et al.* (2006) *Mol. Cancer Ther.* **5**(10), 67.
93. Sesink, A.L., *et al.* (2006) *Mol. Pharmacol.* **67**(6), 1999.
94. van Herwaarden, A.E., *et al.* (2003) *Cancer Res.* **63**(19), 6447.
95. Kruijtzer, C.M., *et al.* (2002) *J. Clin. Oncol.* **20**(13), 2943.
96. Kool, M., *et al.* (1997) *Cancer Res.* **57**(16), 3537.
97. Vander Borght, S., *et al.* (2006) *J. Histochem. Cytochem.* **54**(9), 1051.
98. Ando, T., *et al.* (2007) *Drug Metab. Dispos.* **35**(10), 1873.
99. Enokizono, J., Kusuhara., H., and Sugiyama, Y. (2007) *Drug Metab Dispos.* **35**(2), 209.
100. Hirano, M., *et al.* (2005) *Mol. Pharmacol.* **68**(3), 800.
101. Zamek-Gliszczynski, M.J., *et al.* (2006) *J. Pharmacol. Exp. Ther.* **319**(1), 459.
102. Zamek-Gliszcxynski, M.J., *et al.* (2006) *Mol. Pharmacol.* **70**(6), 2127.
103. Schinkel, A.H., (1997) *Semin. Cancer Biol.* **8**(3), 161.
104. Kusuhara., H. and Sugiyama, Y. (2009) *Drug Metab. Pharcacokinet.* **24**(1), 37.
105. Smit, J.W., *et al.* (1998) *Br. J. Pharmacol.* **124**(2), 416.
106. Schinkel, A.H., *et al.* (1997) *Proc. Natl. Acad. Sci. U.S.A.* **94**(8), 4028.

107. Noonan, K.E., *et al.* (1990) *Proc. Natl. Acad. Sci. U.S.A.* **87**(18), 7160.
108. Huls, M., *et al.* (2008) *Kidney Int.* **73**(2), 220.
109. Fetsch, P.A., *et al.* (2006) *Cancer Lett.* **235**(1), 84.
110. Jonker, J.W., *et al.* (2002) *Proc. Natl. Acad. Sci. U.S.A.* **99**(24), 15649.
111. Okamura, N., *et al.*, (1993) *J. Pharmacol. Exp. Ther.* **266**(3), 1614.
112. van Asperen, J., *et al.* (1999) *Br. J. Cancer* **79**(1), 108.
113. Cisternino, S., *et al.* (2004) *Cancer Res.* **64**(9), 3296.
114. Mizuno, N., *et al.* (2004) *Drug Metab. Dispos.* **32**(9), 898.
115. Mizuno, N., *et al.* (2007) *Drug Metab. Dispos.* **35**(11), 2045.
116. van Herwaarden, A.E. and Schinkel, A.H. (2006) *Trends Pharmacol. Sci.* **27**(1), 10.
117. Jonker, J.W., *et al.* (2005) *Nat. Med.* **11**(2), 127.
118. Vlaming, M.L., Lagas, J.S., and Schinkel, A.H. (2009) *Adv. Drug Deliv. Rev.* **61**(1), 14.
119. Verhasselt, V., *et al.* (2008) *Nat. Med.* **14**(2), 170.
120. Foraker, A.B., Khantwal, C.M., and Swaan, P.W. (2003) *Adv. Drug Deliv. Rev.* **55**(11), 1467.
121. Said, H.M. (2002) *Am. J. Clin. Nutr.* **75**(2), 179.
122. Assaraf, Y.G. (2006) *Drug Resist. Updat.* **9**(4–5), 227.
123. Ganapathy, V., *et al.* (2000) *J. Pharmacol. Exp. Ther.* **294**(2), 413.
124. Cordon-Cardo, C., *et al.* (1990) *J. Histochem. Cytochem.* **38**(9), 1277.
125. Lankas, G.R., *et al.* (1998) *Reprod. Toxicol.* **12**(4), 457.
126. Jun, K., Lee, S.B., and Shin, H.S. (2000) *Mamm. Genome* **11**(10), 843.
127. Ushigome, F., *et al.* (2000) *Eur. J. Pharmacol.* **408**(1), 1.
128. Zhou, L., *et al.* (2008) *Mol. Pharmacol.* **73**(3), 949.
129. Zhang, Y., *et al.* (2007) *Drug Metab. Dispos.* **35**(12), 2154.
130. Enokizono, J., Kushara, H., and Sugiyama, Y. (2007) *Mol. Pharmacol.* **72**(4), 967.
131. Rubin, L.L. and Staddon, J.M. (1999) *Annu. Rev. Neurosci.* **22**, 11.
132. Loscher, W. and Potschka, H. (2005) *Nat. Rev. Neurosci.* **6**(8), 591.
133. Breedveld, P., *et al.* (2005) *Cancer Res.* **65**(7), 2577.
134. Bihorel, S., *et al.* (2007) *J. Neurochem.* **102**(6), 1749.
135. Bihorel, S., *et al.* (2007) *Pharm. Res.* **24**(9), 1720.
136. Enokizono, J., *et al.* (2008) *Drug Metab. Dispos.* **36**(6), 995.
137. Bart, J., *et al.* (2004) *Eur. J. Cancer* **40**(14), 2064.
138. Schwab, M., Eichelbaum, M., Fromm, M.F. (2003) *Annu. Rev. Pharmacol. Toxicol.* **43**, 285.
139. Furuno, T., *et al.* (2002) *Pharmacogenetics* **12**(7), 529.

140. Hoffmeyer, S., *et al.* (2000) *Proc. Natl. Acad. Sci. U.S.A.* **97**(7), 3473.
141. Kimchi-Sarfaty, C., *et al.* (2007) *Science* **315**(5811), 525.
142. Backstrom, G., *et al.* (2003) *Eur. J. Pharm. Sci.* **18**(5), 359.
143. Bosch, T.M., *et al.* (2005) *Am. J. Pharmacogenomics* **5**(2), 123.
144. de Jong, F.A., *et al.* (2004) *Clin. Cancer Res.* **10**(17), 5889.
145. Honjo, Y., *et al.* (2002) *Cancer Biol. Ther.* **1**(6), 696.
146. Lida, A., *et al.* (2002) *J. Hum. Genet.* **47**(6), 285.
147. ltoda, M., *et al.* (2003) *Drug Metab. Pharmacokinet.* **18**(3), 212.
148. Kobayashi, D., *et al.* (2005) *Drug Metab. Dispos.* **33**(1), 94.
149. Zamber, C.P., *et al.* (2003) *Pharmacogenetics* **13**(1), 19.
150. Imai, Y., *et al.* (2002) *Mol. Cancer Ther.* **1**(8), 611.
151. Mizuarai, S., Aozasa, N., and Kotani, H. (2004) *Int. J. Cancer.* **109**(2), 238.
152. Morisaki, K., *et al.* (2005) *Cancer Chemother. Pharmacol.* **56**(2), 161.
153. Tamura, A., *et al.* (2006) *Mol. Pharmacol.* **70**(1), 287.
154. Sharom, F.J., (2008) *Pharmacogenomics* **9**(1), 105.
155. Kondo, C., *et al.* (2004) *Pharm. Res.* **21**(10), 1895.
156. Li, J., *et al.* (2007) *Cancer Biol. Ther.* **6**(3), 432.
157. Sparreboom, A., *et al.* (2004) *Clin. Pharmacol.* **76**(1), 38.
158. Sparreboom, A., *et al.* (2005) *Cancer Biol. Ther.* **4**(6), 650.
159. Cusatis, G., *et al.* (2006) *J. Natl. Cancer. Inst.* **98**(23), 1739.
160. Rudin, C.M., *et al.* (2008) *J. Clin. Oncol.* **26**(7), 1119.
161. Tamura, A., *et al.* (2007) *Cancer Sci.* **98**(2), 231.
162. Honjo, Y., *et al.* (2001) *Cancer Res.* **61**(18), 6635.
163. Selkoe, D. J., (2001) *Physiol. Rev.* **81**(2), 741.
164. Nunan, J. and Small, D.H. (2000) *FEBS Lett.* **483**(1), 6.
165. Lahiri, D.K., *et al.* (2003) *Curr. Drug Targets* **4**(2), 97.
166. Hussain, I., *et al.* (2007) *J. Neurochem.* **100**(3), 802.
167. Cirrito, J.R., *et al.* (2005) *J. Clin. Invest.* **115**(11), 3285.
168. El Ela, A.A., *et al.* (2004) *J. Pharm. Pharmacol.* **56**(8), 967.
169. Ejsing, T.B., Pedersen, A.D., and Linnet, K. (2005) *Hum. Psychopharmacol.* **20**(7), 493.
170. Wang, J.S., *et al.* (2008) *Basic Clin. Pharmacol. Toxicol.* **103**(4), 336.
171. Bazil, C.W. and Pedley, T.A. (1998) *Annu. Rev. Med.* **49**, 135.
172. Tishler, D.M., *et al.* (1995) *Epilepsia* **36**(1), 1.
173. Lazarowski, A., *et al.* (2007) *Epilepsia* **48 Suppl 5**, 140.
174. Potschka, H., Fedrowitz, M., and Loscher, W. (2002) *Neurosci. Lett.* **327**(3), 173.

175. Potschka, H. and Loscher, W. (2001) *Epilepsia.* **42**(10), 1231.
176. Clinkers, R., *et al.* (2005) *J. Pharmacol. Exp. Ther.* **314**(2), 725.
177. Potschka, H., Fedrowitz, M., and Loscher, W. (2001) *Neuroreport* **12**(16), 3557.
178. Lombardo, L., *et al.* (2008) *Curr. Neurocvasc. Res.* **5**(2), 82.
179. Anderson, G.D. and Shen, D.D. (2007) *Epilepsia.* **48**(12), 2372.
180. Summers, M.A., Moore, J.L., and McAuley, J.W. (2004) *Ann. Pharmacother.* **38**(10), 1631.
181. van Vliet, E.A., *et al.* (2006) *Epilepsia* **47**(4), 672.
182. Vogelgesang, S., *et al.* (2004) *Clin. Neuropathol.* **23**(5), 223.
183. Aronica, E., *et al.* (2005) *Epliepsia.* **46**(6), 849.
184. Cerveny, L., *et al.* (2006) *Epilepsia.* **47**(3), 461.
185. Nath, A. and Sacktor, N. (2006) *Curr. Opin. Neurol.* **19**(4), 358.
186. Bachmeier, C.J., *et al.* (2005) *Pharm. Res.* **22**(8), 1259.
187. Shaik, N., *et al.* (2007) *Drug Metab. Dispos.* **35**(11), 2076.
188. Kaddoumi, A., *et al.* (2007) *Drug Metab. Dispos.* **35**(9), 1459.
189. Wang, X. and Baba, M. (2005) *Antivir. Chem. Chemother.* **16**(4), 213.
190. Gupta, A., *et al.* (2004) *J. Pharmacol. Exp. Ther.* **310**(1), 334.
191. Weiss, J., *et al.* (2007) *J. Antimicrob. Chemother.* **59**(2), 238.
192. Hayat, M.J., *et al.* (2007) *Oncologist* **12**(1), 20.
193. Wood, P., *et al.* (1994) *Br. J. Haematol.* **87**(3), 509.
194. Roessner, A., *et al.* (1993) *J. Cancer Res. Clin. Oncol.* **119**(4), 185.
195. Tsukamoto, F., *et al.* (1997) *Breast Cancer* **4**(4), 259.
196. Haber, M., *et al.* (1997) *Eur. J. Cancer* **33**(12), 2031.
197. Niehans, G.A., *et al.* (1992) *Cancer Res.* **52**(13), 3768.
198. Ling, V. (1997) *Cancer Chemother. Pharmacol.* **40 Suppl**, S3.
199. van der Zee, A.G., *et al.* (1995) *J. Clin. Oncol.* **13**(1), 70.
200. Abolhoda, A., *et al.* (1999) *Clin. Cancer Res.* **5**(11), 3352.
201. Mignogna, C., *et al.* (2006) *BMC Cancer* **6**, 293.
202. Raspollini, M.R., *et al.* (2005) *Int. J. Gynecol. Cancer* **15**(2), 255.
203. Schrader, A.J., *et al.* (2007) *Exp. Oncol.* **29**(3), 212.
204. Lu, L., *et al.* (2007) *Oncol. Res.* **16**(8), 395.
205. Brangi, M., *et al.* (1999) *Cancer Res.* **59**(23), 5938.
206. Litman, T., *et al.* (2000) *J. Cell Sci.* **113(Pt 11)**, 2011.
207. Robey, R.W., *et al.* (2001) *Clin. Cancer Res.* **7**(1), 145.
208. Maliepaard, M., *et al.* (1999) *Cancer Res.* **59**(18), 4559.
209. van Herwaarden, A.E., *et al.* (2006) *Carinogenesis* **27**(1), 123.
210. Knutsen, T., *et al.* (2000) *Genes Chromosomes Cancer* **27**(1), 110.
211. Rao, V.K., *et al.* (2005) *Cancer Genet. Cytogenet.* **160**(2), 126.

212. Ross, D.D., *et al.* (2000) *Blood* **96**(1), 365.
213. Benderra, Z., *et al.* (2004) *Clin. Cancer Res.* **10**(23), 7896.
214. Galimberti, S., *et al.* (2004) *Leuk. Res.* **28**(4), 367.
215. Abbott, B.L., *et al.* (2002) *Blood* **100**(13), 4594.
216. van der Kolk, D.M., *et al.* (2002) *Blood* **99**(10), 3763.
217. van der Pol, M.A., *et al.* (2003) *Haematologica* **88**(2), 134.
218. Steinbach, D., *et al.* (2002) *Leukemia* **16**(8), 1443.
219. Uggla, B., *et al.* (2005) *Leuk. Res.* **29**(2), 141.
220. van den Heuvel-Eibrink, M.M., *et al.* (2002) *Leukemia* **16**(5), 833.
221. Wilson, C.S., *et al.* (2006) *Blood* **108**(2), 685.
222. Suvannasankha, A., *et al.* (2004) *Leukemia* **18**(7), 1252.
223. Diestra, J.E., *et al.* (2002) *J. Pathol.* **198**(2), 213.
224. Aust, S., *et al.* (2004) *Lab. Invest.* **84**(8), 1024.
225. Burger, H., *et al.* (2003) *Clin. Cancer Res.* **9**(2), 827.
226. Deichmann, M., *et al.* (2005) *J. Cutan. Pathol.* **32**(7), 467.
227. Diestra, J.E., *et al.* (2003) *J. Urol.* **170**(4 Pt 1), 1383.
228. Faneyte, I.F., *et al.* (2002) *Clin. Cancer Res.* **8**(4), 1068.
229. Friedrich, R.E., Punke, C., and Reymann, A. (2004) *In Vivo* **18**(2), 133.
230. Glasgow, S.C., *et al.* (2005) *Br. J. Cancer* **92**(2), 259.
231. Kanzaki, A., *et al.* (2001) *Jpn. J. Cancer Res.* **92**(4), 452.
232. Kawabata, S., *et al.* (2003) *Clin. Cancer Res.* **9**(8), 3052.
233. Konig, J., *et al.* (2005) *Int. J. Cancer* **115**(3), 359.
234. Scheffer, G.L., *et al.* (2000) *Cancer Res.* **60**(10), 2589.
235. Yoh, K., *et al.* (2004) *Clin. Cancer Res.* **10**(5), 1691.
236. Zurita, A.J., *et al.* (2003) *Br. J. Cancer* **88**(6), 879.
237. Zamboni, W.C. *et al.* (2006) *Invest New Drugs* **24**(5), 393.
238. Hahn, N.M., (2006) *Clin. Cancer Res.* **12**(20 Pt 1), 6094.
239. Schinkel, A.H., *et al.* (1995) *Eur. J. Cancer* **31A**(7–8), 1295.
240. Tsuruo, T., *et al.* (1981) *Cancer Res.* **41**(5), 1967.
241. Ojima, I., *et al.* (1998) *Bioorg. Med. Chem. Lett.* **8**(2), 189.
242. Dantzig, A.H., *et al.* (1996) *Cancer Res.* **56**(18), 4171.
243. Hyafil, F., *et al.* (1993) *Cancer Res.* **53**(19), 4595.
244. Newman, M.J., *et al.* (2000) *Cancer Res.* **60**(11), 2964.
245. Roe, M., *et al.* (1999) *Bioorg. Med. Chem. Lett.* **9**(4), 595.
246. Wang, E., *et al.* (2001) *Cancer Res.* **61**(20), 7525.
247. Sonneveld, P., *et al.* (1996) *Leukemia* **10**(11), 1741.
248. Peng, Z., *et al.* (2004) *Cancer Gene Ther.* **11**(11), 707.
249. Stege, A., *et al.* (2004) *Cancer Gene. Ther.* **11**(11), 699.

250. Dai, C.L., *et al.* (2008) *Cancer Res.* **68**(19), 7905.
251. Hardwick, L.J., Velamakanni, S., and van Veen, H.W., (2007) *Br. J. Pharmacol.* **151**(2), 163.
252. Mao, Q. and Unadkat, J.D. (2005) *AAPS J.* **7**(1), E118.
253. Nicolle, E., *et al.* (2009) *Adv. Drug Deliv. Rev.* **61**(1), 34.
254. Polgar, O., Robey, R.W., and Bates, S.E. (2008) *Expert Opin. Drug Metab. Toxicol.* **4**(1), 1.
255. Shi, Z., *et al.* (2009) *Biochem. Pharmacol.* **77**(5), 781.
256. Shukla, S., *et al.* (2009) *Drug Metab. Dispos.* **37**(2), 359.
257. Takenaka, K., *et al.* (2007) *Cancer Res.* **67**(14), 6965.
258. Kowalski, P., *et al.* (2002) *Cancer Gene. Ther.* **9**(7), 579.

Chapter 5

ABCA4 and Its Role in the Visual Cycle and Inherited Retinal Degenerative Diseases

Robert S. Molday and Laurie L. Molday

1. Introduction

ABCA4, also known as ABCR or the Rim protein, was first identified as a novel photoreceptor ATP binding cassette (ABC) transporter in 1997. Allikmets and coworkers identified the gene responsible for autosomal recessive Stargardt macular degeneration by positional cloning and DNA sequence analysis and showed that it encodes a retinal specific ABC transporter which they named ABCR.[1] At the same time, Illing *et al.* cloned the cDNA for an abundant high molecular weight membrane photoreceptor glycoprotein known as the Rim protein and found that it encodes a novel ABC transporter present in photoreceptor outer segment disc membranes.[2] Sequence comparison of ABCR and the Rim protein revealed that they are the same ABC transporter, which is now commonly known as ABCA4.[3,4]

ABCA4 is the fourth identified member of the ABCA subfamily of ABC transporters.[5–7] In the human genome, there are twelve members of this subfamily designated as ABCA1 through ABCA13. ABCA11 is no longer considered a member since genomic sequence analysis indicates that *ABCA11* is a pseudogene. The ABCA subfamily has been further divided into two subgroups based on phylogenetic analysis and intron structure.[5] One group consists of five genes (*ABCA5, ABCA6, ABCA8, ABCA9, ABCA10*), which map to chromosome 17q24.2-.3 and encode proteins of similar size and sequence.[8] The second group is composed of the seven remaining genes, *ABCA1*, *ABCA2*, *ABCA3*, *ABCA4*, *ABCA7*, *ABCA12* and *ABCA13*, located on different chromosomes.

All members of the human ABCA subfamily are full-length transporters organized in two tandem halves. Each half contains a transmembrane domain consisting of multiple membrane spanning segments followed by a nucleotide binding domain. A characteristic feature of ABCA proteins is the presence of a large extracellular domain that separates the first membrane spanning segment from a cluster of five transmembrane segments in both the N and C halves of the protein.[9] Homologues of mammalian ABCA proteins have been identified in many eukaryotic and prokaryotic organisms suggesting that *ABCA* genes have evolved from a common primordial gene.[10]

Four ABCA transporters have been linked to inherited diseases associated with defective lipid transport.[7] Mutations in *ABCA1* are known to cause Tangier disease and familial high density lipoprotein deficiency associated with defective cholesterol and phospholipid efflux from cells and a deficiency in high-density lipoprotein (HDL) formation.[11–13] Mutations in ABCA3 cause neonatal surfactant deficiency and pediatric interstitial lung disease associated with defective surfactant lipid transport and the formation of abnormal lamellar bodies in lung alveolar type II cells.[12–17] Mutations in ABCA12 are associated with harlequin and lamellar ichthyosis diseases of the skin arising from a defect in the production of long-chain ceramide esters required for the development of normal skin structure and function.[18–21] As discussed in this chapter, mutations in ABCA4 have been linked to a number of retinal degenerative diseases associated with defective clearance of retinal compounds from photoreceptor cells. Other members of the ABCA subfamily have yet to be linked to inherited diseases although ABCA2 has been implicated in sphingolipid metabolism in neural tissue and ABCA7 has been linked to the efflux of phosphatidylcholine and other lipids.[22,23]

This chapter reviews our current knowledge of the cellular localization, structural features and functional properties of ABCA4 and evidence for its proposed role in the visual cycle and inherited retinal degenerative diseases. Prospects for the development of therapeutic treatments for individuals with Stargardt macular degeneration are also discussed.

2. Overview of the Function of ABCA4 in the Photoreceptor Cells and the Mechanism Responsible for Stargardt Macular Degeneration

ABCA4 has been extensively studied at a genetic, biochemical, physiological, molecular and cell biology level.[1,2,24–31] Over 500 mutations in ABCA4 are now known to be responsible for Stargardt disease, an autosomal recessive form of macular degeneration characterized by a severe loss in central vision and the progressive atrophy of photoreceptors and underlying retinal pigment epithelial (RPE) cells in the central or macular region of the retina. Biochemical characterization of purified ABCA4 together with the analysis of *abca4* knockout mice and patients with Stargardt macular degeneration have provided insight into the role of ABCA4 in photoreceptor cells and the molecular and cellular basis for Stargardt disease. These studies implicate ABCA4 in the transport of the Schiff base conjugate of all-*trans* retinal and phosphatidylethanolamine (PE) known as N-retinylidene-phosphatidylethanolamine (N-ret-PE) across photoreceptor disc membranes following photoexcitation. The transport of N-ret-PE coupled with its reversible dissociation into all-*trans* retinal and PE enables all-*trans* retinal to be reduced to all-*trans* retinol by retinol dehydrogenase. This results in the complete clearance of all-*trans* retinal from photoreceptor outer segments through the visual cycle, a series of biochemical reactions that synthesize 11-*cis* retinal from all-*trans* retinal for the regeneration of rhodopsin.

The loss in ABCA4 transport activity leads to a buildup of N-ret-PE and all-*trans* retinal in photoreceptor outer segments.[30,32] These compounds react to form toxic diretinal compounds that accumulate as lipofuscin deposits in RPE cells of patients with Stargardt macular degeneration.[33–35] The progressive loss of RPE cells due to the toxicity of the diretinal pyridinium compound A2E and other diretinal derivatives causes photoreceptor cell death and a loss in vision.

3. Localization of ABCA4 in Outer Segments of Photoreceptor Cells

ABCA4 exhibits a restricted tissue distribution. The gene is expressed at high levels in the vertebrate retina, but not in other tissues including lung, kidney, liver, brain, lung and spleen.[1] ABCA4, however, has been reported to be expressed in rat choroid plexus, but its role in this tissue remains to be defined.[36]

Within the retina, ABCA4 is exclusively expressed in photoreceptors, the light-sensitive cells that line the back of the retina against a layer of cells known as the retinal pigment epithelium (Fig. 1A). Retinal pigment epithelial (RPE) cells are critical for the survival of photoreceptor cells. They function in the removal of "aged" tips of the photoreceptor outer segments through phagocytosis. They also play a crucial role in the visual cycle, a series of biochemical reactions that convert all-*trans* retinal to 11-*cis* retinal for the regeneration of the photopigment rhodopsin in rod photoreceptor cells. Finally, RPE cells provide nutrients and the proper extracellular environment needed to maintain healthy functional photoreceptor cells.[37]

The human retina contains over 120 million rod cells and 6.4 million cone cells. Rod cells function in dim light while cone cells respond to normal light and are responsible for color vision and high visual acuity. Cone photoreceptors in the human retina are further divided into three subtypes based on their sensitivity to light: blue or short wavelength (S) cones, green or middle wavelength (M) cones, and red or long wavelength (L) cones.

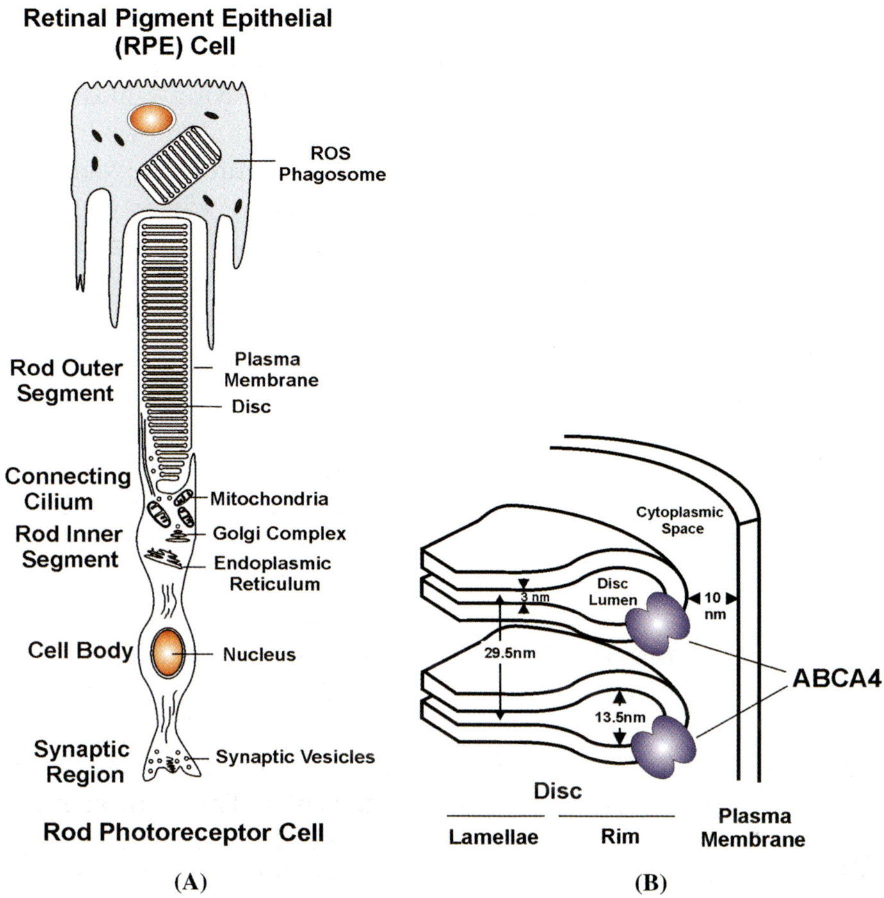

Fig. 1. (**A**) Schematic of the rod photoreceptor cell adjacent to a retinal pigment epithelial (RPE) cell. The five regions of the photoreceptor cell are shown on the left and include the Rod Outer Segment (ROS), Connecting Cilium, Rod Inner Segment, Cell Body, and Synaptic Region. Location of various subcellular organelles is shown on the right. (**B**) Schematic showing a cutout portion of the rod outer segment with the stack of discs surrounded by a plasma membrane. The location of ABCA4 along the rim region of disc membranes is shown.

ABCA4 is present in rod cells and all types of cone photoreceptors in the macular and peripheral regions of the human retina.[29]

Photoreceptors are highly differentiated cells with respect to their structure and function (Fig. 1A). They consist of five principle regions: the outer segment which functions in the detection of light and its conversion into an electrical signal in a process known as phototransduction; a connecting cilium that joins the outer segment to the inner segment and allows for the passage of specific proteins and other molecules between the inner and outer segment; the inner segment which contains the biosynthetic and metabolic machinery of cell including the mitochondria, endoplasmic reticulum (ER), golgi and lysosomes; the cell body which houses the nucleus; and the synaptic region consisting of synaptic vesicles and a ribbon synapse for transmission of electrical signals from the photoreceptors to secondary neurons of the retina.

The outer segment is a unique compartment found only in photoreceptor cells. In the rod cells, it consists of an ordered stack of over 1000 closed discs surrounded by a separate plasma membrane (Fig. 1). Each disc is composed of two flattened membranes or lamellae that are circumscribed by a hairpin loop known as the rim region (Fig. 1B). The continuous disc membrane encloses an intracellular compartment known as the intradiscal space or disc lumen which is topologically equivalent to the lumen of the ER. In cone cells, the discs and plasma membranes are continuous resulting in a folded, disc-like organization often with a tapered "cone-like" appearance. The flattened region of the disc membranes are densely packed with the photopigment protein, rhodopsin in rod cells and cone opsin, also known as iodopsin, in cone cells. The rim regions of the discs in rod and cone outer segments contains a distinct set of membrane proteins.[38]

Immunofluorescence labeling techniques together with subcellular fractionation procedures have localized ABCA4 to rod and cone outer segment disc membranes.[2,29,39] Immunoelectron microscopy has further shown that ABCA4 is confined to the rim region of rod and cone photoreceptor discs (Fig. 1B).[2,40,41] The functional significance of ABCA4 localization is not known. However, the intradiscal space, estimated to be only 2.5–3.0 nm between the two flattened regions of the disc, is not sufficient to accommodate the large exocytoplasmic domains of ABCA4. In contrast the intradiscal space within the disc rim is considerably greater with an axial length of about 13.5 nm.[42] The mechanism by which ABCA4 or other rim specific proteins are targeted to the rim region is not known. The finding that ABCA4 is localized to outer segment disc membranes together with its link to Stargardt disease provided the first clue that ABCA4 may play a central role in photoreceptor outer segment function.

4. Structural Features of ABCA4

The *ABCA4* gene, located on chromosome 1 (position 1p21-p22.1), is most closely related to *ABCA1* on chromosome 9 (position 9q31.1). Both genes consist of 50 exons and code for proteins of similar size, sequence, and structural organization. Human ABCA4 and ABCA1 contain 2,273 and 2,261 amino acids, respectively, and are 52% identical and 60% similar in amino acid sequence. ABCA4 also shares a high degree of sequence similarity to ABCA7 (49%), ABCA2 (40%), and ABCA3 (39%). ABCA4 is highly conserved among vertebrates. Human ABCA4 is 88–89% identical in sequence to bovine and mouse orthologues and 66% identical to *Xenopus laevis* ABCA4. Comparative sequence analysis of ABCA4 from various species supports an evolutionary model in which the full-length ABCA4 transporter evolved from the fusion of two distinct ABCA half-transporter progenitors.[43]

ABCA4 is a relatively abundant membrane glycoprotein in photoreceptor outer segments comprising approximately 2–4% of the total rod outer segment (ROS) membrane protein by weight.[39,41,44] On SDS polyacrylamide gels, it migrates as a single sharp band with an apparent molecular mass of approximately 250 kDa consistent with its molecular weight calculated from its amino acid sequence.[2]

Like P-glycoprotein and CFTR, ABCA4 is organized in two structurally related tandem-arranged halves in which the TMDs precede the NBDs in each half of the polypeptide chain. Computer generated algorithms, together with biochemical studies, have

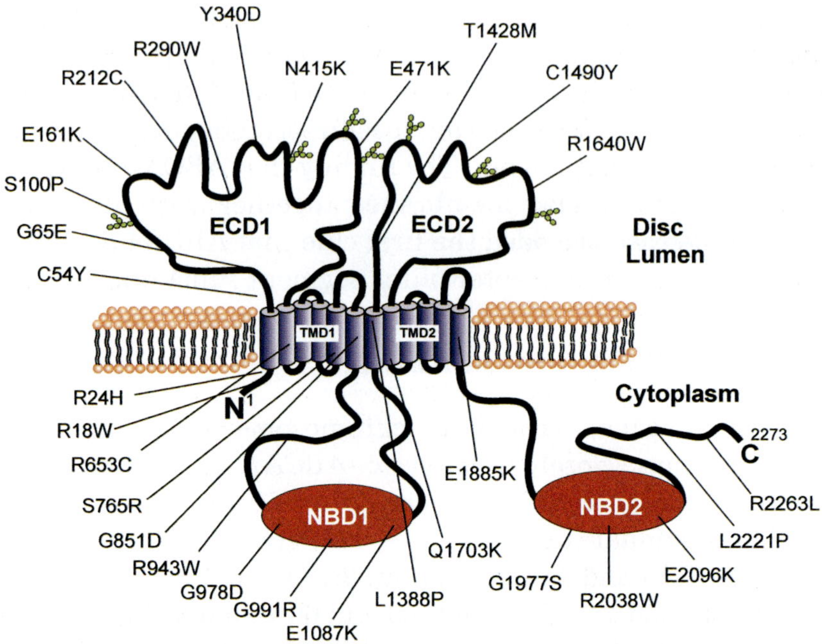

Fig. 2. Topological organization of ABCA4 showing the domain organization (ECD — exocytoplamic domain; NBD — nucleotide binding domain; TMD — transmembrane domain) and selected missense mutations responsible for Stargardt macular degeneration found throughout ABCA4. Some of the N-linked oligosaccharide chains are shown as a series of hexagons in the ECDs. As in other ABC transporters the NBDs are on the cytoplasmic side of the membrane.

been used to predict the topological organization of ABCA4 in membranes (Fig. 2).[2,9] Both the N and C halves are predicted to have a single membrane spanning segment followed by a large exocytoplasmic (extracellular/lumen) domain (ECD), multiple spanning membrane domain, and a nucleotide binding domain (NBD) on the cytoplasmic side of the disc membrane. A similar topological model has been proposed for ABCA1 and other ABCA family members.[9,45]

The small 24 amino acid N-terminal segment preceding the first transmembrane segment of ABCA4 is predicted to reside on the cytoplasmic side of the membrane based on the presence of several positively charged lysine and arginine residues and the positive inside rule.[46] The ECD1 in the N-half contains 602 amino acids. ECD2 in the C-half of ABCA4 is smaller with 289 amino acids. Both ECDs have multiple N-linked oligosaccharide chains as revealed by site-directed mutagenesis together with lectin binding and endoglycosidase studies confirming the location of the ECDs on the lumen side of the disc membrane.[9] In addition there are numerous conserved cysteine residues in the ECDs which most likely form intramolecular disulfide bonds. Trypsin cleavage studies of ABCA4 indicate that one or more disulfide bonds link ECD1 to ECD2.[9] The function of the ECDs is not known although in the case of ABCA1 these domains bind ApoA1, an acceptor of cholesterol and phospholipids.[45,47] To date no ROS proteins have been identified that bind to the ECDs of ABCA4.

NBD1 and NBD2 of ABCA4, containing approximately 200 amino acids each, share a 37% amino acid sequence identity. Both domains display the structural organization found in other ABC transporters including an A-loop, Walker A motif, Q-loop, ABC signature motif, Walker B motif, D-loop, and an H-loop.[48,49] Each NBD has been expressed and purified from *E. coli* and their ATPase activities have been measured.[50,51] NBD1 displays a lower ATPase activity than NBD2, but has broader nucleotide substrate specificity. NBD1 can hydrolyze GTP, CTP and UTP as well as ATP, while NBD2 is specific for ATP *in vitro*.

The transmembrane domains (TMDs) of ABCA4 consist of multiple membrane spanning segments upstream from the NBDs and the initial single membrane spanning segment preceding the ECDs (Fig. 2). The number and identity of the multi-spanning segments remain to be determined experimentally. However, the total number of segments including the initial membrane spanning segment in each TMD is most likely six. The two TMDs presumably form the substrate binding site although the membrane segments which comprise the binding site remain to be identified. ABCA4 contains a stretch of 140 amino acids downstream from NBD2. This C-terminal region is highly conserved between different vertebrates suggesting that it plays an important structural and/or regulatory role. A highly conserved VFVNFA motif near the C-terminus has been shown to play an essential role in the proper folding of ABCA4 into a functional protein.[52] This motif is also present in ABCA1.[53]

5. Evidence for the Role of ABCA4 as a N-Retinylidene-Phosphatidylethanolamine (N-Ret-PE) Transporter

Insight into the role of ABCA4 as a transporter for retinal derivatives has come from two complimentary approaches. In one strategy, ABCA4 was purified from photoreceptor outer segment membranes and its functional properties were investigated.[27,31,54] In another approach an *abca4* knockout mouse was generated and characterized at a molecular, cellular and physiological level.[30,55]

5.1 *Biochemical characterization of ABCA4*

5.1.1 *Binding and hydrolysis of ATP*

The binding of nucleotides to ABCA4 in ROS was first studied by photoaffinity labeling. ABCA4 in ROS membranes was specifically labeled with radiolabeled 8-azido ATP for analysis by SDS gel electrophoresis and autoradiography.[2] Labeling was inhibited by the addition of unlabeled ATP and GTP indicating that either nucleotide could effectively bind ABCA4. Subsequent studies showed that 8-azido-ATP, as well as 8-azido-ADP preferentially labeled NBD2.[56]

The ATPase activity of ABCA4 cannot be readily measured in ROS membranes due to its low activity and the relatively high activity of other ATPases in these preparations. To get around this problem, ABCA4 was purified from ROS by immunoaffinity chromatography and reconstituted into lipid vesicles. Potential transport substrates were identified by their capacity to stimulate the ATPase activity of ABCA4 in these preparations.[27,56] Out of over 40 different compounds tested, two physiologically relevant retinoid compounds, all-*trans* retinal and 11-*cis* retinal, stimulated the ATPase activity of ABCA4 3–4 fold above the basal activity. The 13-*cis* retinal isomer also activated ABCA4, but other retinoids and aliphatic aldehyde derivatives including all-*trans* retinol, retinyl esters, retinoic acid, and nonyl aldehyde showed no stimulation. A number of compounds known to be substrates for P-glycoprotein were also tested. Four of nine such compounds showed ATPase stimulation including amiodarone, digitonin, dehydroabietylamine and 2-tert-butylanthroquinone.[27] Other common drugs that act as substrates for P-glycoprotein including vincristine, verapamil, colchicine, trifluoperazine, and vinblastine were ineffective. These studies provided the first clue that retinal compounds, known to play a central role in the visual response, may serve as transport substrates for ABCA4.[27]

The effect of phospholipids on the ATPase activity of purified and reconstituted ABCA4 was also measured. ABCA4 reconstituted into vesicles consisting of soybean, brain polar, or ROS lipids exhibited both basal and retinal stimulated ATPase activity.[54] The highest activity was observed with vesicles composed of ROS lipids known to have a high content of phosphatidylethanolamine (PE). Addition of exogenous PE to brain polar lipids during reconstitution increased the basal and retinal stimulated ATPase activity of ABCA4 to levels comparable to that measured for ABCA4 reconstituted with ROS lipids. In the absence of PE, the basal ATPase activity of reconstituted ABCA4 was low and activation by all-*trans* retinal was absent. Taken together, these studies indicate that activation of the basal ATPase activity of ABCA4 requires both retinal and PE.

The effect of Walker A mutations (K969M in NBD1 and K1978M in NBD2) on the basal and retinal activated ATPase activities of ABCA4 has been examined by heterologous expression in culture cells. The single and double Walker A mutants expressed at protein levels comparable to wild-type ABCA4 and could be labeled with 8-azido ATP.[28,56] However, the

K969M/K1978M double mutant and the K969M single mutant showed no basal or retinal stimulated ATPase activity, and the K1978M single mutant exhibited basal ATPase activity but no retinal stimulated activity.[28] These studies indicate that NBD1 and NBD2 are both important, but play distinct mechanistic roles.

5.1.2 *N-ret-PE as the substrate for ABCA4*

Rhodopsin in its unbleached state consists of 11-*cis* retinal covalently bound to Lys 296 of opsin through a protonated Schiff base linkage. Light induces the isomerization of 11-*cis* retinal to all-*trans* retinal within the binding pocket of rhodopsin to produce the activated state known as metarhodopsin II. Subsequent decay of the active state of rhodopsin together with the hydrolysis of the Schiff base leads to the release of all-*trans* retinal from opsin. All-*trans* retinal is converted back to 11-*cis* retinal for the regeneration of rhodopsin through the visual cycle (Fig. 4A).

Recent X-ray crystallographic studies suggest that all-*trans* retinal is released into the disc membrane bilayer through an opening between transmembrane helices TM1 and TM7 of opsin.[57,58] Once released from opsin, all-*trans* retinal can freely diffuse between the outer and inner leaflets of the disc lipid bilayer. It has been known for a long time that all-*trans* retinal can react with PE in the disc membrane to form the Schiff base conjugate known as N-retinylidene-phosphatidylethanolamine (N-ret-PE).[59,60] Similarly, addition of all-*trans* retinal to lipid vesicles containing PE results in an equilibrium mixture of free all-*trans* retinal and PE with N-ret-PE (Fig. 3).[54] Accordingly, either all-*trans* retinal or N-ret-PE could be responsible for the stimulation of ATP hydrolysis observed when all-*trans* retinal is added to ABCA4 reconstituted into lipid vesicles containing PE.

A solid-phase binding assay was developed to determine if all-*trans* retinal or N-ret-PE serves as the retinoid substrate for ABCA4.[31] In this assay, selected retinoids in the presence or absence of PE were added to ABCA4 immobilized on an

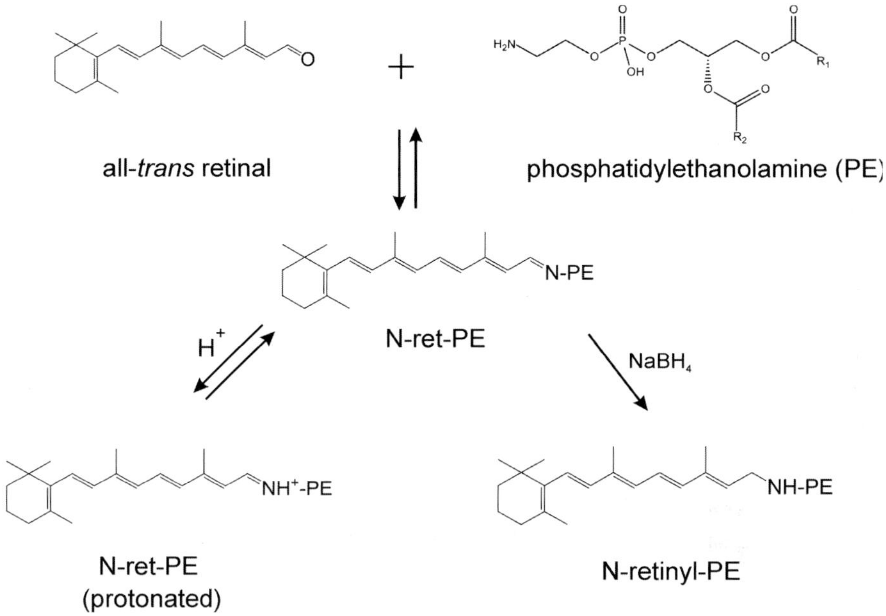

Fig. 3. Formation of N-retinylidene-phosphatidylethanolamine (N-ret-PE) in disc membranes. All-*trans* retinal, released from rhodopsin after photoexcitation, reacts with phosphatidylethanolamine (PE) in disc membranes to produce the Schiff base adduct N-ret-PE. The unprotonated form is in equilibrium with the protonated form. The Schiff base of N-ret-PE can be reduced with sodium borohydride ($NaBH_4$) to form the stabilized derivative N-retinyl-PE. Both N-ret-PE (unprotonated) and N-retinyl-PE bind with high affinity to ABCA4.

immunoaffinity matrix. The bound retinoid was subsequently identified by HPLC and spectrophotometry. When all-*trans* retinal was added to detergent solubilized ABCA4 in the presence of PE, stoichiometric amounts of N-ret-PE bound to ABCA4 with an apparent K_d of 4–5 μM. No retinoid binding was observed, when PE was substituted with PC. This indicates that N-ret-PE formed from all-*trans* retinal and PE, and not free all-*trans* retinal, is the preferred substrate for ABCA4. N-retinyl-PE, a stabilized adduct produced by sodium borohydride reduction of the N-ret-PE Schiff base adduct (Fig. 3), also binds tightly to the same site on ABCA4 as N-ret-PE. Other retinoids, which cannot form a Schiff base with PE due to the absence of an aldehyde group, do not bind to ABCA4.

The Schiff base of N-ret-PE can exist in either a protonated or unprotonated state (Fig. 3). Recent spectrophotometric measurements of ABCA4 containing a bound N-ret-PE indicate that the unprotonated form of N-ret-PE binds to ABCA4 (Zhong and Molday, unpublished results).

The binding of N-ret-PE to ABCA4 is markedly influenced by nucleotides. Addition of ATP (or GTP) causes N-ret-PE to dissociate from ABCA4.[31,52] This is consistent with a mechanism in which in the absence of ATP N-ret-PE binds to a high affinity binding site within the TMDs of ABCA4. The binding of ATP to one or both NBDs induces a dimerization of the NBDs and a protein conformational change which converts the high affinity substrate binding site to a low affinity site resulting in the release of N-ret-PE on the other side of the membrane.[61] Although these studies support N-ret-PE as the substrate for ABCA4, they do not provide information on the direction or mechanism of transport.

5.1.3 *Analysis of the Abca4 knockout mouse*

To investigate the role of ABCA4 in the physiology and pathogenesis of photoreceptors, Travis and coworkers generated and characterized a mouse deficient in the *abca4* gene.[30,55] Analysis of homozygous *abca4* knockout mice up to 11 months old indicated a normal appearance of rod photoreceptor cells with well-preserved outer segments. This indicates that ABCA4 does not play a direct role in the morphogenesis, structure, or turnover of outer segments. In addition these mice displayed a normal photoresponse except for a delay in dark adaptation suggesting that ABCA4 is not directly involved in phototransduction. Finally, the rate of rhodopsin regeneration was not affected indicating that a sufficient amount of all-*trans* retinal is converted to 11-*cis* retinal by the visual cycle in these *abca4* knockout mice.

However, significant light-dependent changes in lipid composition were found.[30] The retina of the *abca4* knockout mice exposed to cyclic or continuous lighting had elevated levels of all-*trans* retinal, protonated N-ret-PE, and PE, and decreased levels of all-*trans* retinol and all-*trans* retinyl esters relative to age-matched wild-type mice or *abca4* knockout mice reared in the dark. Furthermore, RPE cells of these mice were found to progressively accumulate lipofuscin deposits and the diretinal pyridinium compound A2E in RPE cells.[30] Lipofuscin deposits consisting of a mixture of autofluorescent pigments and modified lipids are known to accumulate in high amounts in RPE cells of patients with Stargardt disease and age-related macular degeneration.[62–65] Although initial studies found little or no evidence of photoreceptor degeneration in *abca4* knockout mice, more recent studies indicate that slow progressive photoreceptor degeneration does occur when *abca4* knockout mice are bred onto an albino (balb/c) background possibly due the enhanced generation of A2E epoxides or oxiranes.[66,67] Taken together, these studies indicate that ABCA4 plays a crucial role in the removal of retinal derivatives from photoreceptors following photoexcitation. They further indicate that *abca4* knockout mice serve as a useful animal model for Stargardt disease since they display many of the phenotypic characteristics of this disease.

6. Proposed Role of ABCA4 in the Clearance of N-ret-PE as Part of the Visual Cycle

Following photoexcitation, all-*trans* retinal has to be converted back to 11-*cis* retinal for the regeneration of rhodopsin. This occurs through a series of enzymatic reactions known as the visual or retinoid cycle[68,69] as illustrated in Fig. 4A. A significant fraction of the all-*trans* retinal that is released from rhodopsin is reduced to all-*trans* retinol in a reaction catalyzed by retinol dehydrogenase 8 (RDH8) and other short-chain dehydrogenases using NADPH as a cofactor. This reaction typically occurs on the cytoplasmic surface of the disc membrane.

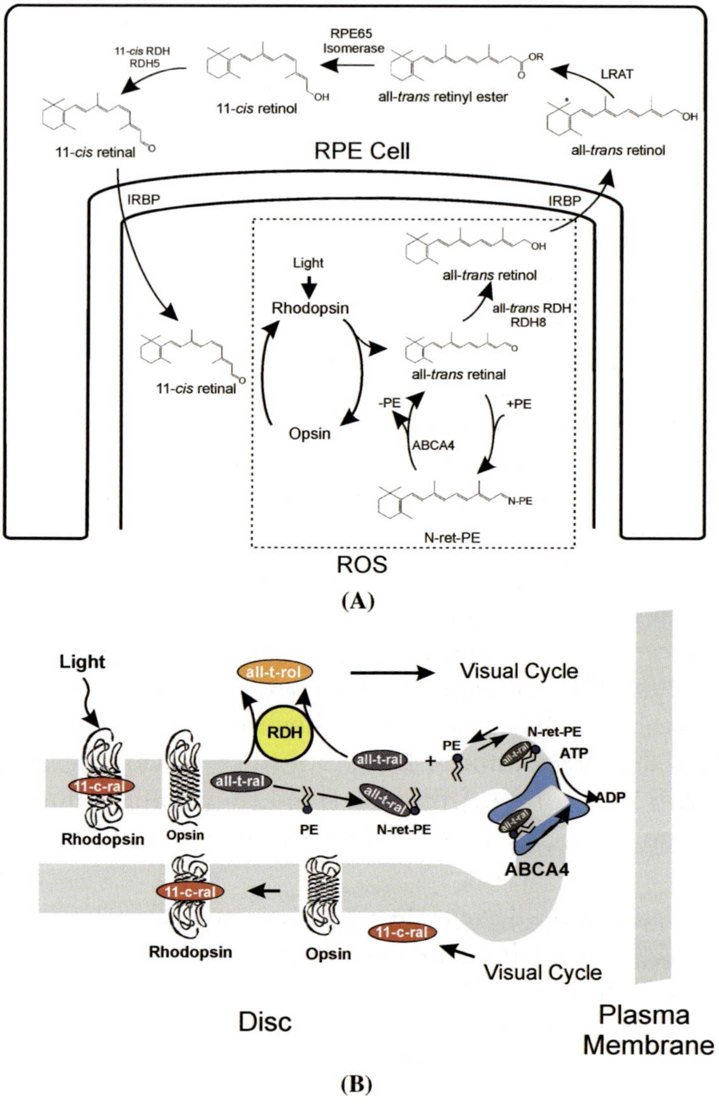

Fig. 4. The role of ABCA4 in the visual cycle. (**A**) All-*trans* retinal released from rhodopsin is reduced to all-*trans* retinol by RDH8 and subsequently shuttled to the RPE cell by IRBP where in a series of reactions it is converted back to 11-*cis* retinal for the regeneration of rhodopsin. Some all-*trans* retinal reacts with phosphatidylethanolamine (PE) to form N-retinylidene-PE (N-ret-PE). Reactions relevant to ABCA4 function are shown in the dashed box and given

All-*trans* retinol is shuttled from photoreceptors to adjacent RPE cells by the interphotoreceptor retinoid binding protein (IRBP). In the RPE cells, all-*trans* retinol is esterified to all-*trans* retinyl ester by the enzyme lecithin:retinol acetyltransferase (LRAT), isomerized to 11-*cis* retinol by the RPE65-isomerohydrolase (RPE65), and oxidized to 11-*cis* retinal by 11-*cis* retinol dehydrogenase (RDH5).[70,71] IRBP then shuttles 11-*cis* retinal back to the photoreceptor outer segment where it recombines with opsin in the disc membrane to regenerate unbleached rhodopsin. All-*trans* retinal produced as a result of the photobleaching in cone photoreceptor cells also has to be converted back to 11-*cis* retinal for regeneration of the cone pigment. Recent studies suggest that in cone dominant retina all-*trans* retinol is converted to 11-*cis* retinal through a separate pathway involving Mueller cells.[69]

Although all-*trans* retinal can be directly reduced to all-*trans* retinol following the photobleaching of rhodopsin or cone opsin, some of the all-*trans* retinal reacts reversibly with PE in disc membranes to form N-ret-PE. The latter can be trapped, possibly in its protonated form, on the lumen side of the disc membrane in rods or the extracellular side in cones (Fig. 4B). In the current model, ABCA4 is proposed to function in the transport or flipping of N-ret-PE from the lumen to the cytoplasmic side of disc membranes.[27,30,32] Once on the cytoplasmic side, N-ret-PE can dissociate to form all-*trans* retinal and PE. All-*trans* retinal can then be reduced to all-*trans* retinol by RDH8 for reentry into the visual cycle. According to this model, the function of ABCA4 is to ensure that all-*trans* retinal and N-ret-PE are efficiently

---

Fig. 4. (*Continued*) in more detail below. (**B**) Diagram depicting the proposed role of ABCA4 in the active transport of N-ret-PE from the lumen to the cytoplasmic side of the disc membrane for re-entry into the visual cycle. RDH — retinol dehydrogenase; LRAT — lecithin:retinol acyl transferase; IRBP — interphotoreceptor binding protein; 11-c-ral — 11-*cis* retinal; all-t-ral — 11-*trans* retinal; all-t-rol — all-*trans* retinol; RDH — retinol dehydrogenase.

removed from photoreceptor outer segment membranes following photoexcitation.

Numerous studies support the role of ABCA4 in the clearance of retinoids from photoreceptors as part of the visual cycle. As discussed above, this includes the light-dependent accumulation of protonated N-ret-PE, all-*trans* retinal and PE in the retina of *abca4* knockout mice, the observed delay in dark adaptation found in *abca4* knockout mice and Stargardt patients suggestive of the slow removal of all-*trans* retinal from discs following photoexcitation, biochemical studies showing binding and ATP-dependent release of N-ret-PE from purified ABCA4, substrate activation of the ATPase activity of ABCA4 and the selective localization of ABCA4 to outer segment disc membranes where all-*trans* retinal and N-ret-PE are generated following the photoactivation of rhodopsin.

7. Direction of Substrate Transport by ABCA4 — An Unresolved Issue

Although there is strong support for the function of ABCA4 as a flippase, the direction of substrate transport remains an unresolved issue. In the current model, ABCA4 is proposed to transport or flip N-ret-PE from the lumen/extracellular to the cytoplasmic side of rod and cone outer segment disc membranes. This direction of transport enables all-*trans* retinal, generated from the dissociation of N-ret-PE to all-*trans* retinal and PE, to be reduced to all-*trans* retinol by RDH8 on the cytoplasmic side of the disc membrane (Fig. 4). However, eukaryotic ABC transporters are typically outward transporters, translocating substrates from the cytoplasmic side to the extracellular or lumen side of the membrane. Accordingly, either ABCA4 differs from other ABC transporters in its direction of transport or the proposed model needs to be revised.

To date there are no assays that directly measure the direction of N-ret-PE transport across membranes by ABCA4. This remains an important but challenging area for future study.

8. Inherited Retinal Degenerative Diseases Associated with Mutations in ABCA4

Inherited retinal degenerative diseases are a major cause of blindness in the world. Typically, these diseases are caused by mutations in genes that encode proteins critical for the structure, function and survival of photoreceptors or RPE cells. To date over 140 genes have been linked to various retinal diseases (http://www.sph.uth.tmc.edu/Retnet/). These genes encode proteins that play key roles in such diverse cellular processes as phototransduction, visual cycle, outer segment membrane structure and morphogenesis, vesicle transport, chaperone function, transcription, phagocytosis, synaptic transmission, retinal development, lipid transport and synthesis, metabolism, ion transport, extracellular matrix, immune response, cell adhesion, RNA splicing, cilium structure and function and others. There has been considerable interest in retinal degenerative diseases caused by mutations in *ABCA4* since this set of disorders represents a significant cause of vision loss in the population.

8.1 *Stargardt macular degeneration and other ABCA4-associated retinal degenerative diseases*

Stargardt disease, first described by the German ophthalmologist Karl Stargardt in 1909, is the most common inherited form of early onset macular degeneration affecting as many as 1 in 10,000 individuals.[72,73] Characteristic features include significant loss in central vision in the first or second decade of life, yellow or white fluorescent lipofuscin deposits in the macula at the level of the RPE cell, progressive bilateral atrophy of the central RPE and photoreceptors, variable decrease in the amplitude of rod and cone electroretinograms (ERGs), and a delay in dark adaptation.[74–77] Samples of retina and RPE from deceased Stargardt patients reveal elevated levels of N-ret-PE in the retina and A2E and related diretinal fluorophores in RPE cells.[33] Fundus flavimaculatus exhibits similar clinical features and is now generally considered as a slow progressive form of Stargardt disease.[76,78]

Over the past decade, an international effort has been undertaken to screen Stargardt patients for mutations in the *ABCA4* gene.[1,3,24–26,79–83] A wide variety of different mutations have been found including missense, nonsense, splice-site, frameshift, and small deletion and insertion mutations. Of these, missense mutations are most common. As shown in Fig. 2, disease-causing missense mutations are distributed throughout ABCA4 including the ECDs, TMDs, NBDs, N- and C-terminus.

Mutations in ABCA4 are also known to cause autosomal recessive cone-rod dystrophy and a recessive form of retinitis pigmentosa (RP), two particularly severe retinal dystrophies.[84–86] Cone-rod dystrophy is a retinal disorder in which cone photoreceptor cells degenerate before rod cells. Affected individuals typically first experience severe loss in visual acuity and color vision with significant degeneration of the macula. Loss in central vision is followed by progressive loss in peripheral vision and nightblindness. Mutations in ABCA4 account for up to 60% of the autosomal recessive forms of cone-rod dystrophy. Many of these are truncation mutations or mutations that are likely to cause abnormal protein folding and a significant loss in function.

RP is a severe retinal degenerative disease associated with night blindness, reduction in peripheral vision and progressive loss in central vision often leading to complete blindness.[87] Over 40 different genes are known to be associated with various genetic forms of RP. Mutations in ABCA4 have been linked to a subset of autosomal recessive forms of RP known as RP19.[88] Affected individuals show characteristic features of RP together with an early loss in central vision and atrophy of the macula. Mutations in ABCA4 that cause RP are often early truncation mutations or mutations which cause the complete loss in ABCA4 activity.

In addition to these disorders, mutations in ABCA4 have been reported to increase one's risk of developing age-related macular degeneration (AMD).[79] In an initial report, individuals heterozygous for selected Stargardt mutations in ABCA4 were found in 16% of patients with AMD. Subsequent genetic screening studies, however, indicate that mutations in ABCA4 are a minor cause of AMD. The involvement of ABCA4 in AMD, however, remains controversial since not all studies have shown that Stargardt-associated mutations in ABCA4 increase one's risk of developing AMD.

Taken together, these studies suggest that mutations in ABCA4 cause a spectrum of related retinal degenerative diseases with variable clinical severity. For the recessive disorders, Stargardt disease has the mildest phenotype, cone-rod dystrophy a moderate to severe phenotype, and RP the most severe phenotype. A model has been proposed in which the severity of the disease phenotype is inversely correlated with the residual functional activity of the mutant ABCA4.[84,89] Individuals harboring mutations in both alleles which result in the absence of a functional transporter display a RP phenotype. Individuals with mutations resulting in only residual ABCA4 function develop cone-rod dystrophy. Finally, individuals with missense mutations which retain some functional activity exhibit the clinical features of Stargardt disease. Although there is evidence that the severity of the disease correlates with the extent of functional activity of ABCA4 as suggested by this model, other factors including the effect of mutations on the trafficking of ABCA4 to photoreceptor outer segment discs, age of diagnosis, progression of the disease, and the general genetic variability of the population may also contribute to the observed phenotype.[83]

8.2 *Biochemical analysis of disease linked mutations in ABCA4*

The effect of disease-causing mutations on the structural and functional properties of ABCA4 has been studied by heterologous expression in HEK293 cells.[28] Deletion of internal amino acids and introduction of charge residues or prolines in the putative transmembrane segments result in significantly reduced level of protein expression. This suggests that such mutations cause severe protein misfolding and rapid degradation by the cell. Many missense mutants express at protein levels similar to wild-type ABCA4, but display reduced or complete loss of ATP binding and hydrolysis. Some mutations cause a reduction in both basal and retinal activated ATPase activities whereas others retain basal ATPase activity, but have reduced or no retinal stimulated activity.[28]

In a more recent study the effect of two disease-associated C-terminal deletion mutants on the biochemical properties of ABCA4 has been examined in considerable detail.[52] The ABCA4Δ30 missing the C-terminal 30 amino acids including the conserved VFVNFA motif is associated with cone-rod dystrophy, whereas the ABCA4Δ24 lacking the C-terminal 24 amino acids, but retaining this motif is linked to a mild form of Stargardt disease.[86,90] The ABCA4Δ30 mutant showed a significant decrease in protein level, complete loss in both N-ret-PE binding and retinal-stimulated ATPase activity, and significant protein misfolding as revealed by its retention in the endoplasmic reticulum and inability to be effectively solubilized by a mild detergent.[52] In contrast, the ABCA4Δ24 mutant displayed normal N-ret-PE binding, only a small reduction in retinal stimulated ATPase activity and intracellular vesicular localization characteristic of wild-type ABCA4 implying proper protein folding. These results highlight the importance of the conserved VFVNFA in correct protein folding and functional activity of ABCA4 and are in general agreement with the model in which the severity of the disease phenotype is inversely proportional to the residual functional activity of the mutant.

The effect of several disease-associated mutations on the trafficking of ABCA4 to outer segments has been studied in transgenic *Xenopus laevis*.[91] Several mutants were retained in the inner segment of the photoreceptors suggesting that protein mislocalization as well as misfolding and diminished function can contribute to the mechanisms underlying ABCA4 associated diseases.

9. Accumulation of Toxic Retinal Compounds is Responsible for Stargardt Disease

Loss in the function of ABCA4 as a retinoid transporter can explain many of the characteristic features found in individuals with Stargardt macular degeneration and *abca4* knockout mice. As discussed above, all-*trans* retinal produced from the photobleaching of rhodopsin and cone opsin reacts with PE to produce an equilibrium mixture of all-*trans* retinal and N-ret-PE in disc membranes. The inability of mutant ABCA4 to efficiently transport N-ret-PE from the lumen to the cytoplasmic side of the disc membrane results in the buildup of all-*trans* retinal and N-ret-PE in disc membranes (Fig. 5A). This leads to the production of diretinal compounds such as A2PE as observed in *abca4* knockout mice and post-mortem samples from Stargardt patients.[33,35,65,92] Upon phagocytosis of ROS, A2PE is hydrolyzed to A2E and phosphatidic acid in the lysosomal compartment of RPE cells. A2E accumulates in RPE cells as fluorescent lipofuscin deposits (Fig. 5B). In addition to A2E, other diretinal compounds have been identified as components of lipofuscin deposits in *abca*4 knockout mice as well as individuals with Stargardt disease.[93] These compounds are also derived from side reactions involving all-*trans* retinal and N-ret-PE.

Normally, the aged disc membranes are efficiently digested by the lysosomal system of the RPE cells. However, RPE cells cannot metabolize diretinal compounds such as A2E and as a result these compounds progressively accumulate in RPE cells in the form of fluorescent lipofuscin deposits seen in Stargardt patients and *abca4* knockout mice. A2E has been shown to have a negative effect on RPE function and survival.[34] It can act as a detergent compromising the membrane integrity of the subcellular organelles and the function of RPE cells.[65,94] In the presence of short wavelength light and oxygen, A2E and other diretinoid compounds can be converted to their corresponding epoxides which are capable of killing RPE cells.[66,95] A2E has also been reported to have a detrimental effect on phagocytosis and

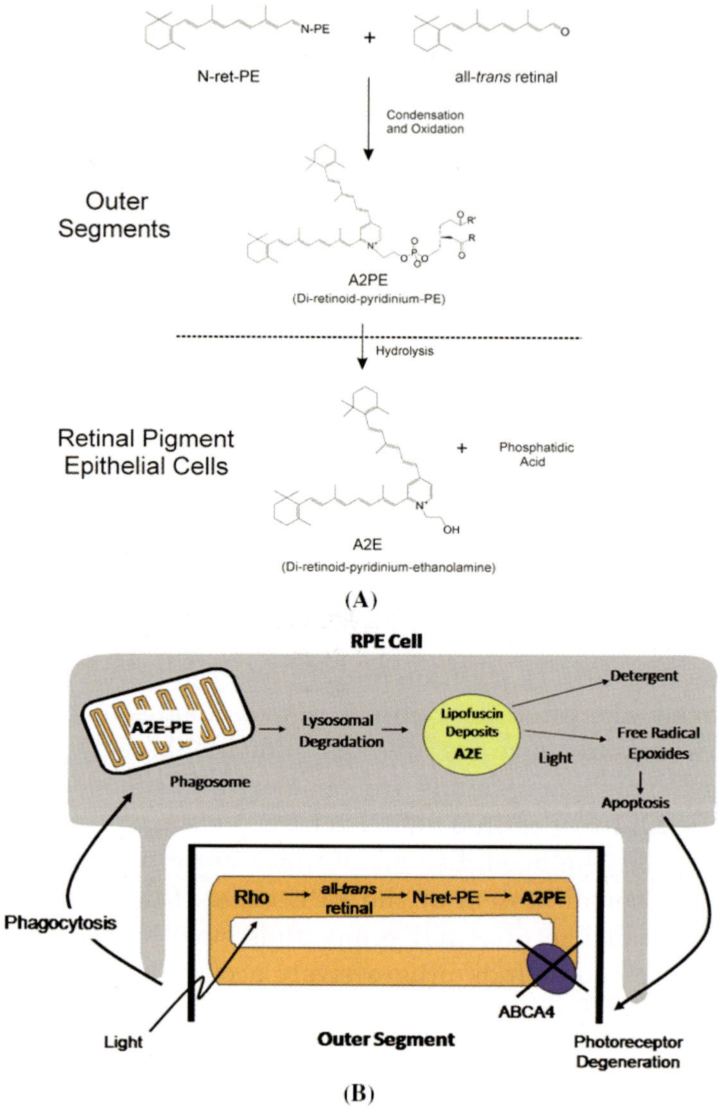

Fig. 5. Mechanism responsible for Stargardt and related retinal degenerative diseases. (**A**) Reactions which lead to the production of A2E, a potential toxic compound. (**B**) Diagram showing how the loss in the function of ABCA4 as an N-ret-PE transporter results in the accumulation of A2PE in disc membranes. After phagocytosis of outer segments by RPE cells, A2PE is hydrolyzed to A2E. This diretinal pyridinium compound together with other fluorophores

mitochondrial function in RPE cells.[96,97] Since RPE cells are crucial for photoreceptor cell survival, the loss in RPE cells leads to photoreceptor degeneration and a loss in vision. Recently, photooxidation products of A2E have been reported to activate the complement system suggesting that A2E and related compounds may also serve as a trigger for age-related macular degeneration.[98] Although A2E appears to be toxic to RPE cells, it is possible that other retinoid compounds that accumulate in RPE cells may also induce RPE cell death. Thus, the delayed removal of N-ret-PE and all-*trans* retinal from outer segments following photoexcitation due to defective ABCA4 can result in the accumulation of a series of toxic compounds which ultimately lead to both RPE and photoreceptor degeneration and a loss in vision.

10. Gene and Drug Based Therapeutic Treatments

Insights into the genetic and molecular basis for Stargardt and related ABCA4-associated retinal degenerative diseases have led to the development of potential treatments for this set of disorders. Gene therapy offers a particularly promising approach to prevent or slow vision loss.[99] Delivery of the normal gene to cells harboring the defective gene can be used to treat recessive diseases such as Stargardt macular degeneration. This strategy has proven successful in animal models for several recessive retinal degenerative diseases caused by mutations in genes expressed in RPE and photoreceptor cells.[100–102] Recombinant adeno-associated viral (rAAV) vectors have been most widely used for the gene delivery to photoreceptor and RPE cells.[99] rAAV vectors are nontoxic and relatively nonimmunogenic. In addition, they can

---

Fig. 5. (*Continued*) progressively accumulates in RPE cells as lipofuscin deposits. A2E and other retinoid derivatives cause RPE apoptosis possibly through their action as detergents or through the light-induced generation of epoxides. Loss of RPE cells leads to photoreceptor degeneration and a loss in vision.

transfect non-dividing cells such as RPE and photoreceptors for lifelong gene and protein expression. For example, a single injection of rAAV containing the *RPE65* gene delivered to RPE cells has restored vision in a canine model for Leber Congenital Amaurosis (LCA2), a severe early onset retinal degenerative disease associated with complete loss in vision for the life of the dog.[101] Success in preclinical animal studies for LCA2 has led to recent Phase I gene therapy clinical trials for this disease. Partial recovery of vision has been demonstrated for most patients enrolled in these initial trials.[103–105]

A potential disadvantage of rAAV for use in Stargardt disease is the limited gene packaging size of 4.7 kb for rAAV. However, a recent study has shown that large genes including the 6.8 kb *ABCA4* gene can be packed into a chimera rAAV (2/5) vector.[106] In one study this vector has been used to deliver the mouse *abca4* gene to photoreceptors of an *abca4* knockout mouse. In another study the human *ABCA4* gene has been delivered to photoreceptors of an *abca4* knockout mouse using lentiviral vector which can accommodate large genes.[107] In both these studies, expression of ABCA4 in photoreceptors correlated with a reduction in A2E levels in RPE cells. These initial studies suggest that gene therapy has potential application as a treatment for Stargardt and other ABCA4 associated diseases.

Another approach has been to reduce the level of 11-*cis* retinal and consequently all-*trans* retinal in photoreceptors through the application of drugs that inhibit enzymes of the visual cycle.[69] In initial studies, isotretinoin or 13-cis-retinoic acid, also known as Accutane, a widely used agent for the treatment of acne, was used to inhibit 11-*cis*-RDH, the enzyme responsible for the formation of 11-*cis* retinal in RPE cell.[108] When *abca4* knockout mice were given isotretinoin, a marked reduction in the light-induced accumulation of A2E containing lipofuscin deposits was observed. Since isotretinoin is associated with side effects, other retinoid-based inhibitors have been developed and tested in animal studies. These include N-(4-hydroxyphenyl)retinamide known to reduce serum levels of vitamin A and retinylamine, an inhibitor of isomerase activity in RPE cells.[109,110] Both these retinoid compounds have reduced A2E accumulation in animal studies pointing to the therapeutic potential of these drugs.

11. Concluding Remarks

Over the past ten years significant progress has been made in characterizing ABCA4 and defining its role in the visual cycle and various retinal degenerative diseases including Stargardt macular degeneration. Like other ABCA family members, ABCA4 is involved in the transport of lipids across biological membranes. The current model implicates ABCA4 in the transport of N-ret-PE across photoreceptor outer segment disc membranes as a mechanism to remove potentially toxic retinal compounds from outer segments. Failure to efficiently remove N-ret-PE and all-*trans* retinal as a result of the loss or decrease in ABCA4 transport activity results in the production of diretinoid compounds that progressively accumulate in RPE cells upon phagocytosis of outer segments. These diretinal compounds induce RPE apoptosis leading to photoreceptor cell death and a loss in vision as found for individuals with Stargardt disease.

Although the biochemical properties of ABCA4 have been investigated in considerable detail, it will be important to obtain a high resolution structure of ABCA4 in order to gain insight into nature of the substrate binding site and the mechanism of energy dependent transport. It will also be critical to develop biochemical assays to define the direction of ABCA4 mediated substrate transport across disc membranes. Another important unresolved issue is the regulation of ABCA4. Is the activity of ABCA4 as a transporter regulated through posttranslational modification or protein-protein interactions? Finally, another active area of research is the development of therapeutic approaches to slow or eliminate RPE and photoreceptor degeneration in individuals with Stargardt disease and related retinal degenerative diseases linked to mutations in ABCA4.

Acknowledgments

This work is supported by grants from the Canadian Institutes for Health Research (MT 5822 and MOP-10667), the National Institutes of Health (EY02422) and the Macula Vision Research Foundation.

References

1. Allikmets, R., Singh, N., Sun, H., Shroyer, N.F., Hutchinson, A., Chidambaram, A., Gerrard, B., Baird, L., Stauffer, D., Peiffer, A., Rattner, A., Smallwood, P., Li, Y., Anderson, K.L., Lewis, R.A., Nathans, J., Leppert, M., Dean, M., and Lupski, J.R. (1997) *Nat. Genet.* **15**, 236.
2. Illing, M., Molday, L.L., and Molday, R.S. (1997) *J. Biol. Chem.* **272**, 10303.
3. Nasonkin, I., Illing, M., Koehler, M.R., Schmid, M., Molday, R.S., and Weber, B.H. (1998) *Hum. Genet.* **102**, 21.
4. Azarian, S.M. and Travis, G.H. (1997) *FEBS Lett.* **409**, 247.
5. Dean, M. and Allikmets, R. (2001) *J. Bioenerg. Biomembr.* **33**, 475.
6. Dean, M. and Annilo, T. (2005) *Annu. Rev. Genomics. Hum. Genet.* **6**, 123.
7. Kaminski, W.E., Piehler, A., and Wenzel, J.J. (2006) *Biochim. Biophys. Acta* **1762**, 510.
8. Schriml, L.M. and Dean, M. (2000) *Genomics* **64**, 24.
9. Bungert, S., Molday, L.L., and Molday, R.S. (2001) *J. Biol. Chem.* **276**, 23539.
10. Peelman, F., Labeur, C., Vanloo, B., Roosbeek, S., Devaud, C., Duverger, N., Denefle, P., Rosier, M., Vandekerckhove, J., and Rosseneu, M. (2003) *J. Mol. Biol.* **325**, 259.
11. Brooks-Wilson, A., Marcil, M., Clee, S.M., Zhang, L.H., Roomp, K., van Dam, M., Yu, L., Brewer, C., Collins, J.A., Molhuizen, H.O., Loubser, O., Ouelette, B.F., Fichter, K., Ashbourne-Excoffon, K.J., Sensen, C.W., Scherer, S., Mott, S., Denis, M., Martindale, D., Frohlich, J., Morgan, K., Koop, B., Pimstone, S.,Kastelein, J.J., Hayden, M.R., *et al.* (1999) *Nat. Genet.* **22**, 336.
12. Bodzioch, M., Orso, E., Klucken, J., Langmann, T., Bottcher, A., Diederich, W., Drobnik, W., Barlage, S., Buchler, C.,

Porsch-Ozcurumez, M., Kaminski, W.E., Hahmann, H.W., Oette, K., Rothe, G., Aslanidis, C., Lackner, K.J., and Schmitz, G. (1999) *Nat. Genet.* **22**, 34.
13. Rust, S., Rosier, M., Funke, H., Real, J., Amoura, Z., Piette, J.C., Deleuze, J.F., Brewer, H.B., Duverger, N., Denefle, P., and Assmann, G. (1999) *Nat. Genet.* **22**, 352.
14. Ban, N., Matsumura, Y., Sakai, H., Takanezawa, Y., Sasaki, M., Arai, H., and Inagaki, N. (2007) *J. Biol. Chem.* **282**, 9628.
15. Shulenin, S., Nogee, L.M., Annilo, T., Wert, S.E., Whitsett, J.A., and Dean, M. (2004) *N. Engl. J. Med.* **350**, 1296.
16. Yamano, G., Funahashi, H., Kawanami, O., Zhao, L.X., Ban, N., Uchida, Y., Morohoshi, T., Ogawa, J., Shioda, S., and Inagaki, N. (2001) *FEBS Lett.* **508**, 221.
17. Fitzgerald, M.L., Xavier, R., Haley, K.J., Welti, R., Goss, J.L., Brown, C.E., Zhuang, D.Z., Bell, S.A., Lu, N., McKee, M., Seed, B., and Freeman, M.W. (2007) *J. Lipid Res.* **48**, 621.
18. Lefevre, C., Audebert, S., Jobard, F., Bouadjar, B., Lakhdar, H., Boughdene-Stambouli, O., Blanchet-Bardon, C., Heilig, R., Foglio, M., Weissenbach, J., Lathrop, M., Prud'homme, J.F., and Fischer, J. (2003) *Hum. Mol. Genet.* **12**, 2369.
19. Akiyama, M., Sugiyama-Nakagiri, Y., Sakai, K., McMillan, J.R., Goto, M., Arita, K., Tsuji-Abe, Y., Tabata, N., Matsuoka, K., Sasaki, R., Sawamura, D., and Shimizu, H. (2005) *J. Clin. Invest.* **115**, 1777.
20. Annilo, T., Shulenin, S., Chen, Z.Q., Arnould, I., Prades, C., Lemoine, C., Maintoux-Larois, C., Devaud, C., Dean, M., Denefle, P., and Rosier, M. (2002) *Cytogenet. Genome Res.* **98**, 169.
21. Zuo, Y., Zhuang, D.Z., Han, R., Isaac, G., Tobin, J.J., McKee, M., Welti, R., Brissette, J.L., Fitzgerald, M.L., and Freeman, M.W. (2008) *J. Biol. Chem.* **283**, 36624.
22. Wang, N., Lan, D., Gerbod-Giannone, M., Linsel-Nitschke, P., Jehle, A.W., Chen, W., Martinez, L.O., and Tall, A.R. (2003) *J. Biol. Chem.* **278**, 42906.
23. Abe-Dohmae, S., Ikeda, Y., Matsuo, M., Hayashi, M., Okuhira, K., Ueda, K., and Yokoyama, S. (2004) *J. Biol. Chem.* **279**, 604.
24. Allikmets, R. (2000) *Am. J. Hum. Genet.* **67**, 793.
25. Rozet, J.M., Gerber, S., Souied, E., Perrault, I., Chatelin, S., Ghazi, I., Leowski, C., Dufier, J.L., Munnich, A., and Kaplan, J. (1998) *Eur. J. Hum. Genet.* **6**, 291.

26. Fishman, G.A., Stone, E.M., Grover, S., Derlacki, D.J., Haines, H.L., and Hockey, R.R. (1999) *Arch. Ophthalmol.* **117**, 504.
27. Sun, H., Molday, R.S., and Nathans, J. (1999) *J. Biol. Chem.* **274**, 8269.
28. Sun, H., Smallwood, P.M., and Nathans, J. (2000) *Nat. Genet.* **26**, 242.
29. Molday, L.L., Rabin, A.R., and Molday, R.S. (2000) *Nat. Genet.* **25**, 257.
30. Weng, J., Mata, N.L., Azarian, S.M., Tzekov, R.T., Birch, D.G., and Travis, G.H. (1999) *Cell* **98**, 13.
31. Beharry, S., Zhong, M., and Molday, R.S. (2004) *J. Biol. Chem.* **279**, 53972.
32. Molday, R.S. (2007) *J. Bioenerg. Biomembr.* **39**, 507.
33. Mata, N.L., Weng, J., and Travis, G.H. (2000) *Proc. Natl. Acad. Sci. U.S.A.* **97**, 7154.
34. Sparrow, J.R. and Boulton, M. (2005) *Exp. Eye Res.* **80**, 595.
35. Ben-Shabat, S., Parish, C.A., Vollmer, H.R., Itagaki, Y., Fishkin N., Nakanishi, K., and Sparrow, J.R. (2002) *J. Biol. Chem.* **277**, 7183.
36. Bhongsatiern, J., Ohtsuki, S., Tachikawa, M., Hori, S., and Terasaki, T. (2005) *J. Neurochem.* **92**, 1277.
37. Bok, D. (1985) *Invest. Ophthalmol. Vis. Sci.* **26**, 1659.
38. Molday, R.S. and Molday, L.L. (1987) *J. Cell Biol.* **105**, 2589.
39. Sun, H. and Nathans, J. (1997) *Nat. Genet.* **17**, 15.
40. Papermaster, D.S., Reilly, P., and Schneider, B.G. (1982) *Vision Res.* **22**, 1417.
41. Papermaster, D.S., Schneider, B.G., Zorn, M.A., and Kraehenbuhl, J.P. (1978) *J. Cell Biol.* **78**, 415.
42. Corless, J.M. and Fetter, R.D. (1987) *J. Comp. Neurol.* **257**, 24.
43. Yatsenko, A.N., Wiszniewski, W., Zaremba, C.M., Jamrich, M., and Lupski, J.R. (2005) *J. Mol. Evol.* **60**, 72.
44. Kwok, M.C., Holopainen, J.M., Molday, L.L., Foster, L.J., and Molday, R.S. (2008) *Mol. Cell Proteomics* **7**, 1053.
45. Fitzgerald, M.L., Morris, A.L., Rhee, J.S., Andersson, L.P., Mendez, A.J., and Freeman, M.W. (2002) *J. Biol. Chem.* **277**, 33178.
46. von Heijne, G. and Gavel, Y. (1988) *Eur. J. Biochem.* **174**, 671.
47. Wang, N., Silver, D.L., Costet, P., and Tall, A.R. (2000) *J. Biol. Chem.* **275**, 33053.
48. Vetter, I.R. and Wittinghofer, A. (1999) *Q. Rev. Biophys.* **32**, 1.

49. Linton, K.J. and Higgins, C.F. (2007) *Pflugers Arch.* **453**, 555.
50. Biswas, E.E. (2001) *Biochemistry* **40**, 8181.
51. Biswas, E.E. and Biswas, S.B. (2000) *Biochemistry* **39**, 15879.
52. Zhong, M., Molday, L.L., and Molday, R.S. (2009). *J. Biol. Chem.* **284**, 3640.
53. Fitzgerald, M.L., Okuhira, K., Short, G.F. 3rd, Manning, J.J., Bell, S.A., and Freeman, M.W. (2004) *J. Biol. Chem.* **279**, 48477.
54. Ahn, J., Wong, J.T., and Molday, R.S. (2000) *J. Biol. Chem.* **275**, 20399.
55. Mata, N.L., Tzekov, R.T., Liu, X., Weng, J., Birch, D.G., and Travis, G.H. (2001) *Invest. Ophthalmol. Vis. Sci.* **42**, 1685.
56. Ahn, J., Beharry, S., Molday, L.L., and Molday, R.S. (2003) *J. Biol. Chem.* **278**, 39600.
57. Hildebrand, P.W., Scheerer, P., Park, J.H., Choe, H.W., Piechnick, R., Ernst, O.P., Hofmann, K.P., and Heck, M. (2009) *PLoS ONE* **4**, e4382.
58. Park, J.H., Scheerer, P., Hofmann, K.P., Choe, H.W., and Ernst, O.P. (2008) *Nature* **454**, 183.
59. Poincelot, R.P., Millar, P.G., Kimbel, R.L. Jr., and Abrahamson, E.W. (1969) *Nature* **221**, 256.
60. Anderson, R.E. and Maude, M.B. (1970) *Biochemistry* **9**, 3624.
61. Molday, R.S., Zhong, M., and Quazi, F. (2009) *Biochim. Biophys. Acta.* **1791**(7), 573.
62. Feeney-Burns, L., Hilderbrand, E.S., and Eldridge, S. (1984) *Invest. Ophthalmol. Vis. Sci.* **25**, 195.
63. Dorey, C.K., Wu, G., Ebenstein, D., Garsd, A., and Weiter, J.J. (1989) *Invest. Ophthalmol. Vis. Sci.* **30**, 1691.
64. Kennedy, C.J., Rakoczy, P.E., and Constable, I.J. (1995) *Eye (Lond)* **9**(6), 763.
65. Eldred, G.E. and Lasky, M.R. (1993) *Nature* **361**, 724.
66. Radu, R.A., Mata, N.L., Bagla, A., and Travis, G.H. (2004) *Proc. Natl. Acad. Sci. U.S.A.* **101**, 5928.
67. Radu, R.A., Yuan, Q., Hu, J., Peng, J.H., Lloyd, M., Nusinowitz, S., Bok, D., and Travis, G.H. (2008) *Invest. Ophthalmol. Vis. Sci.* **49**, 3821.
68. Lamb, T.D. and Pugh, E.N. Jr. (2004) *Prog. Retin. Eye. Res.* **23**, 307.
69. Travis, G.H., Golczak, M., Moise, A.R., and Palczewski, K. (2007) *Annu. Rev. Pharmacol. Toxicol.* **47**, 469.

70. McBee, J.K., Palczewski, K., Baehr, W., and Pepperberg, D.R. (2001) *Prog. Retin. Eye. Res.* **20**, 469.
71. Saari, J.C. (2000) *Invest. Ophthalmol. Vis. Sci.* **41**, 337.
72. Stargardt, K. (1909) *Albrecht von Graefes Arch. Ophthalmol.* **71**, 534.
73. Blacharski, P.A. (1988) Fundu flavimaculatus, in: Newsome, D.A. (ed.) *Retinal Dystrophies and Degenerations.* Raven Press, New York, pp. 135.
74. Weleber, R.G. (1994) *Arch. Ophthalmol.* **112**, 752.
75. Fishman, G.A., Farbman, J.S., and Alexander, K.R. (1991) *Ophthalmology* **98**, 957.
76. Noble, K.G. and Carr, R.E. (1979) *Arch. Ophthalmol.* **97**, 1281.
77. Lois, N., Halfyard, A.S., Bird, A.C., Holder, G.E., and Fitzke, F.W. (2004) *Am. J. Ophthalmol.* **138**, 55.
78. Gelisken, O. and De Laey, J.J. (1985) *Int. Ophthalmol.* **8**, 225.
79. Allikmets, R., Shroyer, N.F., Singh, N., Seddon, J.M., Lewis, R.A., Bernstein, P.S., Peiffer, A., Zabriskie, N.A., Li, Y., Hutchinson, A., Dean, M., Lupski, J.R., and Leppert, M. (1997) *Science* **277**, 1805.
80. Rivera, A., White, K., Stohr, H., Steiner, K., Hemmrich, N., Grimm, T., Jurklies, B., Lorenz, B., Scholl, H.P., Apfelstedt-Sylla, E., and Weber, B.H. (2000) *Am. J. Hum. Genet.* **67**, 800.
81. Webster, A.R., Heon, E., Lotery, A.J., Vandenburgh, K., Casavant, T.L., Oh, K.T., Beck, G., Fishman, G.A., Lam, B.L., Levin, A., Heckenlively, J.R., Jacobson, S.G., Weleber, R.G., Sheffield, V.C., and Stone, E.M. (2001) *Invest. Ophthalmol. Vis. Sci.* **42**, 1179.
82. Maugeri, A., van Driel, M.A., van de Pol, D.J., Klevering, B.J., van Haren, F.J., Tijmes, N., Bergen, A.A., Rohrschneider, K., Blankenagel, A., Pinckers, A.J., Dahl, N., Brunner, H.G., Deutman, A.F., Hoyng, C.B., and Cremers, F.P. (1999) *Am. J. Hum. Genet.* **64**, 1024.
83. Cideciyan, A.V., Swider, M., Aleman, T.S., Tsybovsky, Y., Schwartz, S.B., Windsor, E.A., Roman, A.J., Sumaroka, A., Steinberg, J.D., Jacobson, S.G., Stone, E.M., and Palczewski, K. (2008) *Hum. Mol. Genet.*
84. Maugeri, A., Klevering, B.J., Rohrschneider, K., Blankenagel, A., Brunner, H.G., Deutman, A.F., Hoyng, C.B., and Cremers, F.P. (2000) *Am. J. Hum. Genet.* **67**, 960.

85. Martinez-Mir, A., Paloma, E., Allikmets, R., Ayuso, C., del Rio, T., Dean, M., Vilageliu, L., Gonzalez-Duarte, R., and Balcells, S. (1998) *Nat. Genet.* **18**, 11.
86. Stenirri, S., Battistella, S., Fermo, I., Manitto, M.P., Martina, E., Brancato, R., Ferrari, M., and Cremonesi, L. (2006) *Clin. Chem. Lab. Med.* **44**, 533.
87. Berson, E.L. (1993) *Invest. Ophthalmol. Vis. Sci.* **34**, 1659.
88. Cremers, F.P., van de Pol, D.J., van Driel, M., den Hollander, A.I., van Haren, F.J., Knoers, N.V., Tijmes, N., Bergen, A.A., Rohrschneider, K., Blankenagel, A., Pinckers, A.J., Deutman, A.F., and Hoyng, C.B. (1998) *Hum. Mol. Genet.* **7**, 355.
89. Shroyer, N.F., Lewis, R.A., Allikmets, R., Singh, N., Dean, M., Leppert, M., and Lupski, J.R. (1999) *Vision Res.* **39**, 2537.
90. Fumagalli, A., Ferrari, M., Soriani, N., Gessi, A., Foglieni, B., Martina, E., Manitto, M.P., Brancato, R., Dean, M., Allikmets, R., and Cremonesi, L. (2001) *Hum. Genet.* **109**, 326.
91. Wiszniewski, W., Zaremba, C.M., Yatsenko, A.N., Jamrich, M., Wensel, T.G., Lewis, R.A., and Lupski, J.R. (2005) *Hum. Mol. Genet.* **14**, 2769.
92. Sakai, N., Decatur, J., Nakanishi, K., and Eldred, G.E. (1996) *J. Am. Chem. Soc.* **118**, 1559.
93. Kim, S.R., Jang, Y.P., Jockusch, S., Fishkin, N.E., Turro, N.J., and Sparrow, J.R. (2007) *Proc. Natl. Acad. Sci. U.S.A.* **104**, 19273.
94. Holz, F.G., Schutt, F., Kopitz, J., Eldred, G.E., Kruse, F.E., Volcker, H.E., and Cantz, M. (1999) *Invest. Ophthalmol. Vis. Sci.* **40**, 737.
95. Sparrow, J.R., Nakanishi, K., and Parish, C.A. (2000) *Invest. Ophthalmol. Vis. Sci.* **41**, 1981.
96. Vives-Bauza, C., Anand, M., Shirazi, A.K., Magrane, J., Gao, J., Vollmer-Snarr, H.R., Manfredi, G., and Finnemann, S.C. (2008) *J. Biol. Chem.* **283**, 24770.
97. Finnemann, S.C., Leung, L.W., and Rodriguez-Boulan, E. (2002) *Proc. Natl. Acad. Sci. U.S.A.* **99**, 3842.
98. Zhou, J., Kim, S.R., Westlund, B.S., and Sparrow, J.R. (2009) *Invest. Ophthalmol. Vis. Sci.* **50**, 1392.
99. Dinculescu, A., Glushakova, L., Min, S.H., and Hauswirth, W.W. (2005) *Hum. Gene. Ther.* **16**, 649.

100. Pang, J.J., Chang, B., Kumar, A., Nusinowitz, S., Noorwez, S.M., Li, J., Rani, A., Foster, T.C., Chiodo, V.A., Doyle, T., Li, H., Malhotra, R., Teusner, J.T., McDowell, J.H., Min, S.H., Li, Q., Kaushal, S., and Hauswirth, W.W. (2006) *Mol. Ther.* **13**, 565.
101. Acland, G.M., Aguirre, G.D., Ray, J., Zhang, Q., Aleman, T.S., Cideciyan, A.V., Pearce-Kelling, S.E., Anand, V., Zeng, Y., Maguire, A.M., Jacobson, S.G., Hauswirth, W.W., and Bennett, J. (2001) *Nat. Genet.* **28**, 92.
102. Min, S.H., Molday, L.L., Seeliger, M.W., Dinculescu, A., Timmers, A.M., Janssen, A., Tonagel, F., Tanimoto, N., Weber, B.H., Molday, R.S., and Hauswirth, W.W. (2005) *Mol. Ther.* **12**, 644.
103. Cideciyan, A.V., Aleman, T.S., Boye, S.L., Schwartz, S.B., Kaushal, S., Roman, A.J., Pang, J.J., Sumaroka, A., Windsor, E.A., Wilson, J.M., Flotte, T.R., Fishman, G.A., Heon, E., Stone, E.M., Byrne, B.J., Jacobson, S.G., and Hauswirth, W.W. (2008) *Proc. Natl. Acad. Sci. U.S.A.* **105**, 15112.
104. Maguire, A.M., Simonelli, F., Pierce, E.A., Pugh, E.N., Jr., Mingozzi, F., Bennicelli, J., Banfi, S., Marshall, K.A., Testa, F., Surace, E.M., Rossi, S., Lyubarsky, A., Arruda, V.R., Konkle, B., Stone, E., Sun, J., Jacobs, J., Dell'Osso, L., Hertle, R., Ma, J.X., Redmond, T.M., Zhu, X., Hauck, B., Zelenaia, O., Shindler, K.S., Maguire, M.G., Wright, J.F., Volpe, N.J., McDonnell, J.W., Auricchio, A., High, K.A., and Bennett, J. (2008) *N. Engl. J. Med.* **358**, 2240.
105. Bainbridge, J.W., Smith, A.J., Barker, S.S., Robbie, S., Henderson, R., Balaggan, K., Viswanathan, A., Holder, G.E., Stockman, A., Tyler, N., Petersen-Jones, S., Bhattacharya, S.S., Thrasher, A.J., Fitzke, F.W., Carter, B.J., Rubin, G.S., Moore, A.T., and Ali, R.R. (2008) *N. Engl. J. Med.* **358**, 2231.
106. Allocca, M., Doria, M., Petrillo, M., Colella, P., Garcia-Hoyos, M., Gibbs, D., Kim, S.R., Maguire, A., Rex, T.S., Di Vicino, U., Cutillo, L., Sparrow, J.R., Williams, D.S., Bennett, J., and Auricchio, A. (2008) *J. Clin. Invest.* **118**, 1955.
107. Kong, J., Kim, S.R., Binley, K., Pata, I., Doi, K., Mannik, J., Zernant-Rajang, J., Kan, O., Iqball, S., Naylor, S., Sparrow, J.R., Gouras, P., and Allikmets, R. (2008) *Gene. Ther.* **15**, 1311.
108. Radu, R.A., Mata, N.L., Nusinowitz, S., Liu, X., Sieving, P.A., and Travis, G.H. (2003) *Proc. Natl. Acad. Sci. U.S.A.* **100**, 4742.

109. Radu, R.A., Han, Y., Bui, T.V., Nusinowitz, S., Bok, D., Lichter, J., Widder, K., Travis, G.H., and Mata, N.L. (2005) *Invest. Ophthalmol. Vis. Sci.* **46**, 4393.
110. Golczak, M., Kuksa, V., Maeda, T., Moise, A.R., and Palczewski, K. (2005) *Proc. Natl. Acad. Sci. U.S.A.* **102**, 8162.

Chapter 6

The Role of the Antigen Translocation Machinery TAP in Adaptive Immunity

Daphne Nikles and Robert Tampé

1. Introduction

Tumors and infectious diseases established by bacteria and viruses are among the most common causes of death in the human population. Be it cancer, HIV or tuberculosis, all of these common threats involve an attack on the human immune system, which they sometimes overcome. To shed light on these battles and the consequences of their outcome, molecular medicine needs to understand the processes underlying the adaptive immune response. Indeed, after more than one century of immunological research, we nowadays have a better understanding of this complex interplay. The basic concept of the immune system is the recognition of exogenous structures that have entered the organism. The immune response is generated by a sophisticated machinery that fulfills this function. A complex interplay of different cell types are part of the response generated after infection with a pathogen or the development of cell malignancies. The pathogen invasion and its recognition through the innate immune system is followed by a switch to activate the adaptive immune response. Latter one is even more specific than the first response and usually decreases efficiently the pathogen load in the organism. Each step of the response makes use of certain cell types and signals. Specialized antigen-presenting cells (APCs), such as dendritic cells and macrophages, are capable of presenting extracellular antigens via the so-called Major Histocompatibility Complex (MHC) class II. In addition, these professional APCs display extracellular antigens via MHC class I (MHC I), a process termed cross-presentation. In both cases, the presentation is followed by the recognition of the presented epitope through specialized T cells. Upon stimulation, these help to establish an antibody-based or a cytotoxic immune response against the antigen displayed.

Within the sophisticated process of distinguishing "self" from "non-self", a basic screening mechanism is necessary to recognize all endogenous molecules present in the cells. This is partly accomplished by a macromolecular machinery, the peptide-loading complex (PLC). One major player in this process is the MHC I molecule. As mentioned above, this is also utilized for cross-presentation of exogenous antigens in professional APCs. However, MHC I facilitates the display of the peptidome (the protein fragments mainly produced by proteasomal degradation) on the surface of nearly all nucleated cells of our body. During peptide loading onto MHC I, a representative fraction of the cellular peptidome is shuttled from the cytosol into the endoplasmic reticulum (ER) lumen, via a process carried out by the ABC transporter associated with antigen processing TAP. In infected or in malignant cells, the pathogen-derived and mutated proteins are respectively dissected into peptides and translocated as well as the normal cytosolic library. Subsequently, these cells present "foreign" surface patterns that can then be detected specifically by cytotoxic T cells which in turn eliminate the target cell. Hence, the process of antigen presentation plays a key role in a variety of diseases, which interfere with the main players of antigen presentation. Down-regulation or deficiency of TAP can cause a severe phenotype as it is the case of the Bare Lymphocyte Syndrome I (BLS I) or permit the development of malignant cells subsequently able to escape immune recognition (described below). Importantly, TAP and MHC I molecules are specifically targeted by persistent viruses that have evolved strategies to interfere with peptide translocation and loading as will be described below in more detail.

The following sections will summarize current knowledge on the antigen presentation pathway centered on the ABC transporter TAP and will highlight the processes of immune evasion by cancer cells and viruses.

2. Organization and Function of the PLC

The impact of TAP on MHC I surface expression was first described almost two decades ago.[1–3] It is thus well established that TAP translocates peptides that are then loaded onto MHC I, delivered to the cell surface, and subsequently presented to cytotoxic CD8-positive T cells of the adaptive immune system (for recent review, see Refs. 4 and 5 Consequently, the transporter TAP plays a pivotal role in the complex mechanism of antigen presentation within the immune system of higher vertebrates (see Fig. 1).

The process of peptide loading, however, is tightly regulated and therefore needs not only the presence of the transporter itself but also that of additional molecules, which ensure proper loading of peptides.[6,7] The peptide loading complex (PLC) consists of TAP1 and TAP2, tapasin, the disulfide isomerase ERp57, calreticulin, as well as the heterodimer of MHC I heavy chain (HC) and β_2-microglobulin (β_2m) (see Fig. 2A). Concerning the stoichiometry of this complex, it has been proposed that up to four tapasin-MHC I complexes bind to one TAP subunit[8] and that the tapasin-MHCI complex exists in a 1:1 stoichiometry.[9] Thus, the composition of the PLC was believed to be four molecules of tapasin, four molecules of MHC I (HC and β_2m), four molecules of both ERp57 and calreticulin, that bind to TAP1 and TAP2, respectively. However, recent studies suggest that only two tapasin and ERp57 molecules, one MHC I and one calreticulin molecule associate with TAP1 and TAP2 subunits within the PLC, in which TAP1 and TAP2 alternate in recruiting peptides to one single MHC I molecule.[10] In conclusion, the exact

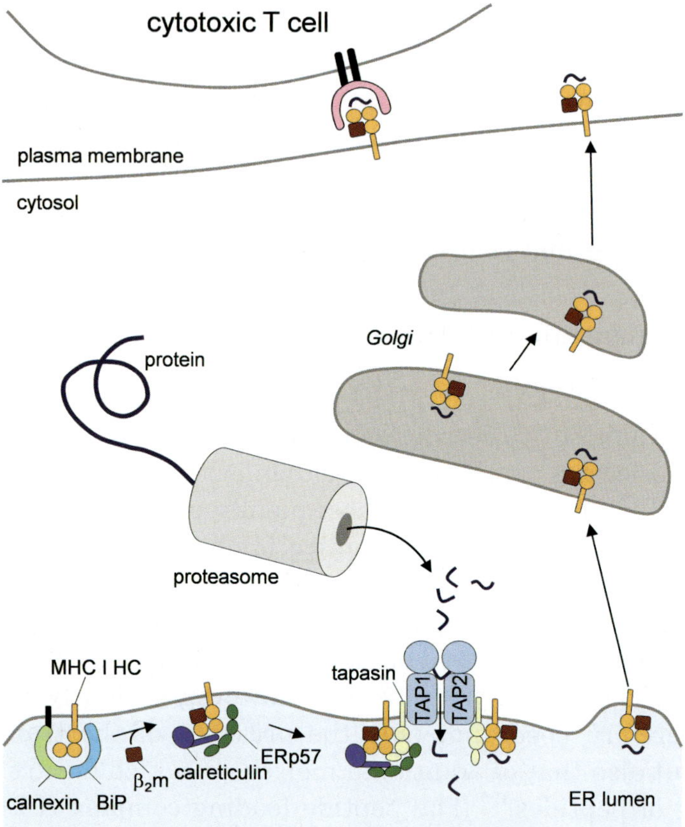

Fig. 1. Principles of the antigen presentation pathway.

The MHC I heavy chain (HC) is cotranslationally translocated into the ER lumen. Subsequently, β_2m binds to HC by help of calnexin and binding protein (BiP). After association with the oxidoreductase ERp57 and calreticulin, the molecule assembles with tapasin and the transporter associated with antigen processing TAP1/TAP2 heterodimer to a fully functional peptide-loading complex (PLC). TAP shuttles peptides generated by the proteasome from the cytosol into the ER lumen, where they are loaded onto MHC I molecules. MHC-peptide complexes equipped with high affinity peptides can overcome the ER quality control and traffic via the *Golgi* network to the cell surface. Here, they are presented to cytotoxic T cells, which recognize the antigenic cargo presented by MHC I molecules via specific T cell receptors. Whereas self-antigens derived from cellular proteins fail to initiate the T cell binding and response, virus- and/or tumor-derived antigens may cause a recognition that leads to a cytotoxic response against the infected or malignantly transformed target cell.

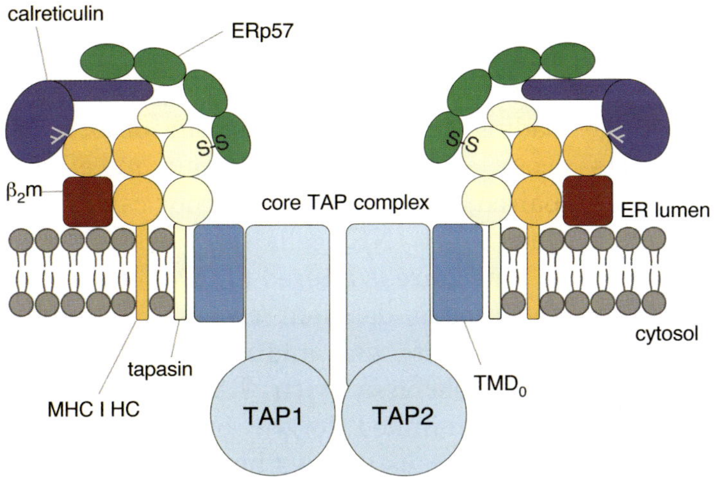

(A)

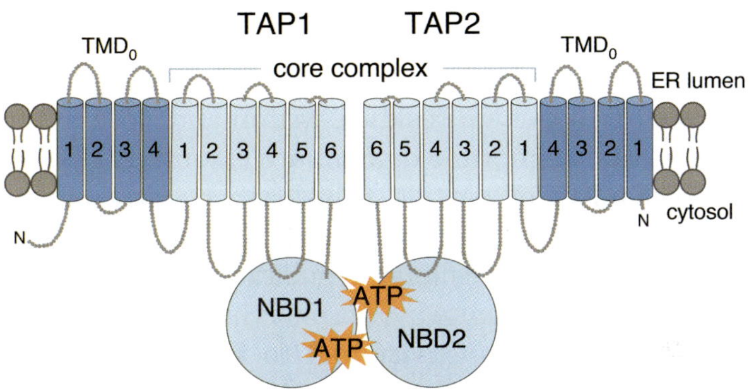

(B)

Fig. 2. Molecular architecture of the PLC and the TAP complex. (**A**) The PLC consists of TAP1/TAP2, tapasin, ERp57, calreticulin, MHC I HC, and β_2m. Tapasin interacts with each TAP subunit and is stably linked to ERp57 by a mixed disulfide bond. In addition, calreticulin and the MHC I molecule bind to the complex that facilitates peptide loading. The exact composition of the PLC, depicted as two molecules of tapasin, two ERp57, two calreticulin and two MHC I molecules per TAP heterodimer, remains to be clarified. (**B**) TAP1 and TAP2 subunits each comprise six putative core transmembrane helices and four N-terminally extended helices (TMD_0). The two nucleotide-binding domains (NBDs) each bind and hydrolyze ATP to drive peptide translocation into the ER lumen.

composition of the PLC remains unclear and apparently represents a challenging future task.

The type I transmembrane glycoprotein tapasin fulfills a key role in peptide loading as underlined by the observation that severe effects occur in tapasin knock-out mice: cytotoxic responses are impaired, dendritic cells are not capable of cross-presentation, natural killer (NK) cells show self-tolerance, and the MHC I peptide repertoire is altered to favor less stably binding peptides in the tapasin-deficient animals.[11] Many studies report that tapasin interacts with and recruits the MHC I heterodimer[9,12–14] and associates with TAP.[8,15,16] Additionally, tapasin is also disulfide-linked to the oxidoreductase ERp57 within the PLC.[17,18] Previous studies have shown that tapasin assists and facilitates peptide selection by MHC I[19–23] and together with ERp57 enables binding of high-affinity peptides to MHC I.[24,25] Importantly, tapasin also stabilizes the TAP complex significantly.[26–30] For the stabilization of TAP2, it has been demonstrated that certain residues within and close to the putative tapasin transmembrane domain are essential.[31]

The mechanism of peptide translocation, although in some respects circumstantial, is a reasonably well understood process while the nature of the cellular regulation of the TAP-mediated translocation event is limited. Phosphorylation of TAP has been reported to regulate its activity and to inhibit TAP function.[32] Nevertheless, this has not yet been confirmed by other groups or linked to a physiological network. More recently, it has been postulated that the heat-shock protein Hsc73 functions in "peptide feeding" of TAP within the cell.[33] However, neither the nature of the regulation of TAP activity within the cell nor the circumstances under which this regulation might be required in relation to immune defense is currently known.

3. Structure and Function of TAP

The TAP complex has a typical blueprint of an ABC exporter as previously described.[34,35] This consists of a heterodimer assembled from two half-transporters named TAP1 (748 aa in human, also termed ABCB2) and TAP2 (686 aa in human, also termed ABCB3) (see Fig. 2B). Six core transmembrane domains (TMDs) per subunit represent the minimal functional unit for transport (termed core TAP).[36,37] However, the core TMDs of both TAP1 and TAP2 are preceded by four additional N-terminal transmembrane helices (TMD_0). These unique domains mediate the binding of tapasin independently from each other.[38] The C-terminal region of TAP1 and TAP2 each consists of ATP-binding domains with the ABC motifs Walker A, Walker B and the C-loop. Both ATP-binding sites of the TAP transporter bind and hydrolyze ATP during the catalytic cycle.[39,40] Nevertheless, it has been shown that the NBDs of TAP1 and TAP2 are functionally non-equivalent.[41–43] The ATP-binding sites I and II consist of Walker A/B TAP1, C-loop TAP2, and Walker A/B TAP2, C-loop TAP1, respectively. In the functional TAP dimer, key residues of the canonical (TAP1) and degenerate (TAP2) C-loops, the Walker B, and the H-loop contribute to an asymmetric behavior of the two NBDs termed non-equivalent functionality. Whereas the canonical ATP binding site II is more important for the catalytic cycle, the degenerate ATP binding site I induces the asymmetric behavior and potentially.[43,44] It has a regulatory role in the ATP hydrolysis cycle.

Previous studies mapped the peptide-binding site that is shared between the two TAP subunits.[45] Peptide binding to TAP stimulates ATP hydrolysis,[40,46] a mechanism that has recently been investigated in more detail. Thus, key residues in TAP1 that upon mutation result in uncoupling of peptide binding from peptide transport have been identified.[47] Recently, the transition interface between the TMDs and the NBDs has been characterized with identification of close contacts between the X-loop of TAP2, which is located close to the C-loop in the α-helical domain of the NBD and the so-called coupling helices 1 and 2, located in the cytosolic loops 1 and 2 where they connect the extended TM helices of TAP1.[48]

Current detailed knowledge of TAP function comes not only from extensive studies of the nature of the transport substrates[45] but also from detailed analysis of the substrate-binding site and binding constraints. The latter have been analyzed by use of combinatorial peptide libraries, which revealed the importance of the three N-terminal and the C-terminal amino acid residues (anchor residues)[49] of peptides that may vary from 8–16 amino acids in length.[50] Subsequently, these analyses resulted in accurate predictions of TAP transport efficiency for certain peptides and antigens.[51] Moreover, the development of specific transport assays greatly increased our understanding of TAP function. The assay has been successfully used in semi-permeabilized cells[52] as well as in isolated microsomes[53] and finally detailed knowledge of the transport mechanism has come from establishment of methods for the isolation and functional reconstitution of the TAP complex in proteoliposomes.[46] Thus, the ATP-dependency, peptide specificity, and the kinetics of peptide transport were demonstrated. Furthermore, the investigation of TAP function has permitted the detailed analysis of TAP inhibitory molecules as described in the following sections.

4. Viral Immune Evasion Strategies Block Peptide-loading

For an immune-relevant molecule such as TAP, one may learn mechanistic details by looking at strategies that have arisen during co-evolution with different pathogens. TAP is indeed a major target for viruses in attempts to evade the immunity system. This is particularly the case for persisting viruses that need to overcome detection by the immune system over the long term. Thus, herpes viruses encode a number of gene products, which interfere with TAP function to escape immune recognition. By inhibiting peptide translocation into the ER lumen, these viruses succeed in repressing the host immune response by remaining unrecognized within the cell. The underlying mechanisms give excellent insight into the fine-tuning and regulation of TAP function and have been investigated in much detail in the past (see Fig. 3).

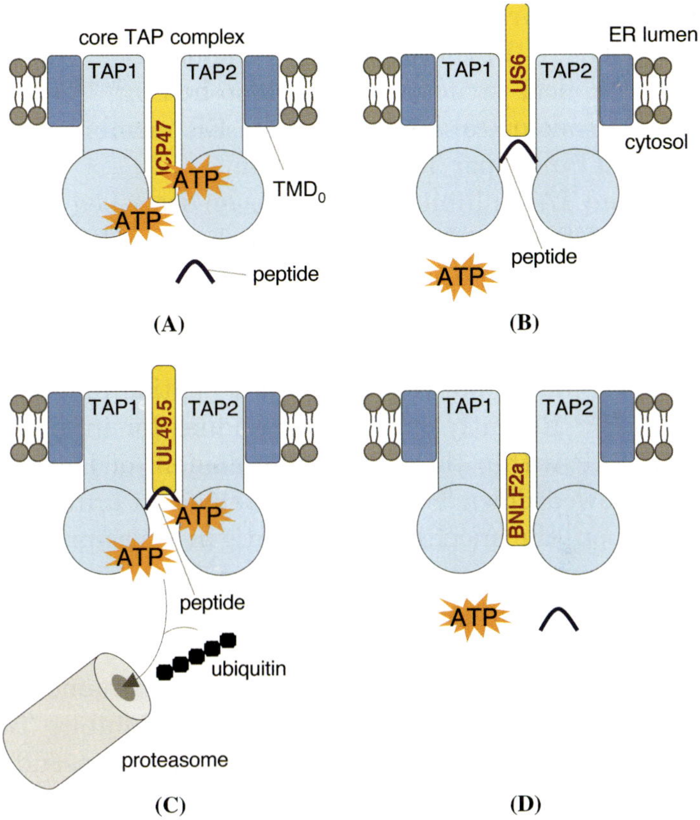

Fig. 3. Viral TAP inhibition mechanisms.

(**A**) The herpes simplex virus protein ICP47 binds to TAP, abolishes peptide binding and subsequent peptide translocation into the ER lumen. The binding of ATP to TAP is not affected. (**B**) US6 from human cytomegalovirus suppresses ATP binding to TAP and acts from the ER lumenal side to inhibit peptide transport. (**C**) The bovine herpes virus factor UL49.5 arrests TAP in a transport-incompetent state and initiates TAP degradation by the proteasome. (**D**) The small factor BNLF2a from Epstein-Barr virus blocks both peptide and ATP binding to TAP and subsequently inhibits peptide translocation into the ER lumen.

The first immune evasion protein to be discovered was ICP47 from herpes simplex.[54,55] This small inhibitor very effectively inhibits peptide binding to TAP with nanomolar affinity without affecting ATP binding.[56,57] The active inhibitory domain of this immediate early viral factor has been mapped to amino acids 3–34 of ICP47.[58,59] In the presence of membranes, this peptide forms an α-helical structure (helix-loop-helix),[60,61] suggesting that under physiological conditions, the ER membrane association is needed for proper TAP inhibition.

The second TAP inhibitor that has been described is the early gene product US6 from human cytomegalovirus HCMV.[62,63] The type I transmembrane protein, consists of an ER-lumenal domain, a transmembrane helix, and a short cytosolic tail. US6 blocks peptide translocation via its ER-luminal domain and inhibits ATP binding but does not affect peptide binding to the transporter.[64,65] Recently, important residues for inhibition and binding to TAP within the ER-luminal domain of US6 could be further narrowed down.[66,67] US6 has helped to unravel important details on the mechanism of antigen cross-presentation. Thus, it transpired that this mechanism is quite independent from endogenous antigen presentation pathways within the cell.[68–70]

More recently, a factor from varicelloviruses named UL49.5 (glycoprotein N) has been characterized that inhibits TAP.[71–74] The small type I transmembrane homologues from pseudorabies virus (PRV) and equine herpesvirus 1 and 4 (EHV-1, EHV-4) inhibit TAP by causing an arrest in a non-functional state. The bovine herpesvirus 1 protein additionally targets TAP for proteasomal degradation.[71]

Targeting for proteasomal degradation of the peptide loading complex (PLC) component MHC I has also previously been shown for the γ-herpesvirus 68 protein mK3. With its RING zinc finger binding domain that contains an E3 ubiquitin ligase activity, mK3 mediates ubiquitination and degradation of MHC I in a TAP- and tapasin-dependent manner.[75,76] A similar activity that controls MHC I expression is present in membrane-associated RING-CH proteins and was possibly adapted by the murine herpesvirus 68 to result in subsequent immune escape. In addition, mK3 induces degradation of TAP and tapasin by a functionally distinct evasion mechanism.[77,78] Nevertheless, the association of mK3 via its cysteine-rich C-terminus with both TAP1 and TAP2 as well as tapasin also provides substrate specificity for mK3 to target MHC I.[79]

Unlike this evasion mechanism of mK3, the UL49.5 protein described above acts in a dual mechanism: UL49.5 arrests TAP in a transport-incompetent state with its N-terminal part and simultaneously targets TAP for proteasomal degradation mediated by its C-terminal domain.[74] The protein thus "signals" across the ER membrane and induces efficient TAP inhibition by acting from both the ER lumenal and the cytosolic side. It has been found very recently that the interactions with TMD_0 of TAP are not required for this inhibition process.[80]

Another inhibitor (BNLF2a) is derived from the Epstein-Barr virus.[81] The small protein of 60 amino acids successfully inhibits both ATP and peptide binding to TAP acting within the cytosol. Recently, it has been shown that BNLF2a is membrane-associated, targeted to TAP and effective during the lytic phase of a productive EBV infection.[82]

As a consequence of all these viral inhibition strategies there is a reduced level of MHC I on the cell surface and therefore a decreased immune recognition through the cytotoxic CTL response.

5. The Bare Lymphocyte Syndrome (BLS)

The Bare lymphocyte syndrome (BLS) has been described in patients exhibiting inherited gene dysregulation of the MHC II or defects in MHC I loci (BLS II or I, respectively). This very rare disease causes a severe combined immunodeficiency (SCID) with symptoms of defective antigen presentation and immune responses. This is particularly important in antigen-presenting cells with the consequence that patients cannot efficiently fight pathogen infections. La Salle and colleagues analyzed cells isolated from BLS type I patients and recognized that the cells lack TAP1 expression.[83,84] In line with the observed instability of TAP2 in the absence of TAP1[85] these cells had also no detectable levels of TAP2. Thus, the antigen presentation pathway was significantly impaired leading to a severe phenotype in these patients.

6. Tumor Escape Mechanisms Target TAP

The escape from recognition by the immune system due to defects or impairment in the MHC I presentation pathway and especially the down-regulation of TAP1 expression has been frequently observed in malignantly transformed cells derived from various origins.[86–88] Upon development, tumor cells benefit from the selection of mutations, which impact MHC I presentation. It is believed, according to numerous reports that TAP1 deficiency, together with the subsequent instability of TAP2 and loss of transport activity, provide an increased survival rate for malignant transformed cells. The shutdown of peptide translocation and presentation leads to a reduced cytotoxic response against malignant cells within the organism. Further investigation of the molecular mechanisms underlying TAP1 deficiency revealed that, in addition to mutations in the *tap1* gene or its promoter, a silencing mechanism involving acetyltransferase and histone H3 deacetylation occurs.[89] More recently, approaches were developed to screen for components inhibiting tumor growth by increasing its immunogenicity, because such agents may be utilized as very long-lasting antitumor drugs (they assist the immune system rather than being toxic to all cells with high proliferation rates). The drug trichostatin A (TSA) for instance, has been shown to increase *tap1* promoter activity and results in enhanced MHC I surface expression and immune recognition *in vivo*.[90]

7. Conclusions and Outlook

Among all molecules involved in the process of antigen presentation, the key players MHC I and in particular the ABC transporter

TAP are major targets for immune evasion by cancer cells and viruses. The inhibition of peptide translocation into the ER is sufficient to repress antigen presentation of either tumor antigens in malignant cells or virus epitopes in productively infected host cells. In the latter case, the herpes virus factors described here (ICP47, US6, UL49.5, and BNLF2a) do not abolish peptide transport completely but reduce the recognition of antigens significantly. Complete loss of loaded MHC I molecules on a cell surface would cause destruction of that cell by natural killer cells of the innate immunity system. Thus, it seems to be important that a basal translocation activity remains also in virus-infected and malignant cells.

Overall, the association of severe disease phenotypes with the deficiency or down-regulation of the ABC transporter TAP described above, underlines the importance of antigen compartmentalization and translocation mechanisms across biological membranes. Intensive efforts have revealed a detailed molecular understanding of the process of antigen processing and the nature of the TAP machinery, which importantly leads to the ability to develop specific drugs as strategies to combat pathogens. However, there still remains some difficult obstacles to full understanding. These mainly concern the precise architecture and stoichiometry as well as the quality control/degradation of the PLC. Can the complex be regulated within the cell or is it constitutively active? And what is the composition of one functional peptide-loading unit? Future goals also clearly concern seeking more detailed knowledge on disease-related processes in antigen presentation within the human immune system.

Acknowledgments

We thank Rupert Abele and David Parcej for critical reading of the manuscript. We also thank Nina Kreissig, Sabine Unger, and Christine Le Gal for helpful comments and corrections.

References

1. Kelly, A., Powis, S.H., Kerr, L.A., Mockridge, I., Elliott, T., Bastin, J., Uchanska-Ziegler, B., Ziegler, A., Trowsdale, J., and Townsend, A. (1992) *Nature* **355**, 641.
2. Powis, S.J., Townsend, A.R., Deverson, E.V., Bastin, J., Butcher, G.W., and Howard, J.C. (1991) *Nature* **354**, 528.
3. Spies, T. and DeMars, R. (1991) *Nature* **351**, 323.
4. Purcell, A.W. and Elliott, T. (2008) *Curr. Opin. Immunol.* **20**, 75.
5. Wearsch, P.A. and Cresswell, P. (2008) *Curr. Opin. Cell Biol.* **20**, 624.
6. Koch, J. and Tampé, R. (2006) *Cell. Mol. Life Sci.* **63**, 653.
7. Schölz, C. and Tampé, R. *Biol Chem.* 2009 May 9. [Epub ahead of print]
8. Ortmann, B., Copeman, J., Lehner, P.J., Sadasivan, B., Herberg, J.A., Grandea, A.G., Riddell, S.R., Tampé, R., Spies, T., Trowsdale, J., and Cresswell, P. (1997) *Science* **277**, 1306.
9. Bangia, N. and Cresswell, P. (2005) *Immunology* **114**, 346.
10. Rufer, E., Leonhardt, R.M., and Knittler, M.R. (2007) *J. Immunol.* **179**, 5717.
11. Garbi, N., Tan, P., Diehl, A.D., Chambers, B.J., Ljunggren, H.G., Momburg, F., and Hammerling, G.J. (2000) *Nat. Immunol.* **1**, 234.
12. Grandea, A.G. 3rd, Lehner, P.J., Cresswell, P., and Spies, T. (1997) *Immunogenetics* **46**, 477.
13. Tan, P., Kropshofer, H., Mandelboim, O., Bulbuc, N., Hammerling, G.J., and Momburg, F. (2002) *J. Immunol.* **168**, 1950.
14. Peh, C.A., Laham, N., Burrows, S.R., Zhu, Y., and McCluskey, J. (2000) *J. Immunol.* **164**, 292.
15. Sadasivan, B., Lehner, P.J., Ortmann, B., Spies, T., and Cresswell, P. (1996) *Immunity* **5**, 103.
16. Li, S., Sjogren, H.O., Hellman, U., Pettersson, R.F., and Wang, P. (1997) *Proc. Natl. Acad. Sci. U.S.A.* **94**, 8708.
17. Dick, T.P., Bangia, N., Peaper, D.R., and Cresswell, P. (2002) *Immunity* **16**, 87.
18. Peaper, D.R., Wearsch, P.A., and Cresswell, P. (2005) *EMBO J.* **24**, 3613.
19. Chen, M. and Bouvier, M. (2007) *EMBO J.* **26**, 1681.
20. Howarth, M., Williams, A., Tolstrup, A.B., and Elliott, T. (2004) *Proc. Natl. Acad. Sci. U.S.A.* **101**, 11737.

21. Williams, A.P., Peh, C.A., Purcell, A.W., McCluskey, J., and Elliott, T. (2002) *Immunity* **16**, 509.
22. Zarling, A.L., Luckey, C.J., Marto, J.A., White, F.M., Brame, C.J., Evans, A.M., Lehner, P.J., Cresswell, P., Shabanowitz, J., Hunt, D.F., and Engelhard, V.H. (2003) *J. Immunol.* **171**, 5287.
23. Rizvi, S.M. and Raghavan, M. (2006) *Proc. Natl. Acad. Sci. U.S.A.* **103**, 18220.
24. Kienast, A., Preuss, M., Winkler, M., and Dick, T.P. (2007) *Nat. Immunol.* **8**, 864.
25. Wearsch, P.A. and Cresswell, P. (2007) *Nat. Immunol.* **8**, 873.
26. Lehner, P.J., Surman, M.J., and Cresswell, P. (1998) *Immunity* **8**, 221.
27. Bangia, N., Lehner, P.J., Hughes, E.A., Surman, M., and Cresswell, P. (1999) *Eur. J. Immunol.* **29**, 1858.
28. Garbi, N., Tiwari, N., Momburg, F., and Hammerling, G.J. (2003) *Eur. J. Immunol.* **33**, 264.
29. Raghuraman, G., Lapinski, P.E., and Raghavan, M. (2002) *J. Biol. Chem.* **277**, 41786.
30. Li, S., Paulsson, K.M., Chen, S., Sjogren, H.O., and Wang, P. (2000) *J. Biol. Chem.* **275**, 1581.
31. Papadopoulos, M. and Momburg, F. (2007) *J. Biol. Chem.* **282**, 9401.
32. Li, Y., Salter-Cid, L., Vitiello, A., Preckel, T., Lee, J.D., Angulo, A., Cai, Z., Peterson, P.A., and Yang, Y. (2000) *J. Biol. Chem.* **275**, 24130.
33. Kamiguchi, K., Torigoe, T., Fujiwara, O., Ohshima, S., Hirohashi, Y., Sahara, H., Hirai, I., Kohgo, Y., and Sato, N. (2008) *Microbiol. Immunol.* **52**, 94.
34. Abele, R. and Tampé, R. (2006) *FEBS Lett.* **580**, 1156.
35. Herget, M. and Tampé, R. (2007) *Pflugers Arch.* **453**, 591.
36. Koch, J., Guntrum, R., Heintke, S., Kyritsis, C., and Tampé, R. (2004) *J. Biol. Chem.* **279**, 10142.
37. Koch, J., Guntrum, R., and Tampé, R. (2005) *FEBS Lett.* **579**, 4413.
38. Koch, J., Guntrum, R., and Tampé, R. (2006) *FEBS Lett.* **580**, 4091.
39. Arora, S., Lapinski, P.E., and Raghavan, M. (2001) *Proc. Natl. Acad. Sci. U.S.A.* **98**, 7241.
40. Chen, M., Abele, R., and Tampé, R. (2003) *J. Biol. Chem.* **278**, 29686.
41. Lapinski, P.E., Neubig, R.R., and Raghavan, M. (2001) *J. Biol. Chem.* **276**, 7526.

42. Karttunen, J.T., Lehner, P.J., Gupta, S.S., Hewitt, E.W., and Cresswell, P. (2001) *Proc. Natl. Acad. Sci. U.S.A.* **98**, 7431.
43. Chen, M., Abele, R., and Tampé, R. (2004) *J. Biol. Chem.* **279**, 46073.
44. Ernst, R., Koch, J., Horn, C., Tampé, R., and Schmitt, L. (2006) *J. Biol. Chem.* **281**, 27471.
45. Nijenhuis, M. and Hammerling, G.J. (1996) *J. Immunol.* **157**, 5467.
46. Gorbulev, S., Abele, R., and Tampé, R. (2001) *Proc. Natl. Acad. Sci. U.S.A.* **98**, 3732.
47. Herget, M., Oancea, G., Schrodt, S., Karas, M., Tampé, R., and Abele, R. (2007) *J. Biol. Chem.* **282**, 3871.
48. Oancea, G., O'Mara, M.L., Bennett, W.F., Tieleman, D.P., Abele, R., and Tampé, R. (2009) *Proc. Natl. Acad. Sci. U.S.A.* **106**, 5551.
49. Uebel, S., Kraas, W., Kienle, S., Wiesmuller, K.H., Jung, G., and Tampé, R. (1997) *Proc. Natl. Acad. Sci. U.S.A.* **94**, 8976.
50. van Endert, P.M., Tampe, R., Meyer, T.H., Tisch, R., Bach, J.F., and McDevitt, H.O. (1994) *Immunity* **1**, 491.
51. Peters, B., Bulik, S., Tampé, R., Van Endert, P.M., and Holzhutter, H.G. (2003) *J. Immunol.* **171**, 1741.
52. Neefjes, J.J., Momburg, F., and Hämmerling, G.J. (1993) *Science* **261**, 769.
53. Shepherd, J.C., Schumacher, T.N., Ashton-Rickardt, P.G., Imaeda, S., Ploegh, H.L., Janeway, C.A. Jr., and Tonegawa, S. (1993) *Cell* **74**, 577.
54. Früh, K., Ahn, K., Djaballah, H., Sempé, P., van Endert, P.M., Tampé, R., Peterson, P.A., and Yang, Y. (1995) *Nature* **375**, 415.
55. Hill, A., Jugovic, P., York, I., Russ, G., Bennink, J., Yewdell, J., Ploegh, H. and Johnson, D. (1995) *Nature* **375**, 411.
56. Tomazin, R., Hill, A.B., Jugovic, P., York, I., van Endert, P., Ploegh, H.L., Andrews, D.W., and Johnson, D.C. (1996) *EMBO J.* **15**, 3256.
57. Ahn, K., Meyer, T.H., Uebel, S., Sempé, P., Djaballah, H., Yang, Y., Peterson, P.A., Früh, K., and Tampé, R. (1996) *EMBO J.* **15**, 3247.
58. Galocha, B., Hill, A., Barnett, B.C., Dolan, A., Raimondi, A., Cook, R.F., Brunner, J., McGeoch, D.J., and Ploegh, H.L. (1997) *J. Exp. Med.* **185**, 1565.
59. Neumann, L., Kraas, W., Uebel, S., Jung, G., and Tampé, R. (1997) *J. Mol. Biol.* **272**, 484.
60. Aisenbrey, C., Sizun, C., Koch, J., Herget, M., Abele, R., Bechinger, B., and Tampé, R. (2006) *J. Biol. Chem.* **281**, 30365.

61. Beinert, D., Neumann, L., Uebel, S., and Tampé, R. (1997) *Biochemistry* **36**, 4694.
62. Hengel, H., Koopmann, J.O., Flohr, T., Muranyi, W., Goulmy, E., Hammerling, G.J., Koszinowski, U.H., and Momburg, F. (1997) *Immunity* **6**, 623.
63. Ahn, K., Gruhler, A., Galocha, B., Jones, T.R., Wiertz, E.J., Ploegh, H.L., Peterson, P.A., Yang, Y., and Früh, K. (1997) *Immunity* **6**, 613.
64. Kyritsis, C., Gorbulev, S., Hutschenreiter, S., Pawlitschko, K., Abele, R., and Tampé, R. (2001) *J. Biol. Chem.* **276**, 48031.
65. Hewitt, E.W., Gupta, S.S., and Lehner, P.J. (2001) *EMBO J.* **20**, 387.
66. Dugan, G.E. and Hewitt, E.W. (2008) *J. Virol.* **82**, 3271.
67. Halenius, A., Momburg, F., Reinhard, H., Bauer, D., Lobigs, M., and Hengel, H. (2006) *J. Biol. Chem.* **281**, 5383.
68. Ackerman, A.L., Kyritsis, C., Tampé, R., and Cresswell, P. (2003) *Proc. Natl. Acad. Sci. U.S.A.* **100**, 12889.
69. Burgdorf, S., Scholz, C., Kautz, A., Tampé, R., and Kurts, C. (2008) *Nat. Immunol.* **9**, 558.
70. Ackerman, A.L., Kyritsis, C., Tampé, R., and Cresswell, P. (2005) *Nat. Immunol.* **6**, 107.
71. Koppers-Lalic, D., Reits, E.A., Ressing, M.E., Lipinska, A.D., Abele, R., Koch, J., Marcondes Rezende, M., Admiraal, P., van Leeuwen, D., Bienkowska-Szewczyk, K., Mettenleiter, T.C., Rijsewijk, F.A., Tampé, R., Neefjes, J., and Wiertz, E.J. (2005) *Proc. Natl. Acad. Sci. U.S.A.* **102**, 5144.
72. Lipinska, A.D., Koppers-Lalic, D., Rychlowski, M., Admiraal, P., Rijsewijk, F.A., Bienkowska-Szewczyk, K., and Wiertz, E.J. (2006) *J. Virol.* **80**, 5822.
73. van Hall, T., Laban, S., Koppers-Lalic, D., Koch, J., Precup, C., Asmawidjaja, P., Offringa, R., and Wiertz, E.J. (2007) *J. Immunol.* **178**, 657.
74. Loch, S., Klauschies, F., Scholz, C., Verweij, M.C., Wiertz, E.J., Koch, J., and Tampé, R. (2008) *J. Biol. Chem.* **283**, 13428.
75. Boname, J.M. and Stevenson, P.G. (2001) *Immunity* **15**, 627.
76. Lybarger, L., Wang, X., Harris, M.R., Virgin, H.W.T., and Hansen, T.H. (2003) *Immunity* **18**, 121.
77. Boname, J.M., de Lima, B.D., Lehner, P.J., and Stevenson, P.G. (2004) *Immunity* **20**, 305.

78. Boname, J.M., May, J.S., and Stevenson, P.G. (2005) *Eur. J. Immunol.* **35**, 171.
79. Wang, X., Lybarger, L., Connors, R., Harris, M.R., and Hansen, T.H. (2004) *J. Virol.* **78**, 8673.
80. Verweij, M.C., Koppers-Lalic, D., Loch, S., Klauschies, F., de la Salle, H., Quinten, E., Lehner, P.J., Mulder, A., Knittler, M.R., Tampé, R., Koch, J., Ressing, M.E., and Wiertz, E.J. (2008) *J. Immunol.* **181**, 4894.
81. Hislop, A.D., Ressing, M.E., van Leeuwen, D., Pudney, V.A., Horst, D., Koppers-Lalic, D., Croft, N.P., Neefjes, J.J., Rickinson, A.B., and Wiertz, E.J. (2007) *J. Exp. Med.* **204**, 1863.
82. Horst, D., van Leeuwen, D., Croft, N.P., Garstka, M.A., Hislop, A.D., Kremmer, E., Rickinson, A.B., Wiertz, E.J., and Ressing, M.E. (2009) *J. Immunol.* **182**, 2313.
83. de la Salle, H., Zimmer, J., Fricker, D., Angenieux, C., Cazenave, J.P., Okubo, M., Maeda, H., Plebani, A., Tongio, M.M., Dormoy, A., and Hanau, D. (1999) *J. Clin. Invest.* **103**, R9.
84. Zimmer, J., Andres, E., Donato, L., Hanau, D., Hentges, F., and de la Salle, H. (2005) *QJM* **98**, 719.
85. Keusekotten, K., Leonhardt, R.M., Ehses, S., and Knittler, M.R. (2006) *J. Biol. Chem.* **281**, 17545.
86. Kageshita, T., Hirai, S., Ono, T., Hicklin, D.J., and Ferrone, S. (1999) *Am. J. Pathol.* **154**, 745.
87. Natali, P.G., Nicotra, M.R., Bigotti, A., Venturo, I., Marcenaro, L., Giacomini, P., and Russo, C. (1989) *Proc. Natl. Acad. Sci. U.S.A.* **86**, 6719.
88. Seliger, B., Ritz, U., Abele, R., Bock, M., Tampé, R., Sutter, G., Drexler, I., Huber, C., and Ferrone, S. (2001) *Cancer Res.* **61**, 8647.
89. Setiadi, A.F., David, M.D., Chen, S.S., Hiscott, J., and Jefferies, W.A. (2005) *Cancer Res.* **65**, 7485.
90. Setiadi, A.F., Omilusik, K., David, M.D., Seipp, R.P., Hartikainen, J., Gopaul, R., Choi, K.B., and Jefferies, W.A. (2008) *Cancer Res.* **68**, 9601.

Chapter 7

The ABCs of Mitochondrial Metabolic Disorders

Yu Fukuda, Dagny Ulrich, John Lynch and John D. Schuetz

1. Introduction

The mitochondria are commonly described as the "powerhouse" of the cell, supplying energy for all of the cellular processes. However, they are also involved in many other activities that are essential for life (Fig. 1). Mitochondria play fundamental roles in control of apoptosis and regulation of cellular redox potential — through control of the production of superoxide anion, hydrogen peroxide, and nitric oxide — and are the site of the initial steps in the urea cycle. In addition, they are involved in the storage and metabolism of iron, notably the production of iron sulfur (Fe-S) clusters and their insertion into apoproteins. Finally, they are the site of steroid synthesis and of the initial and terminal steps in heme biosynthesis.

Mitochondria also hold the key to both life and death. The reactive oxygen species (ROS) generated by the mitochondrion are integral to various cell signaling pathways but also produce cellular damage responsible for the aging process. They also regulate the apoptotic cascade. Disruption of the integrity of mitochondrial processes is linked to several human diseases, such as the lethal neonatal GRACILE (growth retardation, aminoaciduria, cholestasis, iron overload, lactacidosis, and early

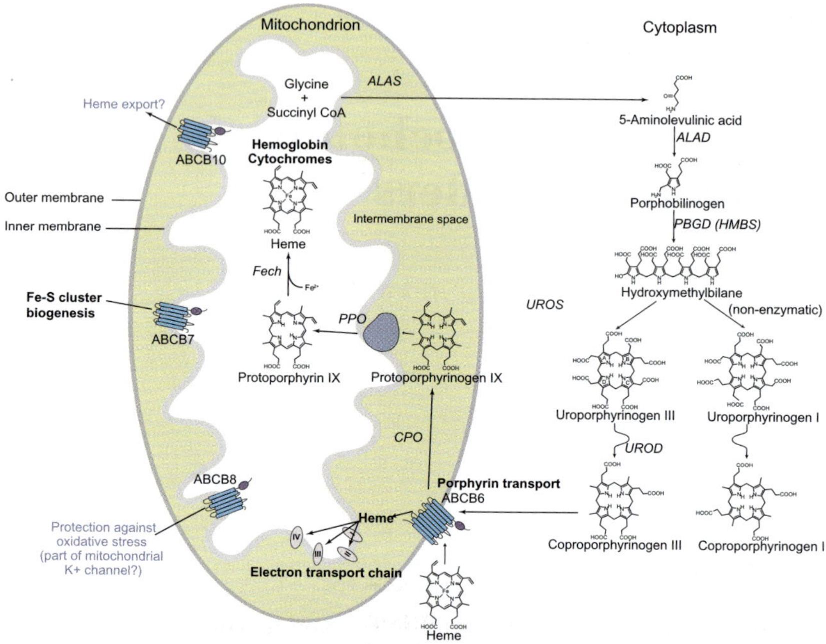

Fig. 1. The location and function of the mammalian mitochondrial ABC transporters. The purple ellipse designates the nucleotide binding domain for each ABC transporter and the blue bars represent transmembrane helices. The enzymes abbreviations and names are as follows: ALAS=Aminolevulinic Acid Synthase; ALAD=Aminolevulinic Acid Dehydratase; PBGD(HMBS)= Porphobilinigen Deaminase (Hydroxymethyl bilane synthase); UROS= Uroporphyrinogen III Synthase; UROD=Uroporphyrinogen Decarboxylase; CPO=Coproporphyrinogen Oxidase; PPO=Protoporphyrinogen Oxidase; Fech= Ferrochelatase.

death) metabolic syndrome.[1] Defects in heme biosynthesis can lead to several diseases, including X-linked sideroblastic anemia with ataxia, and porphyrias.[2–4] The mitochondria are linked by both heme and nonheme pathways to neurological disorders such as Parkinson's and Alzheimer's diseases.[5]

To date, four mitochondrial ABC transport proteins have been identified — ABCB6, ABCB7, ABCB8, and ABCB10 — and are members of subfamily B of the ABC transporter superfamily. These proteins are referred to as "half-transporters" because they

Table 1. Mitochondrial targeting sequence in mitochondrial ATP binding cassette transporters.

Transporter	Localization	Signal sequence	Probability of mitochondrial localization
ABCB6	OM	None	0.0015
ABCB7	IM	MALLAMHSWRWAAAA AAFEK	0.7362
ABCB8	?	MLVHLFRVGIRGGPF PGRLL	0.7279
ABCB10	IM	MRAPSARALLLIPRR GPAVR	0.9962

include only one membrane-spanning domain and a single nucleotide-binding domain. These proteins are localized to either the inner (ABCB7, ABCB10, and ABCB8) or outer (ABCB6) mitochondrial membranes. The inner mitochondrial membrane transporters have their nucleotide binding domains within the mitochondrial matrix, an orientation that would support export from the mitochondria. In contrast, the nuclear binding domain of ABCB6, faces the cytosol, and is believed to be involved in substrate import into the mitochondria.[6] ABCB7, ABCB8, and ABCB10 have a mitochondrial targeting sequence, while ABCB6 does not (Table 1). Defects in ABCB7 are known to cause X-linked sideroblastic anemia with ataxia.[7] ABCB6 and ABCB10 are involved in the regulation of intracellular heme and iron,[7–9] while ABCB8 appears to protect cells against oxidative stress.[10,11] This review summarizes our current understanding of these transporters' function, regulation, and role or potential role in disease.

2. ABCB6

2.1 *Background*

ABCB6 was first cloned from a rat liver cDNA library by using the nucleotide-binding domain of P-glycoprotein as a probe. The resulting partial cDNA sequence was termed P-glycoprotein related protein (PRP).[12] Subsequently, the complete rat cDNA was independently cloned and termed the ubiquitously expressed mammalian ABC half-transporter (UMAT).[13] The human cDNA was then cloned, termed mtABC3, and shown to be localized to chromosome 2q36,[14] the locus initially thought to be responsible for the GRACILE syndrome.[1] However, ABCB6 was subsequently excluded as the primary cause of this syndrome, which is associated with defective iron metabolism. All forms of the protein are now referred to as ABCB6. It is primarily a protein of the outer mitochondrial membrane, a finding confirmed by multiple laboratories,[12,14,15] but it has also been described as localizing to the plasma membrane.[15] Unlike other mitochondrially targeted ABC transporters, however, ABCB6 lacks an inner mitochondrial matrix-targeting sequence (Table 1).[16]

ABCB6 is very similar to the yeast mitochondrial transporters HMT1 (~39% sequence identity) and ATM1p (~46% sequence identity),[14,17] and was initially proposed to be essential for iron homeostasis.[18,19] In complementation studies, ABCB6 rescued yeast harboring a defective ATM1p from the respiratory effects of mitochondrial iron accumulation. While these studies suggested that ABCB6 is a functional orthologue of ATM1p, additional experimentation revealed that, paradoxically, a non-functional ABCB6 mutant provided similar protection.[14] Thus, the physiological function of ABCB6 was unlikely to be a direct efflux of mitochondrial iron. Yeast and mammals differ in their regulation of iron uptake and protoporphyrin synthesis. For example, heme deficiency produces mitochondrial iron accumulation in erythroid cells,[19] whereas the absence of heme does not cause mitochondrial iron accumulation in yeast.[7] The subsequent identification of ABCB7 as the true functional orthologue of ATM1p confirmed the distinction between ABCB6 and ATM1p.[18,20–22] The primary function of ABCB6 appears to be to regulate the heme biosynthesis pathway via the importation of porphyrins (e.g., coproporphyrinogen III), into the mitochondria Fig. 1.[7] We speculate that one reason ABCB6 and its non-functional mutant complement

ATM1p is that the ATM1p mutant was not completely defective,[14] enabling ABCB6 to stabilize it and restore some functionality.

ABCB6 is a highly conserved transporter. Of the four mitochondrial ABC transporters, only ABCB6 and ABCB8 are found in *C. elegans*,[23] suggesting that they have an evolutionarily conserved function. ABCB6 expression in the gut of *C. elegans*,[23] which requires dietary heme, supports its role in heme metabolism. Gene expression profiling of blood formation mutants of Zebrafish provides additional support, revealing ABCB6 to be among the genes of the "erythroid cluster" and suggesting that ABCB6 has a hematopoietic function.[24]

2.2. *ABCB6 regulation and expression*

ABCB6 is ubiquitously expressed in multiple tissues but is expressed most highly in erythroid cells, especially progenitor cells such as fetal liver cells that have a high capacity for heme synthesis. This fact provided an important clue to ABCB6 function.[25] Further, among the non-erythroid tissues it is expressed more highly in those with greater heme requirements because of their metabolic activity (e.g., heart and skeletal muscle). There is evidence that ABCB6 is itself regulated by heme and porphyrins, and its importance to mitochondrial functions implies that it may also be regulated by metabolic changes.[7]

ABCB6 was first identified as a gene that was upregulated in a rodent model of hepatocarcinogenesis; ABCB6 mRNA increased during liver regeneration in a pattern that paralleled the wave of DNA synthesis.[12] The expression of the transcription factor c-myc shows a similar pattern of increase during liver regeneration, suggesting that c-myc may regulate ABCB6. In addition, a genome-wide chromatin immunoprecipitation (ChIP) analysis of a Burkitt's lymphoma cell line showed an association between c-myc and the ABCB6 promoter.[26] C-myc is an oncogenic transcription factor and has a prominent role in regulating proliferative and metabolic genes. Notably, many genes harboring c-myc binding sites have roles in DNA replication, protein synthesis, cell respiration, and energy metabolism.[27–29] A ChIP on ChIP analysis found that the ABCB6 promoter is bound by the transcriptional repressor complex p130-E2F4 under conditions of cell cycle arrest,[30] providing further evidence that ABCB6 plays a regulatory role in the cellular metabolic state. These authors developed a motif-scanning algorithm to identify conserved regions in the promoters bound by p130-E2F4 and discovered that some of these promoters, including the *ABCB6* promoter, also harbored consensus nuclear respiratory factor-1 (NRF1) binding sites. NRF1 contributes to mitochondrial biogenesis and transcriptionally activates nuclear-encoded mitochondrial genes with respiratory functions (e.g., cytochrome c oxidase).[31] These findings fit well with ABCB6's ability to regulate biosynthesis of the oxygen-sensing molecule heme.

Studies indicating that ABCB6 is strongly upregulated by porphyrins and heme agree with findings in publicly available databases (gene profile K562) and with a recent study confirming that ABCB6 is among a subset of genes newly identified as being required for heme synthesis.[32] The molecular signal for ABCB6 upregulation is unknown, but exposure of cells to the transcriptional inhibitor actinomycin D before addition of hemin prevents upregulation of ABCB6, suggesting that transcription is important (unpublished data). It is possible that ABCB6 is regulated by heme-binding transcription factors such as Bach 1 and/or Rev-Erb α and Rev-Erbβ. *ABCB6*, like other erythroid genes, may also be under translational control during erythroid differentiation. The upregulation of ABCB6 mRNA is correlated with expansion of the protoporphyrin IX (PPIX) pool *in vitro* and *in vivo*. The mechanism of ABCB6 upregulation is unknown, but it may reflect the fact that PPIX activates p53, and ABCB6 possesses a p53 response element.[17]

2.3 *ABCB6 transport requirements*

The high expression of ABCB6 in erythroid cells and the regulation of ABCB6 by heme suggests that it plays a role in heme biosynthesis. ABCB6 binds heme and has a specific transmembrane domain required for the binding of hemin, the oxidized form of heme.[7] Transmembrane domains have previously been identified as substrate-binding sites of ABC transporters.[33] For example, P-glycoprotein (Pgp) has defined transmembrane segments that interact with cyclosporin A, a known Pgp substrate, and studies of deletion mutants show that binding to the Pgp transmembrane domain is independent of binding to the nucleotide-binding domain.[34] Traditional heme binding motifs, such as CxxxCH or CPV, that covalently bind heme are absent in ABCB6.[25] The absence of covalent binding sites is not unexpected and is consistent with the anticipated necessity for ABCB6 to capture and rapidly release porphyrins as part of the mitochondrial porphyrin import cycle. The porphyrin binding mechanism of ABCB6 may be analogous to that of hemopexin, a plasma protein secreted by the liver which sequesters free heme. Hemopexin binds and releases heme through the actions of a flexible loop and a more constrained domain.[25] Although ABCB6 has no sequence homology to intracellular heme-binding protein HBP23 or to the general tetrapyrrole binding proteins, such as $p22^{HBP}$,[14] several histidine and arginine residues within its transmembrane domain are commonly seen in heme-binding proteins (*35;* reviewed in (*36*)).[35–36]

The importance of ABCB6 in the heme synthetic pathway has been demonstrated in several ways. The binding of ABCB6 to a heme-affinity resin could be specifically disrupted only by certain tetrapyrrole-containing molecules such as hemin, but not by monopyrroles such as porphobilinogen. Moreover, coproporphyrin III (the oxidized form of coproporphyrinogen III, the final cytosolic protoporphyrin precursor) is highly effective in disrupting ABCB6 interaction with hemin-agarose.[7]

The ability of ABCB6 to recognize hemin implies that the iron oxidation state is unimportant for ABCB6-mediated transport of heme, although hemin may be reduced prior to transport. Nevertheless, the strong competitive displacement of ABCB6 by non-iron containing coproporphyrin III in binding to hemin-agarose and the ability of coproporphyrin III to reduce heme transport indicate that heme coordination with iron is not essential to ABCB6 tetrapyrrole recognition. This finding is consistent with early transport studies showing that mitochondria can import protoporphyrin IX, which does not contain iron.[37]

2.4 *Relation between mitochondrial iron, heme biosynthesis, and ABCB6*

In addition to binding and transporting porphyrins, ABCB6 is a regulator of heme and porphyrin biosynthesis.[7] ABCB6 function is a prerequisite for the elevation of intracellular porphyrins, as PPIX concentrations increase in cell lines overexpressing ABCB6 but not in those expressing non-functional ABCB6,[7] suggesting that it is the transport of substrate which promotes porphyrin synthesis. Further, increased porphyrin levels are directly correlated with enhanced *de novo* porphyrin biosynthesis, which is abrogated by silencing ABCB6 gene expression. ABCB6 overexpression also activates genes important for *de novo* heme biosynthesis, suggesting that ABCB6 may represent a previously unrecognized regulator of porphyrin biosynthesis. One might envision a positive regulatory loop where ABCB6 increases heme levels leading to increased gene transcription. This would require a transcription factor that represses gene expression when heme levels are low. One example is BACH1, a transcription factor regulated by heme. Heme binding promotes BACH1 dissociation from the promoter it is repressing.

ABCB6 regulates the pool of heme available for hemoglobin biosynthesis in erythroid cells by increasing the uptake of heme precursors (e.g., coproporphyrinogen III). This level of regulation affects the physiology of the cell substantially, as heme is involved in many different cell processes, and as changes are seen in the activity and expression of a number of hemoproteins, both inside and outside the mitochondria, in response to changes in ABCB6 expression.[8]

2.5 *Functional implications of ABCB6*

ABCB6 also plays a role in the survival of cancer cells. For example, ABCB6 is upregulated in many tumor cell lines and in liver tumors.[12,17] The *ABCB6* gene is amplified in tumor cells that acquire resistance to chemotherapy drugs, suggesting that it has a dominant effect in tumor cells.[38,39]. In addition, microarray analysis shows an inverse relationship between ABCB6 and sensitivity to doxorubicin and cisplatin.[40] Cells that have acquired resistance to arsenite exhibit profound increases in ABCB6 expression and are cross-resistant to cisplatin.[41] Treatment of cancer cells with artesunate resulted in increased ABCB6 expression and, curiously, downregulation of ABCB7,[42] possibly due to the importance of ferrous iron levels in the mitochondria. One explanation for the effectiveness of ABCB6 in drug resistance requires plasma membrane transport, and ABCB6 has been reported to reside in the plasma membrane in some cell types overexpressing ABCB6.[15]

However, an analysis of resistance conferred by ABCB6 orthologues in lower phyla may offer an alternative explanation of resistance. The *S. pombe* ABCB6 orthologue Hmt1p was localized to an intracellular vacuole and reported to compartmentalize xenobiotic substances by vacuolar import.[43] However, recent findings of studies of ABCB6 orthologues in drosophila (D_MHMT-1) and green algae (CrCDS1, which localizes to the mitochondria) indicate that resistance to xenobiotics (e.g., cadmium) is independent of factors known to facilitate vacuolar compartmentation. We speculate that these ABCB6 orthologues may protect from xenobiotics by activating heme biosynthesis.

As mentioned previously, ABCB6 is expressed in non-erythroid tissues such as liver.[12,14,17] In the liver, heme biosynthesis is regulated by feedback inhibition of non-erythroid ALAS1 (5- Aminolevulinate synthetase — see Fig. 1, the first and rate-limiting first step in porphyrin synthesis.[44] Heme blocks the import of ALAS1 into the mitochondrial matrix by binding to an ALAS1 amino-terminal domain called the heme-regulatory motif.[45,46]

However, the source of the mitochondrial heme is unclear. It is possible that in liver cells, ABCB6 regulates heme biosynthesis by re-importing cytoplasmic heme into the mitochondria, where it binds to the heme regulatory motif in ALAS1 located in the inner mitochondrial membrane and thus prevents ALAS1 passage in the inner mitochondrial matrix (Fig. 1). This hypothesis is consistent with the *in vitro* findings that heme does not block ALAS1 binding to outer mitochondrial membranes but does block its passage into the mitochondrial matrix.[45] Moreover, given the recent finding that ABCB6 overexpression increases the activity of complex IV of the electron transport chain, it appears likely that ABCB6 can supply heme to the mitochondria via re-importation.[9]

3. ABCB7

3.1 *Background*

ABCB7 is highly conserved from yeast to man, with the yeast orthologue, Atm1p, sharing 49% identity with the human protein. ATM1 deletion causes mitochondrial iron accumulation in yeast, suggesting that ATM1 has an important role in mitochondrial iron homeostasis. Expression of ABCB7 in Atm1p-deficient yeast provided evidence that ATM1 and ABCB7 have similar functions with ABCB7 able to restore cytochrome levels and reverse mitochondrial iron over-accumulation.[18,20–22]

ABCB7 is a half transporter that homodimerizes[21,47,48] and is located in the mitochondrial inner membrane with its nucleotide-binding domain (NBD) facing the matrix, a position that suggests it transports its substrate into the cytosol inter membrane space.[19,49] ABCB7 is found in many tissues and is most highly expressed in muscle.[47] Several studies have suggested that the transport substrate of ABCB7 is an iron-sulfur (Fe-S) cluster or that ABCB7 has a role in general Fe-S cluster biogenesis[19,49] However, recent studies (see below) suggest that the role of ABCB7 in Fe-S cluster biogenesis may be restricted to a subset of Fe-S cluster proteins.

3.2 *ABCB7 in human disease*

ABCB7 is best known as the causative (defective) gene in X-linked sideroblastic anemia with cerebellar ataxia (XLSA/A).[20,50,51] XLSA/A is a rare disease that results from iron accumulation within the mitochondria, leading to ring sideroblast formation. Affected males have a mild anemia (slightly decreased or low-normal hemoglobin with microcytosis). The neurologic defects in patients with XLSA/A are the distinguishing feature of this disease, presenting as a non-progressive ataxia associated with coordination difficulties. Erythrocytes are hypochromic and have a dramatically elevated protoporphyrin concentration. While XLSA/A results in iron overload only within the mitochondria (visible as sideroblasts), ALAS2 deficiency (see Fig. 1) in common sideroblastic anemia results in both systemic and mitochondrial iron overload without observable ataxia.[52] The ataxia in XLSA/A may be caused by damage from the mitochondrial iron overload or possibly by secondary effects of the disruption of iron homeostasis (concurrent with changes in protoporphyrin levels) within the neural cells. In 1991, XLSA/A was linked to chromosome Xq13[53] and mutational analysis of several XLSA/A-affected families showed that missense point mutations within *ABCB7* exon 9 or 10 are responsible for the defective protein. One study found a point mutation (T1200G) that changes the encoded amino acid from Ile (ATT) to Met (ATG) at position 400 in the protein (I400M). This residue is located within the predicted fifth transmembrane helix of ABCB7.[20] Other familial studies have found a mutation at G1299C (V411L), in the putative sixth transmembrane helix,[51] and at G1305A (E433K).[50] While these findings are consistent with the view that the transmembrane domains are the substrate binding sites in ABC transporters it is just as likely that these mutations affect ABCB7 folding, which could affect inter-domain communication and/or dimerization.

3.3 *ABCB7 function*

The physiological manifestations of ABCB7 defects are consistent with the role of the mitochondrion in regulating iron-sulfur (Fe-S) clusters and iron homeostasis. Mitochondria play a central role in iron metabolism, as evidenced by both the heme biosynthetic pathway and the biosynthesis of Fe-S clusters. Iron that is transported to the mitochondrion is used in heme biosynthesis, stored as mitochondrial ferritin, or used in the biogenesis of Fe-S clusters.[54,55] Fe-S clusters are integral to the structure and function of both mitochondrial and extra-mitochondrial enzymes. The mitochondrion is the synthetic site for both 2Fe-2S and 4Fe-4S clusters that are crucial to the function of enzymes involved in a variety of pathways, including the regulation of iron itself. Mitochondrial Fe-S containing proteins include respiratory enzymes (complexes I and III), ferrochelatase, and enzymes of the citric acid cycle (aconitase and succinate dehydrogenase). Cytosolic and nuclear Fe-S containing proteins include iron regulatory protein 1 (IRP1 — whose holo-form is c-aconitase) and human endonuclease III homolog (hNTH1).[19,56–58]

Several of these Fe-S cluster-containing proteins are directly involved in the regulation of iron homeostasis. IRP1 contains a 4Fe-4S cluster when in the active state. However, during states of iron deficiency, IRP1 lacks its 4Fe-4S cluster and is able to bind to the iron response element (IRE) within the mRNA of transferrin receptor 1 (TfR1) (causing increased expression) and ferritin (causing decreased expression)[59] with the net result producing enhanced iron uptake by TfR1 and decreased storage/sequestration by the iron scavenger, ferritin.

The mitochondrial enzyme ferrochelatase is the final enzyme in the heme biosynthetic pathway (Fig. 1), inserting a ferrous iron ($Fe^{2}+$) into protoporphyrin IX to form heme. The enzyme contains a 2Fe-2S cluster at its C-terminus and is necessary for enzyme activity, as C-terminal truncations of the enzyme are inactive.[60,61]

3.4 *ABCB7 knockout and knockdown*

Several recent studies in mammalian models of ABCB7 deficiency provide insight into its physiological role in Fe-S cluster regulation. Pondarre *et al.*[62] looked at the tissue-specific responses of a conditional *Abcb7*-null mouse. Complete absence of ABCB7 resulted in prenatal death, indicating that ABCB7 is essential for normal development during early gestation; death was caused by the tissue-specific deletion of Abcb7 in most organs, including the CNS and bone marrow. These findings support the attribution of human XLSA/A to a partially defective ABCB7, as opposed to a complete loss of ABCB7 function. Surprisingly Abcb7 was not required for the survival of hepatocytes and endothelial cells. There are several possible explanations for this finding: incomplete deletion of the conditional allele within these cells, the possibility of redundancy between Abcb7 and other regulators of mitochondrial iron or Fe-S clusters in certain tissues, or compensation mechanisms for the loss of Abcb7 in certain tissues. It is likely that their role in iron metabolism renders hepatocytes uniquely able to adapt to metabolic fluctuations caused by the loss of Abcb7.[62]

Loss of ABCB7 produces Fe-S cluster deficiency and activation of IRP1 and IRP2. This result is consistent with the disruption of iron homeostasis, as evidenced by increased iron uptake (likely due to increased TfR1 expression). Paradoxically, ferritin expression also increased, possibly because of increased iron influx from the constitutive activation of TfR1. Consistent with the elevated ferritin was the finding of cytoplasmic iron inclusions that were morphologically similar (as imaged by electron microscopy) to structures seen in murine neurons lacking frataxin. This observation suggests that iron sequestration may be a generalized response to cytosolic Fe-S–cluster deficiency.[63]

In contrast to the clinical presentation of XLSA/A, Abcb7-deficient mice did not show iron overload, and mitochondrial Fe-S proteins such as ferrochelatase, succinate dehydrogenase, and m-aconitase were only mildly affected. However, the levels of some cytosolic Fe-S cluster proteins were reduced; these included c-aconitase (the holoform of IRP1) and xanthine oxidase. This discrepancy between levels of mitochondrial and cytosolic Fe-S cluster proteins may be an indication of compensation by another mitochondrial ABC transporter in the absence of Abcb7. While these results suggest that Abcb7 plays a role in cytosolic Fe-S cluster formation, they do not specifically address the role of ABCB7 in cellular iron movement or the cause of the sideroblastic anemia seen in XLSA/A patients.

Cavadini *et al.* used siRNA knockdown of ABCB7 in HeLa cells to address some of these questions about the role of ABCB7 in iron storage. When the protein was suppressed for seven days by sequential siRNA transfections, cytosolic iron deprivation in the cells was followed by a six-fold increase in mitochondrial iron that was not available to mitochondrial ferritin. Markers of mitochondrial activity (membrane potential, succinate dehydrogenase activity and ATP content) were not affected. As with the *ABCB7* conditional knockout, there was increased activation of the IRE-binding activity of IRP1, leading to an increase in TfR1. There was also a decrease in ferritin and frataxin levels, possibly due to reduced stability of the apo-form of the enzyme.[64]

The findings of the above studies did not fully explain why the defective ABCB7 renders XLSA/A patients unable to synthesize sufficient heme despite the presence of sufficient PPIX and iron. One explanation comes from the finding that ABCB7 expression upregulates the expression of ferrochelatase, the final enzyme in the heme biosynthetic pathway, which catalyzes the insertion of iron into protoporphyrin IX.[65] Without ABCB7, the catalytic activity of ferrochelatase is insufficient for heme biosynthesis. However, this explanation does not account for the fact that most of these patients' erythrocyte protoporphyrin is in the form of Zn-protoporphyrin, which is formed by ferrochelatase.[66] Moreover, the conditional knockout mouse study showed no inhibition of ferrochelatase activity in the absence of Abcb7.[62] Another explanation is that Abcb7 deficiency activates IRP1, which binds to the IRE in ALAS2, thereby disrupting expression of this critical heme synthesis protein. However, this explanation is incomplete because XLSA/A patients have sufficient PPIX and iron. Nonetheless, the role of iron regulatory protein IRP1 in heme synthesis is implied by the elevated erythrocyte protoporphyrin in *IRP2*-null mice, which may result from localized iron deficiency.[24] A final and more plausible explanation is that the iron that accumulates as a consequence of ABCB7 deficiency is not directly usable by enzymes such as ferrochelatase in heme synthesis. This possibility is borne out by the results of siRNA knockdown in HeLa cells, in which the iron excess in the *ABCB7*-silenced cells was not available to mitochondrial ferritin (MtF). In this case, loss of ABCB7 apparently reduces the pool of Fe (II) ions available to MtF and ferrochelatase, thereby increasing the level of free protoporphyrin.

3.5 *ABCB7 in refractory anemia*

Myelodysplastic syndrome (MDS) comprises a group of hematopoietic malignancies characterized by peripheral-blood cytopenias and ineffective hematopoiesis. Refractory anemia with ringed sideroblasts (RARS) is a subtype of MDS in which iron accumulates in the mitochondria, but the molecular and genetic basis of RARS is unknown.[6] Because defective ABCB7 results in iron accumulation in the mitochondria of XLSA/A patients, a recent study tested the involvement of ABCB7 in RARS. While patients with RARS had a normal *ABCB7* gene, *ABCB7* mRNA expression was reduced, supporting the role of ABCB7 deficiency in RARS. Further, a strong inverse correlation was shown between ABCB7 expression and the percentage of ringed sideroblasts.[6]

4. ABCB8

4.1 *Background*

While ATM1p was the first mitochondrial ABC transporter identified in yeast, ABCB8 was the first to be identified in mammals and was initially named M-ABC1 (mitochondrial ABC1).[67] ABCB8 was first discovered when Ling and colleagues identified an expressed sequence tag (EST) clone that contained a putative ABC transporter sequence by using the conserved NBD domain of P-glycoprotein (ABCB1) to screen the EST database. This EST clone was subsequently used to obtain the full sequence of ABCB8 mRNA. The 2417-bp cDNA had an open reading frame of 2154 bp, a deduced amino acid sequence of 718 amino acids, and a domain organization of a single N-terminal membrane-spanning domain followed by a C-terminal nucleotide-binding domain (NBD).[67] Northern blot analysis revealed that ABCB8 mRNA is ubiquitously expressed in human tissues,[67] and the *ABCB8* gene was mapped to chromosome 7q35–q36.[68]

4.2 *Mitochondrial localization and targeting of ABCB8*

Under immunofluorescence microscopy, ABCB8 co-localized with the mitochondrial marker MitoTracker Red.[68] However, it appeared to undergo post-translational processing, as the size of ABCB8 produced by *in vitro* transcription/translation (72 kDa) differed from that of ABCB8 produced by transcription *in cellulo* (60 kDa). One possibility was a cleavable mitochondrial targeting sequence. Mitochondrial targeting sequences are characteristically α-helices enriched with positively charged amino acids that reside in the N-terminus of the protein (Table 1).[69–71] These positively charged α-helices are recognized by chaperones and translocons (TIM/TOM). After mitochondrial translocation is completed the signal sequence is cleaved. To test the role of such a targeting sequence in the N-terminus, the first 55 amino acids of ABCB8 were fused to enhanced green fluorescent protein (EGFP) and expressed in SKVO3 cells. The green fluorescence co-localized with MitoTracker Red.[67] The ability of the N-terminal 55 amino acids of ABCB8 to target a non-mitochondrial protein to the mitochondria suggests that this segment is sufficient to act as a mitochondrial targeting sequence. In a separate study, ABCB8 protein was detected in purified mitochondrial inner membranes.[72] Typically, mitochondrial inner membrane proteins contain N-terminal targeting sequences that are cleaved upon reaching their destination.[69] Although there is no formal evidence, it is likely that the ABCB8 N-terminal mitochondrial targeting sequence is cleaved, a finding that would be consistent with the decreased molecular mass and inner mitochondrial localization of ABCB8.[67,72]

4.3 *Biochemistry and function of ABCB8*

ABC half transporters must homo- or hetero-dimerize to meet the functional requirement for two nucleotide-binding domains. Previous studies demonstrated that ABCB6, Atm1p (ABCB7 orthologue), and ABCB10 homodimerize;[7,73,74] therefore, it is likely that ABCB8 forms a homodimer. While homodimerization has not been formally demonstrated, it is supported by cross-linking studies that show an ABCB8 band of 120–130 kDa, consistent with an ABCB8 homodimer (monomer = 60 kDa).[67]

The conserved motifs in the nucleotide-binding domain (NBD) are crucial for ATP-dependent transport by ABC transporters; mutation of conserved residues within this domain disrupts transport function. Ling and colleagues hypothesized that an NBD mutant (Walker A motif) of ABCB8 would exert a dominant negative effect on wild-type cellular ABCB8.[67] Unexpectedly, the mutant protein was undetectable despite an abundance of mutant RNA transcripts. Paradoxically, *in vitro* transcription and translation produced comparable amounts of wild-type and mutant ABCB8 proteins. This finding suggested that an intact Walker A motif is important for the stability of ABCB8 *in vivo*.

The substrates transported by ABCB8 are currently unknown, but recent studies have closely linked ABCB8 function to mitochondrial ATP-sensitive K^+ (mitoK_{ATP}) channels whose molecular identity is unknown.[10,72] Activation of mitoK_{ATP} channels in myocardial cells protected the cells against ischemia and oxidant-induced apoptosis.[75–77] Functional and pharmacological studies revealed overlaps between mitoK_{ATP} channels and succinate dehydrogenase (SDH) function, although SDH had been suspected to be a part of the channel. Notably, SDH inhibition activated the mitoK_{ATP} channel. Immunoprecipitation from the mitochondrial inner membrane using antibodies against putative mitoK_{ATP} channels (e.g., SDH and ABCB8) revealed four mitochondrial inner membrane proteins in the immunocomplex: ABCB8, phosphate carrier (PIC), adenine nucleotide translocator, and ATP synthase.[72] The interaction between ABCB8, SDH, and PIC was further supported by yeast two-hybrid analysis, suggesting that the proteins directly contacted each other in this multiprotein complex.[72] This possibility was supported by the fact that isolated mitochondrial inner membranes enriched for ATP synthase, PIC, and ATP/DP carrier were re-constituted into proteoliposomes and exhibited activatable K^+ channel activity. However, we do not know to what extent ABCB8 contributes to this ATP-sensitive K^+ channel. To address this issue, ABCB8 was targeted to the plasma membrane by removing its N-terminal mitochondrial targeting sequence. Cells expressing ABCB8 at the plasma membrane exhibited passive ion movement, supporting the idea that the movement is mediated by ABCB8.[78] However, this interpretation assumes that removal of the mitochondrial targeting sequence does not affect the function of ABCB8 by altering its folding/topology. Thus, further elucidation of the role of ABCB8 as a mitoK_{ATP} channel is required.

While the exact contribution of ABCB8 to the mitoK_{ATP} channel is unclear, recent studies have examined its role in protection against oxidative stress. Marban *et al.* showed that ABCB8 knockdown in neonatal rat cardiomyocytes disrupts mitochondrial membrane potential and cell viability.[10] Conversely, overexpression of ABCB8 in these cells protected against oxidant-induced reduction of mitochondrial membrane potential. Paradoxically, the pattern of mitochondrial membrane potential was not affected by treating the ABCB8-overexpressing cells with a mitoK_{ATP} channel inhibitor, 5-hydroxydecanoate, before H_2O_2 exposure. This finding suggests that the protective effect of ABCB8 is independent of K^+ channel activity. However, an alternative explanation is that ABCB8 overexpression altered the stoichiometry of the proteins in the mitoK_{ATP} channel multiprotein complex and rendered the K^+ channel insensitive to the inhibitor. Additional studies are needed to reconcile the relation of ABCB8, mitoK_{ATP} channel, and ABCB8-mediated protection against oxidative damage.

Multidrug resistance conferred by ABC transporter subfamily members has typically been ascribed to plasma membrane ABC transporters that pump drugs out of cells. However, the ABC transporters that reside in the intracellular organelles, such as ABCB8, can also confer drug resistance through indirect mechanisms. For example, ABCB8 protects against oxidants. A screen of doxorubicin-resistant melanoma cell lines suggested that ABCB8 contributes to resistance. Knock-down of ABCB8 by shRNA increased the doxorubicin sensitivity of the cells. This effect was specific to doxorubicin, as ABCB8 deficiency did not alter sensitivity to taxol, 5-fluorouracil, or cisplatin.[11] Significantly greater DNA damage was caused by doxorubicin when the cells expressed *ABCB8* shRNA. Because doxorubicin causes mitochondrial DNA damage, the mechanism by which ABCB8 protects mitochondrial DNA from doxorubicin requires further investigation, although it may be related to the ability of ABCB8 to protect against oxidants.

Although ABCB8 has no known substrates, its role in protecting against oxidant-induced impairment of mitochondrial function suggests a physiological pathway that may be coupled to its role in mitoK_{ATP} channels. Development of animal models that lack ABCB8 may provide valuable insights into its role in basal and stressed/diseased states.

5. ABCB10

5.1 *Background*

ABCB10 was identified as a gene that is upregulated during erythroid differentiation and induced by the erythroid transcription factor GATA-1. GATA-1 is essential for erythropoiesis and regulates many erythroid-specific genes. The proerythroblast cell line G1ER expresses a conditionally regulated GATA-1 and terminally differentiates after GATA-1 activation. PCR-based subtractive gene analysis was used to identify mRNAs selectively increased by GATA-1. While a majority of the transcripts identified were known targets of GATA-1, such as α- and β-globins, some transcripts (6 of 212) had little homology to known sequences in databases. One such clone displayed homology to members of the ABC transporter superfamily. Northern blot analysis using this fragment as a probe revealed a 4.1-kb mRNA that was strongly upregulated by GATA-1 and induced in differentiating murine proerythroblasts. This gene was named ABC-me (mitochondrial localization and erythroid expression).[79]

The full-length ABC-me cDNA contained an open reading frame encoding 716 amino acids. The domain arrangement was the same as that of other mitochondrial ABC half-transporters: the N-terminal membrane-spanning domain had six α-helices (predicted from hydrophilicity plot) followed by an NBD. The *ABCB10* gene was mapped to human chromosome 1q32 and is highly expressed at sites of definitive hematopoiesis, such as adult bone marrow and fetal liver.[79]

5.2 *Mitochondrial targeting of ABCB10*

The substrates of ABCB10 are unknown, although ABCB10 is upregulated by erythroid differentiation and multiple studies have posited a role in heme export (Fig. 1). Biochemical studies have provided some clues to the molecular requirements for transport. Mass spectrometry of ABCB10 immunoprecipitated from cross-linked mitochondria showed that ABCB10 forms a homodimer that does not appear to interact with other proteins.[74] In addition, the C-terminal NBD resides in the mitochondrial matrix. Taken together, these findings suggest that ABCB10 resides in the inner membrane as a homodimer that is not closely associated with other proteins. Biochemical studies place the NBD of ABCB10 in the mitochondria matrix, which suggests ABCB10 transports substrates out of the matrix into the intermembrane space. However, additional studies are needed to confirm the direction of the transport.

To date, the ABCB10 domains required for mitochondrial targeting are among those best characterized within the mitochondrial ABC transporter group. Mitochondrial inner membrane localization of ABCB10 was first demonstrated by immunofluorescence microscopy and by subcellular fractionation (removal of the outer mitochondrial membrane) coupled with density gradient centrifugation.[79] The mitochondrial inner membrane/matrix localization of the C-terminus was also demonstrated by protection from proteinase K digestion. Deletion of the N-terminal 69 amino acids produced cytosolic distribution of ABCB10, suggesting that this region is required for mitochondrial targeting.

Further studies revealed a long mitochondrial targeting sequence in ABCB10. Unexpectedly, immunoblot analysis produced a single band of 65kDa rather than the expected 77.5 kDa.[74] N-terminal sequencing of immunopurified ABCB10 was performed to determine whether this finding reflected the removal of a mitochondrial targeting sequence. The first 17 amino acids identified in the immunopurified ABCB10 corresponded to amino acids 106–123 in full length ABCB10, suggesting that the first 105 amino (a non-transmembrane helix acids act as a cleavable (presequence) mitochondrial targeting sequence. A chimeric protein created by appending this putative presequence (amino acids 1–105) to EGFP was localized to the mitochondria, while the deletion of these residues from full-length ABCB10 prevented mitochondrial localization.[74] These results indicated that the first 105 amino acid residues are both necessary and sufficient for ABCB10 mitochondrial localization. However, because this sequence is much longer than the typical mitochondrial targeting sequence (usually 34 ± 17 amino acids), it was further characterized.

To determine whether the mitochondrial targeting sequence of ABCB10 could be subdivided, three distinct subdomains (1 = residues 1 – 35; 2 = 36 – 70; and 3 = 71 – 105) were selected. Each contained both a predicted α-helix and positively charged residues, which are typically required for mitochondrial targeting. Each subdomain was fused to EGFP. The second subdomain was the most effective in mitochondrial targeting of EGFP, but the first subdomain was required for efficient targeting of ABCB10 to the inner membrane. Unexpectedly, neither positively charged amino acids nor the putative α-helix were crucial for mitochondrial targeting of the second subdomain; instead, reduced hydrophobicity resulted in reduced mitochondrial localization of this domain. Together, these findings show that although the first 105 amino acids are sufficient for ABCB10 mitochondrial localization, internal subdomains appear essential for the correct positioning of ABCB10 within the mitochondria.[74]

5.3 *The ABCB10 orthologue Mdl1*

A phylogenetic analysis suggested that yeast Mdl1 is an ABCB10 homolog,[80] and it was speculated that analysis of its substrates might facilitate identification of ABCB10 substrates. Mdl1 was identified by screening yeast databases for mitochondrial ABC transporters with homology to ABCB2/3 (transporters associated with antigen processing; TAPs)(see Chapter 6),[81] which reside in the endoplasmic reticulum and transport peptides into the ER lumen. Mitochondria from yeast cells deficient in Mdl1 showed decreased release of long peptides, whereas release of short peptides or free amino acids was unaffected. Therefore, it was hypothesized that Mdl1 is a component of mitochondrial protein quality control. Mdl1 exports peptides from the mitochondrial matrix in an ATP-dependent manner, and Mdl1 peptide export is then coupled to the mitochondrial M-AAA protease.[81] However, although Mdl1 is an inner membrane protein with a cleavable mitochondrial targeting sequence, it is unknown whether Mdl1 is a functional ortholog of ABCB10, as no direct evidence of peptide transport by ABCB10 (e.g., ABCB10 complementing the delta Md11 yeast phenotype) has been reported to date.

5.4 *Role of ABCB10 in hemoglobin biosynthesis*

Induction of ABCB10 during erythroid differentiation and the mitochondrial matrix localization of ABCB10 suggest that ABCB10 exports mitochondrial heme. When ABCB10 was overexpressed by stable transfection of proerythroblasts undergoing differentiation, intracellular hemoglobin content increased.[79] If ABCB10 exports heme, one might expect an increase in hemoglobin production as heme is co-translationally inserted into hemoglobin. ABCB10 overexpression did not accelerate the onset of hemoglobin production, which suggests that it does not directly export heme. Moreover, ABCB10 overexpression did not induce spontaneous differentiation. However, heme biosynthesis and erythroid differentiation can be uncoupled. Because heme is generated in the mitochondrial matrix, it could be speculated that ABCB10 transports heme into the intermembrane space, where it is then released into the cytosol via an unknown carrier or chaperone. To date, no alterations in mitochondrial heme content have been demonstrated in ABCB10-expressing cells. Further, direct binding or efflux of heme by ABCB10 has yet to be determined. At this time, there is no direct evidence that heme is a substrate of ABCB10. Future studies using knockdown of ABCB10 in proerythroblasts and animal models lacking ABCB10 will certainly provide insight into its role in erythroid differentiation. However, elucidating this role may be complex, as ABCB10 stabilizes mitoferrin, a gene essential for heme and Fe-S cluster synthesis in erythroblasts.[9,82] Because most of the current studies have used murine ABCB10, additional studies are needed to determine whether human ABCB10 has a role in hemoglobin synthesis. Nonetheless, it is worth noting that three of the four mitochondrial ABC transporters — ABCB6, ABCB7, and ABCB10 — are involved in erythroid biology.

5.5 *Regulation of ABCB10*

The initial identification of ABCB10 indicated that ABCB10 expression increases in proerythroblasts programmed to differentiate by GATA-1.[79] Moreover, GATA-1-deficient erythroid cells contain little to no ABCB10 mRNA, supporting the role of GATA-1 in regulating ABCB10 expression. Indeed, the promoter region of Abcb10 contains a GATA-1 binding motif; however, evidence of direct transcriptional regulation by GATA-1 is lacking.

Interestingly, addition of heme to a differentiating proerythroblast cell line significantly reduced ABCB10 transcript levels. Because iron, heme, and heme precursors regulate several steps in the heme biosynthetic pathway, heme may activate a negative feedback mechanism in which elevated heme concentrations downregulate ABCB10. Because it is still unclear whether heme or its precursor(s) are substrates of ABCB10, this possible feedback mechanism warrants further investigation.

6. Conclusions and Perspectives

There are several common themes that emerge for the mitochondrial transporters. The first is that three of them (ABCB6, ABCB7 and ABCB10) have significant roles in heme/iron homeostasis. Deficiency of ABCB7 produces mitochondrial iron accumulation and defects in heme synthesis with a sideroblastic anemia that is uniquely characterized by the cerebellar ataxia it causes. While the transport substrate for ABCB10 is unknown it appears to be important in regulating mitochondrial iron import by interacting with mitoferrin 1. However, it is unknown if this interaction regulates heme synthesis. ABCB6 expression is increased by activitation of heme synthesis which fits with its capability to import porphyrins and regulate heme synthesis. Notably, the capability of ABCB6 to increase heme synthesis provides cells with the capability to survive certain types of oxidative stress.[8] ABCB8 also provides a survival advantage against oxidative stress challenge.

Another theme is the finding that the mitochondrial ABC transporters, when targeted to the mitochondrial matrix, do not use a mitochondrial targeting sequence of typical length. In the case of ABCB8 and ABCB10 the cleavable presequence is 2–3 times longer than expected. This suggests a unique pathway to the matrix that might require novel protein interactions and chaperones. Finally, it is unknown how ABCB6 targets the outer mitochondrial membrane.

Acknowledgments

This work was supported by NIH grants ES058571, P30 CA21745, and CA21865 and by the American Lebanese Syrian Associated Charities (ALSAC). We thank Sharon Naron for excellent editoral assistance.

References

1. Visapaa, I., *et al.* (2002) *Am. J. Hum. Genet.* **71**, 863.
2. Napier, I., Ponka, P., and Richardson, D.R. (2005) *Blood* **105**, 1867.
3. Fontenay, M., Cathelin, S., Amiot, M., Gyan, E., and Solary, E. (2006) *Oncogene* **25**, 4745.
4. Moore, M.R. (1993) *Int. J. Biochem.* **25**, 1353.
5. Beal, M.F. (2004) *J. Bioenerg. Biomembr.* **36**, 381.
6. Boultwood, J., *et al.* (2008) *PLoS One* **3**, e1970.
7. Krishnamurthy, P.C., *et al.* (2006) *Nature* **443**, 586.
8. Lynch, J., Fukuda, Y., Krishnamurthy, P., Du, G., and Schuetz, J.D. (2009) *Cancer Res.* **69**, 5560.
9. Chen, W., *et al.* (2009) *Proc. Natl. Acad. Sci. U.S.A.* **106**, 16263.
10. Ardehali, H., O'Rourke, B., and Marban, E. (2005) *Circ. Res.* **97**, 740.
11. Elliott, A.M. and Al-Hajj, M.A. (2009) *Mol. Cancer Res.* **7**, 79.

12. Furuya, K.N., Bradley, G., Sun, D., Schuetz, E.G., and Schuetz, J.D. (1997) *Cancer Res.* **57**, 3708.
13. Hirsch-Ernst, K.I., *et al.* (1998) *Biochem. Biophys. Res. Commun.* **249**, 151.
14. Mitshhashi, N., *et al.* (2000) *J. Biol. Chem.* **275**, 17536.
15. Paterson, J.K., *et al.* (2007) *Biochemistry* **46**, 9443.
16. Claros, M.G., Brunak, S., and von Heijne, G. (1997) *Curr. Opin. Struct. Biol.* **7**, 394.
17. Emadi-Konjin, H.P., *et al.* (2002) *Biochim. Biophys. Acta.* **1574**, 117.
18. Kispal, G., Csere, P., Guiard, B., and Lill, R. (1997) *FEBS Lett.* **418**, 346.
19. Kispal, G., Csere, P., Prohl, C., and Lill, R. (1999) *EMBO J.* **18**, 3981.
20. Allikmets, R., *et al.* (1999) *Hum. Mol. Genet.* **8**, 743.
21. Csere, P., Lill, R., and Kispal, G. (1998) *FEBS Lett.* **441**, 266.
22. Senbongi, H., Ling, F., and Shibata, T. (1999) *Mol. Gen. Genet.* **262**, 426.
23. Sheps, J.A., Ralph, S., Zhao, Z., Baillie, D.L., and Ling, V. (2004) *Genome. Biol.* **5**, R15.
24. Weber, G.J., *et al.* (2005) *Blood* **106**, 521.
25. Paoli, M., Marles-Wright, J., and Smith, A. (2002) *DNA Cell Biol.* **21**, 271.
26. Li, Z., *et al.* (2003) *Proc. Natl. Acad. Sci. U.S.A.* **100**, 8164.
27. Dominguez-Sola, D., *et al.* (2007) *Nature* **448**, 445.
28. Iritani, B.M. and Eisenman, R.N. (1999) *Proc. Natl. Acad. Sci. U.S.A.* **96**, 13180.
29. Sutiphin, P.D., Giaccia, A.J., and Chan, D.A. (2007) *Dev. Cell.* **12**, 845.
30. Cam, H., *et al.* (2004) *Mol. Cell* **16**, 399.
31. Virbasius, J.V. and Scarpulla, R.C. (1994) *Proc. Natl. Sci. U.S.A.* **91**, 1309.
32. Nilsson, R., *et al.* (2009) *Cell Metab.* **10**, 119.
33. Sauna, Z.E. and Ambudhar, S.V. (2001) *J. Biol. Chem.* **276**, 11653.
34. Dey, S., Ramachandra, M., Pastan, I., Gottesman, M., and Ambudkar, S.V. (1997) *Proc. Natl. Acad. Sci. U.S.A.* **94**, 10594.
35. Allen, J.W., Leach, N., and Ferguson, S.J. (2005) *Biochem. J.* **389**, 587.
36. Chan, M.K. (2001) *Curr. Opin. Chem. Biol.* **5**, 216.

37. Koller, M.E. and Romslo, I. (1980) *Biochem. J.* **188**, 329.
38. Yasui, K., *et al.* (2004) *Cancer Res.* **64**, 1403.
39. Gillet, J.P., *et al.* (2004) *Cancer Res.* **64**, 8987.
40. Ju, Y.S.J., Fodstad, O., and Bruheim, S. (2005) *ASCO Annual Meeting Proceedings.*
41. Paterson, J.K., Tachiwada, T., Akiyama, S., and Gottesman, M. (2005) *Proc. Am. Cancer Res.* **46**, 122.
42. Kelter, G., *et al.* (2007) *PLoS ONE* **2**, e798.
43. Ortiz, D.F., *et al.* (1992) *EMBO J.* **11**, 3491.
44. Srivastava, G., *et al.* (1983) *Biochem. Biophys. Res. Commun.* **117**, 344.
45. Lathrop, J.T. and Timko, M.P. (1993) *Science* **259**, 522.
46. Munakata, H., *et al.* (2004) *J. Biochem.* **136**, 233.
47. Savary, S., *et al.* (1997) *Genomics* **41**, 275.
48. Shimada, Y., *et al.* (1998) *J. Hum. Genet.* **43**, 115.
49. Lill, R. and Muhlenhoff, U. (2005) Trends *Biochem. Sci.* **30**, 133.
50. Bekri, S., *et al.* (2000) *Blood* **96**, 3256.
51. Maguire, A., Hellier, K., and Hammans, S. (2001) *Br. J. Haematol.* **115**, 910.
52. Camaschella, C. (2008) *Br. J. Haematol.* **143**, 27.
53. Raskind, W.H., *et al.* (1991) *Am. J. Hum. Genet.* **48**, 335.
54. Muhlenhoff, U. and Lill, R. (2000) *Biochim. Biophys. Acta.* **1459**, 370.
55. Ponka, P. (1997) *Blood* **89**, 1.
56. Leighton, J. and Schatz, G. (1995) *EMBO J.* **14**, 188.
57. Rouault, T.A. and Tong, W.H. (2005) *Nat. Rev. Mol. Cell. Biol.* **6**, 345.
58. Beinert, H., Holm, R.H., and Munck, E. (1997) *Science* **277**, 653.
59. Eisenstein, R.S. (2000) *Annu. Rev. Nutr.* **20**, 627.
60. Furukawa, T., Kohno, H., Tokunaga, R., and Taketani, S. (1995) *Biochem. J.* **310**, 533.
61. Dailey, H.A., Finnegan, M.G., and Johnson, M.K. (1994) *Biochemistry* **33**, 403.
62. Pondarre, C., *et al.* (2006) *Hum. Mol. Genet.* **15**, 953.
63. Puccio, H., *et al.* (2001) *Nat. Genet.* **27**, 181.
64. Cavadini, P., *et al.* (2007) *Blood* **109**, 3552.
65. Taketani, S., Kakimoto, K., Ueta, H., and Masaki, R. (2003) *Blood* **101**, 3274.

66. Jacobs, J.M., *et al.* (1998) *Toxicology* **125**, 95.
67. Hogue, D.L., Liu, L., and Ling, V. (1999) *J. Mol. Biol.* **285**, 379.
68. Allikmets, R., Gerrad, B., Hutchinson, A., and Dean, M. (1996) *Hum. Mol. Genet.* **5**, 1649.
69. Pfanner, N. (2000) *Curr. Biol.* **10**, R412.
70. Roise, D. and Schatz, G. (1998) *J. Biol. Chem.* **263**, 4509.
71. von Heijne, G., Steppuhn, J., and Herrmann, G. (1989) *Eur. J. Biochem.* **180**, 535.
72. Ardehali, H., Chen, Z., Ko, Y., Meijia-Alvarez, R., and Marban, E. (2004) *Proc. Natl. Acad. Sci. U.S.A.* **101**, 11880.
73. Chloupkova, M., Reaves, S.K., LeBard, L.M., and Koeller, D.M. (2004) *FEBS Lett.* **569**, 65
74. Graf, S.A., Haigh, S.E., Corson, E.D., and Shirihai, O.S. (2004) *J. Biol. Chem.* **279**, 42954.
75. Grover, G.J. and Garlid, K.D. (2000) *J. Mol. Cell. Cardiol.* **32**, 677.
76. Liu, Y., Sato, T., O'Rourke, B., and Marban, E. (1998) *Circulation* **97**, 2463.
77. Murata, M., Akao, M., O'Rourke, B., and Marban, E. (2001) *Circ. Res.* **89**, 891.
78. Ardehali, H., Xue, T., Dong, P., and Machamer, C. (2005) *Biochem. Biophys. Res. Commun.* **338,** 1143.
79. Shirihai, O.S., Gregory, T., Yu, C., Orkin, S.H., and Weiss, M.J. (2000) *EMBO J.* **19**, 2492.
80. Herget, M. and Tampe, R. (2007) *Pflugers Arch.* **453**, 591.
81. Young, L., Leonhard, K., Tatsuta. T., Trowsdale, J., and Langer, T. (2001) *Science* **291**, 2135.
82. Shaw, G.C., *et al.* (2006) *Nature* **440**, 96.

Chapter 8

The Role of ABCC Family Members in the Disposition of Endogenous Compounds and Drugs

Jörg König, Martin F. Fromm and Hartmut Glaeser

1. MRP1 (Gene Symbol *ABCC1*)

MRP1 is the founding member of the ABCC subfamily cloned in 1992 from a drug-selected human lung cancer cell line.[1] Although selected by a single drug this H69AR cell line displays cross-resistance to a variety of structurally unrelated compounds and drugs but shows no overexpression of the well studied drug efflux transporter MDR1 P-glycoprotein.[2] A cytogenic analysis of this cell line indicated that an mRNA was overexpressed which was located on chromosome 16p13.1 (Table 1). This led to the identification and subsequent cloning of a protein initially named multidrug resistance-associated protein (MRP) and now termed MRP1. Detailed analysis of the deduced amino acid sequence of this isolated mRNA demonstrated that the protein has two cytoplasmic domains containing motifs typical of ATP-binding domains of known ABC transporters, including MDR1 (P-glycoprotein). This suggested that MRP1 was the cause of the drug-resistance phenotype in the H69AR cell line. Since then 48 different ABC transporters have been identified in the human

Table 1. **Molecular properties of the human ABCC subfamily members.**

Gene symbol	Protein	Chromosomal localization	Amino acids	NCBI nucleotide accession number
ABCC1	MRP1	16p13.1	1531	NM_004996
ABCC2	MRP2	10q24	1545	NM_000392
ABCC3	MRP3	17q22	1527	NM_003786
ABCC4	MRP4	13q32	1325	NM_005845
ABCC5	MRP5	3q27	1437	NM_005688
ABCC6	MRP6	16p13.1	1503	NM_001171
ABCC7	CFTR	7q31.2	1480	NM_000492
ABCC8	SUR1	11p15.1	1581	NM_000352
ABCC9	SUR2A	12p12.1	1549	NM_005691
	SUR2B		1549	NM_020297
ABCC10	MRP7	6p21.1	1464	NM_033450
ABCC11	MRP8	16q12.1	1382	NM_032583
ABCC12	MRP9	16q12.1	1359	NM_033226

genome[3,4] and based on their sequence similarities these have been classified into the seven subfamilies ABCA–ABCG. Starting with MRP1 as the first member of the ABCC family, 12 human ABCC family members have been identified (Table 1). These are nine MRPs (MRP1–MRP9/*ABCC1–ABCC6* and *ABCC11–ABCC12*),[5,6] the CFTR protein (*ABCC7*)[7] and both SUR proteins (*ABCC9* and *ABCC10*).[8]

Based on the predicted topology of their membrane spanning regions the ABCC family can be divided into two subgroups. MRP1, MRP2, MRP3, MRP6, together with the two SUR proteins, form the so-called "long" MRPs, containing three predicted transmembrane domains (TMDs) and two ATP-binding domains (NBDs) with a structure TMD0–TMD1–NBD1–TMD2–NBD2, (Fig. 1), MRP4, MRP5, MRP8 (the "short" MRPs) and CFTR have the more typical ABC transporter structure with two transmembrane domains and two nucleotide-binding domains — TMD1–NBD1–TMD2–NBD2 — Fig. 1. The NH_2-terminal extension

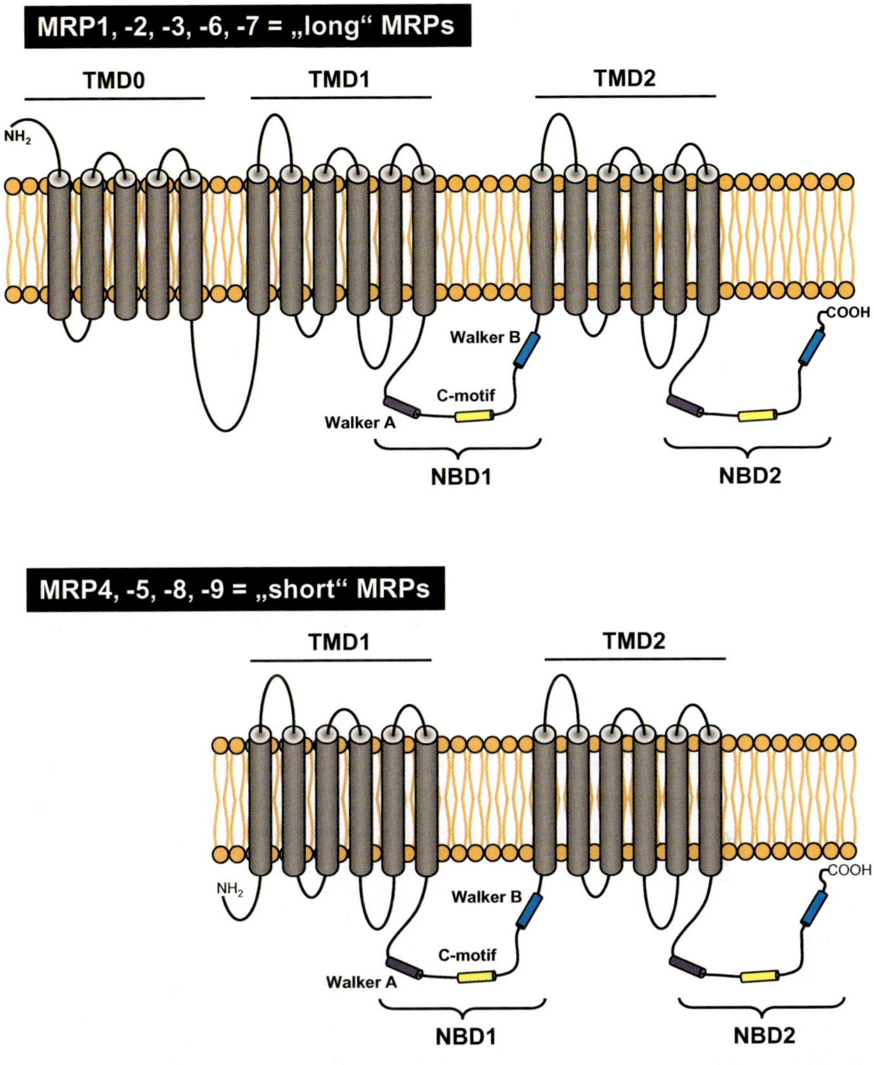

Fig. 1. Membrane topology of human MRP family members. All MRPs contain two membrane domains (TMD1 and TMD2) and two nucleotide-binding domains (NBD1 and NBD2). Each of the two TMDs likely consists of six transmembrane helices, both nucleotide-binding domains contain the C-motif and the Walker A and B motifs typical for ABC transporters. MRP1, -2, -3, -6 and -7 contain an additional aminoterminal transmembrane-spanning domain (TMD0) and are therefore designated as "long" MRPs whereas MRP4, -5, -8 and -9 have the typical ABC transporter structure and are therefore called the "short" MRPs.

(TMD0) of the so-called "long" ABCC family members is composed of five transmembrane helices but the amino acid sequence of this extension is relatively poorly conserved. Interestingly, based on algorithms for a computer-assisted transmembrane-analysis, the NH_2-terminal extension of the "long" MRP family members contain five transmembrane helices and therefore, the NH_2-termini of these proteins are predicted to be extracellular (Fig. 1).[9] Indeed this topology was experimentally confirmed for MRP1 by epitope insertion studies.[10,11]

MRP1 is expressed in several tissues with high expression levels in lung, kidney, skeletal and cardiac muscles, testis and placenta,[12,13] but is barely detectable in human liver. It is also expressed in the blood-brain barrier, the cerebrospinal fluid barrier and the brain.[14,15] In polarized cells, MRP1 is localized mostly in the basolateral membrane,[16] which is in contrast to the exclusive apical localization of the ABCC family member MRP2.[5] However, in some cell types, such as endothelial cells of brain microvessels,[17] MRP1 localizes to the apical membrane.

Following the observation that MRP1 functions as an ATP-dependent export pump for the endogenous glutathione *S*-conjugate leukotriene C_4,[18–20] its substrate specificity was studied in more detail using inside-out oriented membrane vesicles or stably-transfected whole cell approaches. MRP1 displays a broad substrate spectrum including structurally diverse amphiphilic organic anions, many of which conjugated with glutathione, sulphate or glucuronate.[12,21–23] Endogenous compounds include bilirubin-glucuronosides, 17β-estradiol glucuronide ($E_2$17βG), sulfatolithocholyl-taurine and the glutathione conjugate leukotriene C_4 (LTC_4).[12,21–23] LTC_4 is the substrate with the highest affinity for MRP1 identified to date, with a K_m value of 0.1 μM.[19] In contrast, oxidized glutathione (GSSG) has a relatively low affinity with a K_m value of 93 μM.[24] The estrogen conjugate 17β-estradiol glucuronide ($E_2$17βG) is a major physiological substrate for MRP1 with a K_m value of 1.5 μM.[19] The other MRP family members MRP2, MRP3 and MRP4 also transport $E_2$17βG, although with 10-fold higher K_m values.

Interestingly, Loe and coworkers[25] demonstrated that substrates could be transported by MRP1 without being covalently linked to glutathione. Using membrane vesicles prepared from MRP1-overexpressing cells they have shown that the vinca alkaloid vincristine is a poor inhibitor of MRP1-mediated LTC_4 transport but this inhibition was enhanced by adding glutathione. Furthermore, they found that vincristine is transported by MRP1 (ATP dependent) only in the presence of glutathione[25] without being covalently linked to it. The detailed substrate spectrum of MRP1 (and subsequently of other MRP family members) has been established by several methods. The gold standard for analyzing a substance as a substrate for a MRP protein are experiments using so-called inside-out oriented membrane vesicles. These are prepared from stably-transfected cells overexpressing the respective transport protein. Using these MRP-containing vesicles compared to vesicles prepared from respective control cell lines, it is possible not only to identify a substance as a substrate for the respective transport protein but also to determine kinetic constants (K_m values and V_{max} values) for the given substrate. Most of the drug substrates listed in Table 2 have been identified by this method. Another method for the analysis of possible substrates is drug accumulation using stably-transfected cells or drug-selected cells overexpressing an MRP family member. Using these methods, several other drugs have been identified as MRP1 substrates which are either conjugated or transported in a glutathione-substrate complex. MRP1 confers resistance to anthracyclines (e.g. daunorubicin), to vinca alkaloids (e.g., vincristine), epipodophyllotoxins (e.g. etoposide) and low-level resistance to taxanes.[2,16] Interestingly, the drug resistance profiles of MRP1 and MRP2 are very similar and relatively broad. Selected drug substrates are summarized in Table 2.

Several studies have investigated the clinical relevance of this export pump. MRP1-specific monoclonal antibodies,[26] together with RT-PCR and immunoblot approaches, have provided convincing evidence of elevated expression levels of

Table 2. Selected drug and drug metabolite substrates of MRP family members.

MRP1		
Daunorubicin	Antineoplastic agent	[174]
Vincristine	Antineoplastic agent	[175]
Vinblastine	Antineoplastic agent	[2]
Etoposide	Antineoplastic agent	[2]
Paclitaxel	Antineoplastic agent	[2]
Epirubicine	Antineoplastic agent	[2]
Melphalan-SG	Antineoplastic agent	[176]
Methotrexate	Antineoplastic agent	[177]
Chlorambucil	Antineoplastic agent	[178]
Irinotecan	Antineoplastic agent	[179]
Ciprofloxacin	Antibiotic	[180]
Saquinavir	HIV protease inhibitor	[181]
Ritonavir	HIV protease inhibitor	[181]
Indinavir	HIV protease inhibitor	[182]
MRP2		
Chlorambucil	Antineoplastic agent	[183]
Cisplatin	Antineoplastic agent	[40]
Doxorubicin	Antineoplastic agent	[40]
Etoposide	Antineoplastic agent	[40]
Epirubicin	Antineoplastic agent	[40]
Methotrexate	Antineoplastic agent	[177]
Vincristine	Antineoplastic agent	[184]
Vinblastine	Antineoplastic agent	[184]
Cerivastatin	HMG-Co-reductase inhibitor	[185]
Pravastatin	HMG-Co-reductase inhibitor	[185]
Indinavir	HIV protease inhibitor	[73]
Ritonavir	HIV protease inhibitor	[73]
Saquinavir	HIV protease inhibitor	[73]
Morphine-3-glucuronide	Conjugate of the narcotic morphine	[186]
Olmesartan	Angiotensin II receptor antagonist	[187]

(*Continued*)

Table 2. (*Continued*).

Rifampicin	Antibiotic	[71]
Fexofenadine	Antiallergic, H1-receptor antagonist	[72]
Ethinyl-estradiol-glucuronide	Conjugate of oral contraceptive	[111]
MRP3		
Etoposide glucuronide	Antineoplastic agent, conjugate of etop.	[115]
Methotrexate	Antineoplastic agent	[188]
Morphine-3-glucuronide	Conjugate of the narcotic morphine	[189]
Acetaminophen-glucuronide	Conjugate of the analgetic/ antipyretic drug	[190]
Ethinyl-estradiol-glucuronide	Conjugate of oral contraceptive	[111]
MRP4		
Adefovir (PMEA)	Antiviral agent	[117]
Irinotecan	Antineoplastic agent	[155]
Leucovorin	Antineoplastic agent	[147]
Methotrexate	Antineoplastic agent	[147]
PMEG	Antiviral agent	[117]
Thiopurine metabolites	Antineoplastic agents	[191]
Topotecan	Antineoplastic agent	[153]
Vincristine	Antineoplastic agent	[155]
3TC (Lamivudin)	Protease inhibitor	[117]
Ganciclovir	Nucleoside analog	[145]
Cefmetazole	Antibiotic	[192]
Ceftizoxime	Antibiotic	[192]
Enalapril	Angiotensin-converting enzyme inhibitor	[193]
Furosemide	Loop diuretic	[194]
Hydrochlorothiazide	Loop diuretic	[194]
Olmesartan	Angiotensin II receptor antagonist	[187]
Pravastatin	HMG-CoA-reductase inhibitor	[193]

(*Continued*)

Table 2. (*Continued*).

MRP5		
5-FDUMP, 5-FUMP, dUMP	Antineoplastic agent	[152]
Methotrexate	Antineoplastic agent	[148]
Pemetrexed	Antineoplastic agent	[152]
Thiopurine metabolites	Antineoplastic agent	[146]
MRP6		
Etoposide	Antineoplastic agent	[171]
Daunorubicin	Antineoplastic agent	[171]
Teniposide	Antineoplastic agent	[171]
BQ-123	Endothelin-A receptor antagonist	[170]

Antineoplastic agent, a drug that inhibits or combats the development of neoplasms; PMEG: 9-(2-Phosphonylmethoxyethyl)guanine; 5-FDUMP: 5-Fluorodeoxy-uridine monophosphate; 5-FUMP: 5-fluoro-uridine-monophosphate; dUMP: deoxy-uridine monophosphate.

MRP1 in a variety of solid tumors, including breast, prostate and lung cancer. High expression levels of MRP1 have been found in non-small cell lung cancer (NSCLC) and this high expression correlates with a higher grade — tumors are classified from G1 (well differentiated) to G4 (undifferentiated) — with respect to differentiation of NSCLC.[27] A recently published study of NSCLC confirmed these analyses and demonstrated that the level of MRP1 expression was a highly significant negative predictor of the response to chemotherapy and overall survival.[28]

Furthermore, several studies have demonstrated that MRP1 expression is a negative prognostic factor for some types of breast cancer. When MRP1 was detected in early stages of breast cancer, patients have a shorter time to relapse after postsurgical adjuvant chemotherapy.[29,30] This has been confirmed in a large study with over 500 premenopausal women with early-stage breast cancer.[31] In relapsed breast cancer patients who received presurgical adjuvant chemotherapy the expression levels of MRP1 were higher than in the initial tumor.[32,33] In prostate cancer cell lines MRP1 expression has been found to be elevated after treatment of these cells with doxorubicin.[34] Furthermore, MRP1 is expressed at high levels in prostatic intraepithelial neoplasis and prostatic adenocarcinoma but only moderately detectable in normal prostatic epithelia.[35] The level of MRP1 expression increases with invasiveness and cancer stage and is positively associated with mutant p53 status of the tumor.[36] In neuroblastoma, the most common extracranial solid tumor in children, MRP1 expression is positively correlated with the amplification of the NMYC gene, which is a strong negative molecular indicator of outcome.[37]

2. MRP2 (Gene Symbol *ABCC2*)

The first cDNA fragment of rat Mrp2 (formerly termed cMrp or cMoat) was identified in 1995 by comparing normal and transport-deficient GY/TR$^-$ (Groningen yellow/transport-deficient) mutant rats,[38] following a search for the ATP-dependent LTC_4 transporter localized in the hepatocyte canalicular membrane.[39] Based on this sequence information the full-length cDNAs of rat and human MRP2 were cloned and further analyzed.[9,40,41] Since then, the full-length cDNA sequences from several mammalian species have been cloned including orthologes from rabbit, dog, horse and chimpanzee.[9,41–44] Furthermore, MRP2-related sequences from other organisms including Caenorhabditis elegans, zebrafish and the plant Arabidopsis thaliana[45,46] have been identified and characterized. Within the human MRP subfamily, MRP2 shows the highest degree of identity to MRP1 (ABCC1: 48%) followed by MRP3 (ABCC3: 47%). The lowest degree of similarity was found between MRP2 and MRP8 (29%) and CFTR (26%). Interes-tingly, the level of identity of human MRP2 to the MDR1 P-glycoprotein[47] (ABCB subfamily) is only 18%, emphasizing the major differences between the ABCB and ABCC transporter subfamilies. The human *ABCC2* gene has been localized to chromosome 10q24 (Table 1),[48] spanning approximately 65 kbp and contains 32 exons with a high proportion of class 0 introns.[49] Interestingly, a comparison of the genomic organization of human *ABCC1* and *ABCC2* shows remarkable similar-ities as indicated by number and size of exons,[50] suggesting that both genes share a close evolutionary relationship perhaps originating from gene duplication.

Initially, MRP2 was localized to the apical membrane of human hepatocytes[41,51] and was, therefore, termed canalicular MRP (cMRP)[9] or cMOAT (canalicular multispecific organic anion transporter).[41] Since then, MRP2 has been detected in several tissues including kidney[52,53] (Fig. 2), small intestine (Fig. 2),[54] lung,[55] gallbladder[56] and placenta.[57] Recently, MRP2 was detected in the apical membrane of the non-pigmented ciliary epithelium of the eye where it perhaps protects intraocular tissues from potentially harmful substances.[58] Localization of MRP2 in the apical membrane of hepatocytes, kidney proximal tubules and epithelial cells of the small intestine, the lung and the gallbladder was confirmed by immunofluorescence microscopy or immunohistochemistry. So far, a localization of MRP2 in other than the apical membrane domains has not been demonstrated. Interestingly, the distribution of MRP2 within an organ may change during different pathophysiological conditions. Thus, rat Mrp2 is homogeneously distributed throughout a lobule in normal rat livers, whereas after cholestasis (blockage of bile flow from liver to duodenum) Mrp2 concentrates near the perivenous area of the lobule.[59] On the subcellular level, retrieval of Mrp2 from the canalicular membrane into perinuclear vesicles of rat hepatocytes was demonstrated to be an early event of cholestasis.[60,61]

In early studies, the substrate spectrum of the apical export pump MRP2 was elucidated using inside-out oriented membrane vesicles from normal and transport-deficient GY/TR$^-$ (Groningen Yellow/transport deficient) or EHBR (Eisai hyperbilirubinemic mutant) rats, without knowing the molecular identity of the export pump.[39,62] However, as cell lines stably expressing human MRP2 became available[40,63] the substrate spectrum of MRP2 has

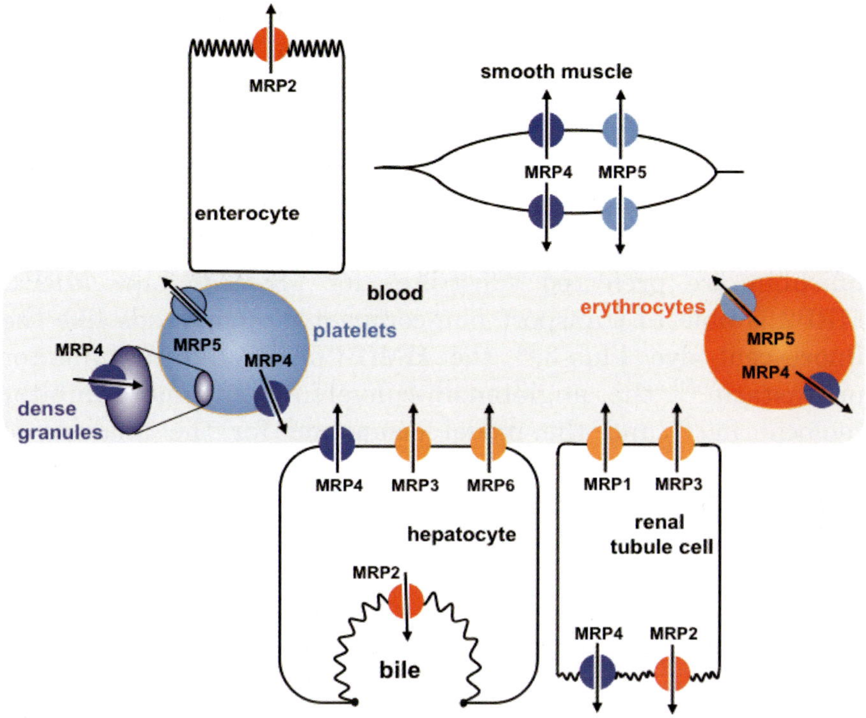

Fig. 2. Localization of human MRP family members in different cell types important for the absorption, distribution and excretion of drug substrates. MRP2 is localized to the apical membrane of enterocytes, hepatocytes and kidney tubule cells. MRP3 is localized to the basolateral membrane of hepatocytes and tubule cells. MRP4 shows an apical localization in tubule cells whereas in hepatocytes, MRP4 is localized to the basolateral membrane. Furthermore, MRP4 and MRP5 are also expressed in smooth muscle cells, erythrocytes, and platelets.

been studied under more defined conditions. Human MRP2 has also been purified and shown to exhibit transport activity when reconstituted in proteoliposomes.[64] As in the case of MRP1, leukotriene C_4 (LTC_4) was identified as the preferred substrate for MRP2. ATP-dependent transport of radioactively labeled substrates into membrane vesicles was measured by incubating the vesicles with the labeled compound for a defined time period followed by separating the vesicles from extravesicular labeled substrate.[65] These analyses demonstrated that MRP2 shares a very similar substrate spectrum with MRP1. High-affinity substrates include amphiphilic anions, particularly conjugated with glutathione, sulphate or glucuronate.[5,6,66] The comparison of both substrate spectra shows that MRP1 has a 10-fold higher affinity for LTC_4 and a 5-fold higher affinity for 17ß-estradiol glucuronide than MRP1.[40] In contrast, mono- and bisglucuronosyl bilirubin are preferred substrates for MRP2.[67] Like MRP1, MRP2 is able to transport non-conjugated compounds like the fluorescent dye Fluo-3,[68] the HMG-CoA-Reduktase inhibitor pravastatin,[69] the angiotensin-converting enzyme inhibitor temocaprilat[70] and the model compound for the analysis of hepatic transport processes, bromosulfophthalein.[71] Several other widely prescribed drugs have been identified as substrates for MRP2. These include antibiotics such as rifampicin,[71] antiallergics like fexofenadine[72] and HIV protease inhibitors like indinavir, ritonavir and saquinavir.[73] Interestingly, besides being substrates for drug transporting export pumps, drugs may also have an influence on the expression of these transporters. This has been demonstrated for the MRP2 substrate rifampin. In human duodenal biopsies, obtained after oral treatment with 600 mg rifampin/day, an elevated expression of MRP2 mRNA and protein could be detected[54] suggesting that the drug modulates transporter expression. Other drugs which appear to regulate the expression of MRP2 are cisplatin, phenobarbital and ritonavir[74] and it was shown that this regulation is mediated via the nuclear hormone receptor PXR (pregnane X receptor).[74] Selected drug substrates for the family members MRP1–MRP6 are summarized in Table 2.

The strategic localization of MRP2 to the apical membranes of hepatocytes or renal tubular cells and the epithelia of the small intestine, the colon and the lung suggests that a physiological function of this export pump is to excrete endogenous metabolites and xenobiotics and to prevent tissues from being entered by toxic compounds. Since MRP1 mediates resistance to several anticancer drugs and since MRP1 and MRP2 have a very similar substrate spectrum it has been proposed that MRP2 may also confer drug resistance. The first direct evidence for MRP2-mediated drug resistance was obtained by analyzing stably transfected MDCK and HEK cells expressing human recombinant MRP2.[40] Expression of MRP2 in these cells leads to significant resistance to cisplatin, etoposide, epirubicin, doxorubicin and vincristine. In addition, MRP2 mediates resistance to methotrexate, an antifolate drug frequently used in the treatment of childhood acute leukaemia and rheumatoid arthritis.[75] Furthermore, a mutation in the *ABCC2* gene, leading to loss of transport activity was shown to be associated with impaired methotrexate elimination.[76] The possible clinical relevance of MRP2-mediated drug resistance was further suggested by the detection of MRP2 in various carcinoma cells and tissue biopsy samples. Several studies have shown that in cisplatin-resistant carcinoma cells overexpression of MRP2 mRNA and protein could be detected.[48,77,78]

More than 200 naturally occurring sequence variations in the *ABCC2* gene encoding human MRP2 have been discovered.[49,79–81] Most of these are single-nucleotide polymorphisms (SNPs) located in introns or in exons without changing the amino acid sequence of the MRP2 protein (silent mutations) and therefore, probably, without functional consequences.[82–84] However, some mutations are associated with the absence of the MRP2 protein from the hepatocyte canalicular membrane leading to a phenotype designated as Dubin-Johnson syndrome.[85] In humans, Dubin-Johnson syndrome, originally described in 1954[86] is an autosomal, recessively inherited disorder, characterized by conjugated hyperbilirubinemia.[86,87] This is due to deficient transport of anionic conjugates, including monoglucuronosyl- and bisglucuronosyl-bilirubin into bile, associated with mutations in the *ABCC2* gene and the consequent absence of functional MRP2 protein from the canalicular membrane.[80,85] Mutations in the *ABCC2* gene identified in patients with Dubin-Johnson syndrome include missense mutations,[88–90] splice site mutations, leading to exon deletions with subsequent premature termination codons,[91–93] nonsense mutations,[49,80] a deletion mutation leading to the loss of two amino acids[49] and a deletion/in-frame insertion.[94] Although all mutations leading to Dubin-Johnson syndrome result in the absence of a functionally active MRP2 protein from the canalicular membrane, their effects on synthesis and function of the conjugate export pump are different. It has been demonstrated that premature termination codons may cause rapid *ABCC2* mRNA degradation and a truncated MRP2 protein could not be detected.[49,80] Other mutations are associated with deficient protein maturation, impaired sorting of the protein to the canalicular membrane or with an apically localized, but inactive MRP2 protein.[79,88,95,96] Recently, Mrp2-knock-out mice have been generated and characterized by several groups.[97,98] These knock-out mice are fertile and apparently healthy and it will be of interest in the future to crossbreed these mutant mice with strains lacking other ABC transporters like Mrp3, Mrp4, Mdr1a/Mdr1b or Abcg2. Recently, the first results on such double-knock-out models have been published analyzing an MRP2 + MRP3-deficient mouse strain [Abcc2; Abcc3(–/–)].[99] The authors found that Mrp2 is important for the biliary excretion of methotrexate and its metabolite 7-hydroxy-methotrexate and that this metabolite accumulated in liver in Abcc2 (–/–) mice. When Mrp2 is absent, Mrp3 transports both substances back from liver into blood leading to increased plasma levels and urinary excretion. Together with other studies showing a compensatory upregulation of Mrp3 under cholestatic conditions or in *Abcc2* mutant mice, the complex interplay of these transporters for the elimination of endogenous compounds such as bilirubin and bilirubin-glucuronide and for drugs such as methotrexate became clearer.

3. MRP3 (Gene Symbol *ABCC3*)

The first MRP3 sequence was identified in 1997 together with the first partial sequences of MRP4 and MRP5 by analyzing EST databases.[100] MRP3 together with MRP1, MRP2, MRP6 and MRP7 belong to the so-called long MRP group, with three membrane-spanning domains (MSDs) and two nucleotide-binding domains (NBDs) (Fig. 1). The *ABCC3* gene encoding human MRP3 is located on chromosome 17q22 (Table 1), the protein has an identity of 56% to MRP1 and is, therefore, more similar to MRP1 than the apical export pump MRP2. MRP3 is expressed in human liver, gut, gall bladder, placenta and adrenal gland.[101–104] Lower expression was detected in kidney and prostate.[104,105] In contrast to the apically localized export pump MRP2, MRP3 is present in basolateral membranes of hepatocytes,[103] close to bile ducts and in cholangiocytes.[106] In rat cholestatic models it has been demonstrated that under cholestatic conditions Mrp2 in the canalicular membrane is downregulated,[107] whereas in compensation Mrp3 in the basolateral membrane is upregulated.[108] This inverse regulation between the two major hepatic MRP family members is presumed to protect the hepatocyte from the accumulation of potentially toxic bile constituents in hepatocytes under conditions normal cellular export across the canalicular membrane into bile is impaired. In the kidney, MRP3 is present in the distal tubules[109] and has also been detected in enterocytes in the ileum and colon.[110]

Whereas MRP1 and MRP2 have nearly identical substrate spectra MRP3 differs with respect to the ability to transport glutathione conjugates. GSH-conjugates are poor substrates for MRP3, which shows a preference for glucuronidated compounds[111,112] consistent with its role in protecting the hepatocyte from the accumulation of potential harmful compounds synthesized intracellularly (e.g., conjugated bile salts). In contrast to MRP2, MRP3 transports also monovalent bile salts like taurocholate and glycocholate.[111–113] The localization of MRP3 in the basolateral hepatocyte membrane, together with its ability to mediate bilirubin-glucuronide transport out of the hepatocyte back into blood support the hypothesis that MRP3 constitutes for an alternative transport route when biliary excretion via MRP2 is impaired (e.g., cholestasis or Dubin-Johnson syndrome — see MRP2).

MRP3, like MRP1 and MRP2, confers resistance to epipodophyllotoxins (etoposide), but not to vinca alkaloids or anthracyclines.[114,115] Furthermore, the transport of etoposide, mediated by MRP3 is not glutathione dependent.

In summary, MRP3 may contribute to the detoxification of glucuronidated drug conjugates, a fact that is supported by studies with Mrp3 (–/–) knock out mice. These mice lack the transport of acetaminophen glucuronide or morphine-3-glucuronide from hepatocytes into blood.[116]

4. MRP4 and MRP5 (Gene Symbols *ABCC4* and *ABCC5*)

In contrast to ABCB1 (P-glycoprotein) and MRP1, which were identified by characterization of drug resistant cell lines, the first description of MRP4 (*ABCC4*) and MRP5 (*ABCC5*) was the consequence of an EST library search and an expression analysis in different cancer cell lines.[100] *ABCC4* mRNA (encoding MRP4) was not overexpressed in any of the investigated cell lines whereas mRNA encoding MRP5 was overexpressed in some of these cell lines. MRP4 and MRP5 belong to the C branch of the ABC transporter superfamily and the respective genes are located on chromosome 13q32.1 and 3q27 (Table 1).[100,101,117] The *ABCC4* gene encodes 1,325 amino acids and is the shortest member of subfamily C, while *ABCC5* encodes a protein of 1,437 amino acids. Like other members of the MRP subfamily MRP4 and MRP5 are typical organic anion transporters and since both share a similar substrate spectrum, we have combined the discussion of these short MRPs in one section.

A major difference in the structure of MRP4 and MRP5 compared to other members of subfamily C is the absence of the N-terminal transmembrane domain 0 (TMD0), which is predicted to contain four to six transmembrane helices.[118–120] Therefore, MRP4 and MRP5 are characterized, together with MRP8 (*ABCC11*) and MRP9 (*ABCC12*) by a typical ABC transporter structure comprised of two TMDs (TMD1 and TMD2) and two typical ABC NBDs (Fig. 1). Each of these TMDs is predicted to consist of six transmembrane helices.[118] Regarding expression, both *ABCC4* and *ABCC5* mRNA are expressed in various tissues. *ABCC4* mRNA is expressed in human lung, kidney, bladder, gall bladder and tonsils.[100] In an additional study different transcript levels of MRP4 were detected in several tissues: high levels in prostate, testis, spleen, thymus, ovary, small intestine, colon, lung, skeletal muscle, and pancreas, whereas heart, brain, liver, and blood leukocytes showed low transcript levels of *ABCC4* mRNA.[121] Furthermore, *ABCC4* mRNA was also detected in adrenal glands.[122] In contrast, expression analysis of MRP5 revealed that *ABCC5* mRNA is ubiquitously expressed.[100] Interestingly, in most tissues and cell types such as tubuloacinar cells in prostate[123] and hepatocytes in liver, MRP4 is localized to the basolateral membrane,[124,125] whereas in proximal tubule cells of kidney MRP4 is localized to the apical membrane.[126] For MRP5, the localization in smooth muscle cells of the corpus cavernosum, ureter, urethra, bladder, in epithelial cells of the mucosa in ureter and urethra, and in blood vessels within the lamina propria of a male urethra was observed.[127] In addition, human cardiomyocytes were positively stained for MRP5.[128] In human brain MRP5 is highly expressed in pyramidal neurons, astrocytes, and in endothelial cells of brain.[129] MRP5 is also preferentially localized to the basal membrane of syncytiotrophoblasts in human placenta, as well as in and around fetal blood vessels. Furthermore, levels of *ABCC5* mRNA decreased with increasing gestational age.[130] In addition to other members of the subfamily C such as MRP1, MRP4 and MRP5 are also expressed in blood cells, for example erythrocytes (Fig. 2), leukocytes and platelets. The protein expression of both transporters was detected in human erythrocytes using immunoblot analyses.[131–133] *ABCC5* mRNA was also detected in monocytes and different subsets of peripheral blood mononuclear cells.[134] Finally, MRP4 shows high expression in human platelets, which as demonstrated by immunoblot and immunofluorescence analyses. MRP4 was also localized to the plasma membrane and is enriched in dense granules of human platelets (Fig. 2).[135] Interestingly, it was also shown that the

ABCC4 mRNA expression changes during the differentiation of blood cells.[136] Thus, *ABCC4* mRNA expression increased during differentiation from CD34 positive cells to megakaryocytes (progenitor cells of platelets), whereas mRNA expression decreased during maturation to leukocytes. In contrast, *ABCC5* mRNA expression was not affected during these differentiation processes. This study also revealed that MRP4 protein was localized to granular structures and to the plasma membrane in differentiated progenitors and bone marrow megakaryocytes.[136]

The application of uptake studies employing MRP4- and MRP5-enriched membrane vesicles and/or the use of MRP4- and MRP5-overexpressing cell lines for efflux studies led to the identification of numerous MRP4 and MRP5 substrates and inhibitors. Both transporters are characterized by the ability to mediate the transport of the second messengers 3′,5′-cyclic AMP and 3′,5′-cyclic GMP. However, the observed affinities of cAMP and cGMP for MRP4 and MRP5 are quite different. The affinity of cAMP for MRP4 is about nine-fold higher than that for MRP5 whereas, for cGMP the reported affinity is higher for MRP5 compared to MRP4.[131,137] The affinity of MRP4 and MRP5 for cGMP is higher compared to cAMP. From the observed expression pattern of MRP4 and MRP5 and their ability to transport cyclic nucleotides, different physiological roles may be envisaged for these ABC transporters. One possible function could be the modulation of cyclic nucleotide signaling by decreasing the intracellular excess of such messengers. A further possible physiological function is the extracellular allocation of cyclic nucleotides for intercellular signaling. Sildenafil (Viagra™), which modulates the intracellular concentration of cyclic nucleotides by inhibition of phosphodiesterase is of interest since this compound inhibits the transport activity of both MRP4 and MRP5.[131,138] ADP is the major signaling molecule in dense granules of human platelets with concentrations above 0.6 mM inside the vesicles.[139,140] The abundant expression of MRP4 in dense granules has suggested its possible physiological role in transporting ADP. Although it is experimentally challenging to demonstrate ADP-mediated transport by an ABC transporter, time dependent uptake of [^{3}H]ADP into dense granule inside-out vesicles was obtained in the presence of ATP. Moreover, transport was sensitive to the ATP-hydrolysis inhibitor orthovanadate and was inhibited by the more specific MRP4 inhibitors dipyridamole and MK-571, indicating an important role for platelet MRP4 in mediator storage in intracellular vesicles.[135]

As found for other members of the subfamily C such as MRP1-3, MRP4 also transports steroid conjugates such as 17β-estradiol glucuronide and dehydroepiandrosterone-3-sulfate.[122,137] It was also suggested that MRP4 expressed on the basolateral membrane of human hepatocytes could function as a compensatory bile salt efflux pump.[124] This was confirmed in a later study showing that MRP4 is up regulated in the liver of cholestatic patients.[125] Further physiological substrates of MRP4 are prostaglandins such as PGE_1 and PGE_2 suggesting an important role for inflammatory processes and pain.[141,142] For example, mice lacking Mrp4 showed a decrease in prostaglandin metabolites and an increase in the inflammatory pain threshold compared to wild type mice.[143] Interestingly, the MRP4-mediated transport of 17β-estradiol glucuronide can be inhibited by non-steroidal anti-inflammatory drugs.[141,142]

Until now, no selective inhibitors of MRP4 and MRP5 were found. Both transporters are inhibited to different extents, for example, by probenecid, sulfinpyrazone, dipyridamole, zaprinast, trequensin, and sildenafil.[124,126,135,138,144]

In addition to the transport of endogenous substrates, MRP4 and MRP5 are also involved in the efflux of therapeutically used drugs. For MRP4, the transport of such substrates was first shown using the PMEA [Phosphonic acid, (9-(2-phosphonylmethoxyethyl)adenine)]-resistant human T-lymphoid cell line CEMr1.[117] PMEA (adefovir) is an antiviral drug used for the treatment of chronic hepatitis B. The depletion of intracellular ATP by azide and 2-deoxyglucose showed that the efflux of PMEA was considerably decreased in CEMr1 cells but not in the parental CEMss cells, indicating that an ATP-binding cassette transporter might participate in the efflux of PMEA.[117] The same result was also observed for monophosphorylated azidothymidine (AZT). Subsequent analyses revealed an amplification of the *ABCC4* gene and increased expression of MRP4, indicating that MRP4 plays a role in resistance to nucleoside-based antiviral drugs. A higher resistance of CEMr1 cells against other antiviral drugs such as 2′,3′-dideoxy-3′-thiacytidine (3TC, lamivudine) and PMEG [(9-(2-phosphonylmethoxyethyl)guanine)] was observed.

The role of MRP5 in cellular resistance against PMEA was also investigated using MRP5 transfected HEK293 cells.[144] Thus, MRP5-overexpressing cells were characterized by a 2.5-fold increase in resistance against PMEA compared to parental cells, providing further evidence that MRP5 also contributes to the efflux of PMEA. In relation to transport of antiviral drugs it should be noted that ganciclovir is also a substrate of MRP4.[145] Following treatment with ganciclovir, a dose dependent reduction in cell viability and increased rate of apoptosis was observed in CEMss cells compared to the MRP4 overexpressing CEMr1 cells. Moreover, CEMr1 cells accumulated considerably less drug than the parental CEMss cells.

Resistance to thiopurines such as 6-mercaptopurine and 6-thioguanine was observed in MRP4 overexpressing NIH3T3 cells.[137] Subsequently, it was demonstrated that thiopurine metabolites were transported by MRP4 and MRP5. However, the substrate specificity regarding these metabolites differs between MRP4 and MRP5.[146] It was found that thionucleoside monophosphates (tNMPs) are exported by both MRP4 and MRP5. However, only MRP5-overexpressing cells transported thioxanthosine monophosphate (tXMP), whereas MRP4-overexpressing cells showed higher rates of transport of thio-guanosine 5′-monophosphate (tGMP) and 6-methyl-thio-inosine 5′-monophosphate (MetIMP).[146] Resistance to methotrexate, a widely used anticancer agent is, in addition to MRP1, 2, 3, also mediated by MRP4 after short term exposure (4 h) but not with continuous exposure.[123] In this regard, it should be noted that MRP4 is able to mediate ATP dependent transport of not only methotrexate, but also of folic acid and N(5)-formyltetrahydrofolic acid (leucovorin).[147] However, mono- or polyglutamylated methotrexate metabolites were not transported.

In another study, MRP5 was identified as efflux pump for methotrexate and folic acid using membrane vesicles HEK293 cells overexpressing MRP5.[148] MRP5 was also able to transport the diglutamylated form of methotrexate but not the methotrexate-triglutamate. In addition it was also found that the accumulation of total [^{3}H]methotrexate and methotrexate polyglutamates were significantly reduced in MRP5 overexpressing cells.[148] Despite the experimental advances in identifying substrates of MRP transporters, for some compounds such as gemcitabine and cisplatin the current data is still controversial. For example, it was assumed, that gemcitabine is a substrate of MRP5. Although, experimental confirmation using membrane vesicles is currently still lacking, a further study regarding the sensitivity and resistance to gemcitabine in non-small lung cancer cell lines showed a correlation between gemcitabine sensitivity and the level of *ABCC5* mRNA expression. This indicated that MRP5 might be involved in the transport and/or sensitivity of gemcitabine.[149] Furthermore, there is some evidence that MRP4 and MRP5 confer resistance to the anticancer agent cisplatin.[150,151] However, experimental proof of MRP4 and MRP5 mediated cisplatin transport is still awaited. The contribution of MRP5 to the resistance of 5′-fluorouracil, a drug used for the treatment of colon and breast cancer, was demonstrated in an *in vitro* study,[152] which showed that the observed resistance was a result of MRP5-mediated transport of the monophosphorylated 5′-FU metabolites 5′-fluoro-2′-deoxyuridinemonophosphate (5-FdUMP), 5′-fluoro-uridinemonophosphate (5-FUMP) and deoxyuridinemonophosphate (dUMP). In contrast, the metabolite 5′-fluoro-2′-deoxyuridine and 5′-Fluorouracil were not transported by MRP5.[152] The preferred transport of monophosphorylated nucleoside analog drugs[117,144,146] seems to be a functional characteristic of MRP4 and MRP5 since, as already discussed, the monophosphorylated thiopurine metabolites are also transported by both transporters.[146] MRP4 also transports the anticancer agents such as topotecan, irinotecan and its active metabolite SN-38.[153–155] Topotecan and irinotecan are used in the therapy of bronchial carcinoma and colon carcinoma, respectively.

The clinical relevance of MRP4 and MRP5 for drug therapy is still under debate since so far only a few studies have indicated the importance of these two transporters *in vivo*. For example, in the case of neuroblastomas a high level of *ABCC4* mRNA expression was associated with a poor clinical outcome.[155] In childhood and adult acute lymphoblastic leukemias (ALL)[156] the expression of MRP1, -2, -3, -5, and -6 encoding mRNA in ALL blasts was higher in patients who relapsed than in patients who remained in complete response. In contrast, *ABCC4* mRNA expression was similar in both groups of patients. Taken together, MRP4 plays a different role in relation to the prognosis depending upon the type of malignancy. ALL-patients with a high expression of MRP1, -2, -3, -5, and -6, except MRP4, had a shorter relapse-free survival than patients with low MRP expression. The same effect of mRNA expression was observed when compared to overall survival.[156] In contrast, in a study investigating the influence of MRP transporter expression on clinical outcome in childhood acute myeloid leukemia (AML), *ABCC4* and *ABCC5* mRNA expression was not associated with a failure to achieve remission or overall survival.[157] The notion of disease dependent influence of ABC transporters on the clinical outcome was further confirmed in a study, using gene expression profiling of ABC transporters as a predictor of pathological response to neoadjuvant chemotherapy in breast cancer.[158] Using pretreatment samples and microarray analysis it was shown that ABC transporters such as ABCA12, ABCA1, MRP5, MRP8, and MRP9 showed a higher expression in patients with residual disease compared to patients with a pathological complete response. Furthermore, patients with residual disease showed a decreased responsiveness to sequential weekly paclitaxel/FEC (5′-fluorouracil, epirubicin, and cyclophosphamide) neoadjuvant chemotherapy. MRP5 was identified as having the highest statistically significant increase in expression levels in patients with residual disease.[158] These result suggest that MRP5 may contribute to the resistance in the neoadjuvant chemotherapy (5′-fluorouracil, epirubicin, and cyclophosphamide), since it was already known that MRP5 confers resistance to 5-FU.[152] This study showed that gene expression profiling of certain ABC transporters in breast cancer patients might be useful in order to predict the pathological response to sequential weekly paclitaxel/FEC therapy. Further studies support the involvement of MRP5 in resistance to chemotherapy of pancreatic carcinoma. An expression analysis of MRP transporters in pancreas carcinoma tissue, compared to normal pancreatic tissue, revealed that *ABCC5* mRNA expression was higher in pancreatic carcinoma tissue.[159] A later study showed that a pancreatic cell line (Capan-1), with acquired chemoresistance against 5-FU was characterized by an increased expression of mRNAs encoding MRP3, -4, and -5.[160] Since chemotherapeutic protocols containing 5-FU and gemcitabine are widely used in the treatment of advanced pancreatic cancer, the data presented strongly suggest that MRP5 is at least involved in chemoresistance to 5-FU and perhaps also to gemcitabine.

Thiopurine induced hematopoietic toxicity (toxic effects related to the production of all type of blood cells) is one of the limitations in the immunosuppressant therapy with thiopurines such as azathiopurine and 6-mercaptopurine. As already discussed above mercaptopurine metabolites are substrates of MRP4 and MRP5.[146] Abundant Mrp4 expression was detected in myeloid progenitors in mice[161] and using Mrp4 deficient mice, an *Abcc4* gene dose dependent toxicity caused by accumulation of 6-thioguanine nucleotides in myelopoietic cells was demonstrated.[161] This study clearly highlights the protective function of Mrp4 against thiopurine-induced hematopoietic toxicity. Patients with a lower expression of MRP4 or a decreased function may have a higher risk for thiopurine-mediated hematopoietic toxicity.

5. MRP6 (Gene Symbol *ABCC6*)

Chromosomal sequencing in the course of the human genome project has revealed the existence of this family member. The

ABCC6 gene encoding human MRP6 is located immediately next to the *ABCC1* gene (encoding MRP1) on chromosome 16p13 (Table 1).[162] Despite the likely origin of both genes originating by gene duplication, the MRP6 protein is less similar to MRP1 (only 45% amino acid identity) than to MRP2 (47% identity) or MRP3 (56% identity). Since mutations in the *ABCC6* gene are associated with the degenerative connective tissue disease pseudoxanthoma elasticum (PXE).[163,164] the tissue distribution of MRP6 is of particular interest. Earlier investigations have shown that MRP6 protein expression seems to be restricted to kidney and liver.[165] This suggests that MRP6 normally is involved in the excretion of metabolites responsible for the degeneration observed in the affected tissues.[166,167] Subsequently, immunohistochemistry and *in situ* hybridization studies have detected murine Mrp6 also in epithelial cells from many tissues.[168] This study also showed *Abcc6* mRNA and/or protein expression in keratinocytes of the skin, intestinal mucosa, corneal epithelium and bronchial and tracheal epithelium as well as in endothelial and smooth muscle cells of the cardiovascular system.[168]

Like MRP1 and MRP2, MRP6 mediates the ATP-dependent transport of a number of glutathione conjugates including leukotriene C_4, N-ethylmaleimide S-glutathione (NEM-SG) and S-(2,4-dinitrophenyl)glutathione (DNP-SG).[169] In addition, the endothelin receptor antagonist BQ-123 but not endothelin-1 itself is transported by MRP6.[170] Furthermore, transport inhibition by known non-specific transport inhibitors like probenecid and indomethacin has been demonstrated.[171,172] In contrast, glucuronidated substances appear not to be MRP6 substrates but MRP6 confers only low level of resistance to various antineoplastic drugs like teniposide, etoposide, doxorubicin and daunorubicin.[171] Interestingly, to this day it is unclear how tissue expression and substrate spectrum relate to the pathology of PXE, which is characterized by calcification of elastic fibers in the skin, retina and arteries.[173]

6. Summary and Perspective

The export pumps MRP1 to MRP6 (gene symbols *ABCC1* to *ABCC6*) belong to the ATP-binding cassette transporter subfamily C (ABCC) within the ABC transporter superfamily. Like all ABC family members, they are primary active transporters for a variety of different endogenous compounds and xenobiotics including several widely prescribed drugs. All six MRP family members are expressed in a variety of different tissues including gut, brain, kidney and liver. In polarized cells MRP2 is localized in apical membrane domains, e.g., in the canalicular membrane of hepatocytes or the apical membrane of renal proximal tubule cells. All other MRPs are predominantly localized in basolateral membrane domains except MRP4 that is localized in the apical membrane of renal proximal tubule cells. MRP1 and MRP2 share a very similar and overlapping substrate spectrum including structurally diverse amphiphilic organic anions conjugated with glutathione, glucuronate or sulphate. Whereas glutathione conjugates are poor substrates for MRP3, glucuronidated compounds are preferred substrates for this family member. MRP4 and MRP5 are characterized by the ability to mediate the transport of the second messengers cAMP and cGMP. Additional physiological substrates for MRP4 are prostaglandins (PGE_1 and PGE_2) and steroid conjugates. Due to their broad substrate spectrum including xenobiotics and chemotherapeutic agents and their ability to transport structurally related substances, the six human MRP family members MRP1–MRP6 are important for conferring resistance to several antineoplastic agents including folate antimetabolites, platinum compounds and nucleoside- and nucleotide analogs. To date, detailed and large-scale *in vivo* studies elucidating the role of a certain MRP family member in a specific type of cancer are lacking. In the future, such studies are necessary to gain more insights into MRP-mediated drug resistance and to correlate *in vitro* expression and transport data with the role of the export pumps in cancer treatment.

References

1. Cole, S.P., Bhardwaj, G., Gerlach, J.H., Mackie, J.E., Grant, C.E., Almquist, K.C., Stewart, A.J., Kurz, E.U., Duncan, A.M., and Deeley, R.G. (1992) *Science* **258**, 1650.
2. Cole, S.P., Sparks, K.E., Fraser, K., Loe, D.W., Grant, C.E., Willson, G.M., and Deeley, R.G. (1994) *Cancer Res.* **54**, 5902.
3. Allikmets, R., Gerrad, B., Hutchinson, A., and Dean, M. (1996) *Hum. Mol. Genet.* **5**, 1649.
4. Klein, I., Sarkardi, B., and Varadi, A. (1999) *Biochim. Biophys. Acta* **1461**, 237.
5. König, J., Nies, A.T., Cui, Y., Leier, I., and Keppler, D. (1999) *Biochim. Biophys. Acta.* **1461**, 377.
6. Kruh, G.D. and Belinsky, M.G. (2003) *Oncogene.* **22**, 7537.
7. Morales, M.M., Capella, M.A., and Lopes, A.G. (1999) *Braz. J. Med. Res.* **32**, 1021.
8. Bryan, J. and Aguilar-Bryan, L. (1999) *Biochim. Biophys. Acta* **1461**, 285.
9. Büchler, M., König, J., Brom, M., Kartenbeck, J., Spring, H., Horie, T., and Keppler, D. (1996) *J. Biol. Chem.* **271**, 15091.
10. Hipfner, D.R., Almquist, K.C., Leslie, E.M., Gerlach, J.H., Grant, C.E., Deeley, R.G., and Cole, S.P. (1997) *J. Biol. Chem.* **272**, 23623.
11. Hipfner, D.R., Gao, M., Scheffer, G., Scheper, R.J., Deeley, R.G., and Cole, S.P. (1998) *Br. J. Cancer* **78**, 1134.
12. Cole, S.P. and Deeley, R.G. (1998) *Bioessays* **20**, 931.
13. Flens, M.J., Zaman, G.J., van der Valk, P., Izquierdo, M.A., Schroeijers, A.B., Scheffer, G.L., van der Groep, P., de Haas, M., Miejer, C.J., and Scheper, R.J. (1996) *Am. J. Pathol.* **148**, 1237.
14. Mercier, C., Masseguin, C., Roux, F., Gabrion, J., and Scherrmann, J.M. (2004) *Brain Res.* **1021**, 32.
15. Wijnholds, J., deLange, E.C., Scheffer, G.L., van den Berg, D.J., Mol, C.A., van der Valk, M., Schinkel, A.H., Scheper, R.J., Breimer, D.D., and Borst, P. (2000) *J. Clin. Invest.* **105**, 279.
16. Evers, R., Zaman, G.J., van Deemter, L., Jansen, H., Calafat, J., Oomen, L.C., Oude Elferink, R.P., Borst, P., and Schinkel, A.H. (1996) *J. Clin. Invest.* **97**, 1211.
17. Zhang, Y., Schuetz, J.D., Elmquist, W.F., and Miller, D.W. (2004) *J. Pharmacol. Exp. Ther.* **311**, 449.

18. Jedlitschky, G., Leier, I., Buchholz, U., Center, M., and Keppler, D. (1994) *Cancer Res.* **54**, 4833.
19. Leier, I., Jedlitschky, G., Buchholz, U., and Keppler, D. (1994) *Eur. J. Biochem.* **220**, 599.
20. Müller, M., Meijer, C., Zaman, G.J., Borst, P., Scheper, R.J., Mulder, N.H., de Vries, E.G., and Jansen, P.L. (1994) *Proc. Natl. Acad. Sci. U.S.A.* **91**, 13033.
21. Jedlitschky, G., Leier, I., Buchholz, U., Barnouin, K., Kurz, G., and Keppler, D. (1996) *Cancer Res.* **56**, 988.
22. Keppler, D., Kamisako, T., Leier, I., Cui, Y., Nies, A.T., Tusjii, H., and König, J. (2000) *Adv. Enzyme Regul.* **40**, 339.
23. Leslie, E.M., Deeley, R.G., and Cole, S.P. (2005) *Toxicol. Appl. Pharmacol.* **204**, 216.
24. Leier, I., Jedlitschky, G., Buchholz, U., Center, M., Cole, S.P., Deely, R.G., and Keppler, D. (1996) *Biochem. J.* **314 Pt 2**, 433.
25. Loe, D.W., Deeley, R.G., and Cole, S.P. (1998) *Cancer Res.* **58**, 5130.
26. Flens, M.J., Izquierdo, M.A., Scheffer, G.L., Fritz, J.M., Meijer, C.J., Scheper, R.J., and Zaman, G.J. (1994) *Cancer Res.* **54**, 4557.
27. Nooter, K., Burger, H., and Stoter, G. (1996) *Leuk. Lymphoma* **20**, 381.
28. Berger, W., Setinek, U., Hollaus, P., Zidek, T., Steiner, E., Elbling, L., Cantonati, H., Attems, J., Gsur, A., and Micksche, M. (2005) *J. Cancer Res. Clin. Oncol.* **131**, 355.
29. Nooter, K., Brutel de la Riviere, G., Look, M.P., van Wingerden, K.E., Henzen-Logmans, S.C., Scheper, R.J., Flens, M.J., Klijn, J.G., Stoter, G., and Foekens, J.A. (1997) *Br. J. Cancer* **76**, 486.
30. Nooter, K., de la Riviere, G.B., Klijn, J., Stoter, G., and Foekens, J. (1997) *Lancet* **349**, 1885.
31. Filipits, M., Pohl, G., Rudas, M., Dietze, O., Lax, S., Grill, R., Pirker, R., Zielinski, C.C., Hausmaninger, H., Kubista, E., Samoningg, H., and Jakesz, R. (2005) *J. Clin. Oncol.* **23**, 1161.
32. Filipits, M., Suchomel, R.W., Dekan, G., Haider, K., Valdimarsson, G., Depisch, D., and Pirker, R. (1996) *Clin. Cancer Res.* **2**, 1231.
33. Filipits, M., Suchomel, R.W., Zochbauer, S., Malayeri, R., and Pirker, R. (1996) *Leukemia* **10 Suppl 3**, S10.
34. Zalcberg, J., Hu, X.F., Slater, A., Parisot, J., El-Osta, S., Kantharidis, P., Chou, S.T., and Parkin, J.D. (2000) *Prostate Cancer Prostatic Dis.* **3**, 66.

35. Hinoshita, E., Uchiumi, T., Taguchi, K., Kinukawa, N., Tsuneyoshi, M., Maehara, Y., Sugimachi, K., and Kuwano, M. (2000) *Clin. Cancer Res.* **6**, 2401.
36. Sullivan, G.F., Amenta, P.S., Villanueva, J.D., Alvarez, C.J., Yang, J.M., and Hait, W.N. (1998) *Clin. Cancer Res.* **4**, 1393.
37. Manohar, C.F., Bray, J.A., Salwen, H.R., Madafiglio, J., Cheng, A., Flemming, C., Marshall, G.M., Norris, M.D., Haber, M., and Cohn, S.L. (2004) *Oncogene* **23**, 753.
38. Mayer, R., Kartenbeck, J., Buchler, M., Jedlitschky, G., Leier, I., and Keppler, D. (1995) *J. Cell Biol.* **131**, 137.
39. Ishikawa, T., Müller, M., Klunemann, C., Schaub, T., and Keppler, J. (1990) *J. Biol. Chem.* **265**, 19279.
40. Cui, Y., König, J., Buchholz, J.K., Spring, H., Leier, I., and Keppler, D. (1999) *Mol. Pharmacol.* **55**, 929.
41. Paulusma, C.C., Bosma, P.J., Zaman, G.J., Bakker, C.T., Otter, M., Scheffer, G.L., Scheper, R.J., Borst, P., and Oude Elferink, R.P. (1996) *Science* **271**, 1126.
42. Conrad, S., Viertelhaus, A., Orzechowski, A., Hoogstraate, J., Gjellan, K., Schrenk, D., and Kauffmann, H.M. (2001) *Toxicology* **156**, 81.
43. Fritz, F., Chen, J., Hayes, P., and Sirotnak, F.M. (2000) *Biochim. Biophys. Acta.* **1492**, 531.
44. Van Aubel, R.A., Koenderink, J.B., Peters, J.G., Van Os, CH., and Russel, F.G. (1999) *Mol. Pharmacol.* **56**, 714.
45. Broeks, A., Gerrad, B., Allikmets, R., Dean, M., and Plasterk, R.H. (1996) *EMBO. J.* **15**, 6132.
46. Lu, Y.P., Li, Z.S., Drozdowicz, Y.M., Hortensteiner, S., Martinoia, E., and Rea, P.A. (1998) *Plant Cell* **10**, 267.
47. Ambudkar, S.V. (1995) *J. Bioenerg. Biomembr.* **27**, 23.
48. Taniguchi, K., Wada, M., Kohno, K., Nakamura, T., Kawabe, T., Kawakami, M., Kagotani, K., Okumura, K., Akiyama, S., and Kuwano, M. (1996) *Cancer Res.* **56**, 4124.
49. Tsujii, H., König, J., Rost, D., Stockel, B., Leuschner, U., and Keppler, D. (1999) *Gastroenterology* **117**, 653.
50. Grant, C.E., Kurz, E.U., Cole, S.P., and Deeley, R.G. (1997) *Genomics* **45**, 368.
51. Keppler, D. and Kartenbeck, J. (1996) *Prog. Liver Dis.* **14**, 55.
52. Schaub, T.P., Kartenbeck, J., König, J., Spring, H., Dorsam, J., Staehler, G., and Storkel, S. (1999) *J. Am. Soc. Nephrol.* **10**, 1159.

53. Schaub, T.P., Kartenbeck, J., König, J., Vogel, O., Witzgall, R., Kriz, W., and Keppler, D. (1997) *J. Am. Soc. Nephrol.* **8**, 1213.
54. Fromm, M.F., Kauffmann, H.M., Fritz, P., Burk, O., Kroemer, H.K., Warzok, R.W., Eichelbaum, M., Seigmund, W., and Schrenk, D. (2000) *Am. J. Pathol.* **157**, 1575.
55. Young, L.C., Campling, B.G., Cole, S.P., Deeley, R.G., and Gerlach, J.H. (2001) *Clin. Cancer Res.* **7**, 1798.
56. Rost, D., König, J., Weiss, G., Klar, E., Stremmel, W., and Keppler, D. (2001) *Gastroenterology* **121**, 1203.
57. St-Pierre, M.V., Serrano, M.A., Macias, R.I., Dubs, U., Hoechli, M., Lauper, U., Meier, P.J., and Marin, J.J. (2000) *Am. J. Physiol. Regul. Integr. Comp. Physiol.* **279**, R1495.
58. Pelis, R.M., Shahidullah, M., Ghosh, S., Coca-Prados, M., Wright, S.H., and Delamere, N.A. (2009) *J. Pharmacol. Exp. Ther.* **329**, 479.
59. Paulsuma, C.C., Kothe, M.J., Bakker, C.T., Bosma, P.J., van Bokhoven, I., van Marle, J., Bolder, U., Tytgat, G.N., and Oude Elferink, R.P. (2000) *Hepatology* **31**, 684.
60. Dombrowski, F., Kubitz, R., Chittattu, A., Wettstein, M., Saha, N., and Haussinger, D. (2000) *Biochem. J.* **348 Pt 1**, 183.
61. Rost, D., Kartenbeck, J., and Keppler, D. (1999) *Hepatology* **29**, 814.
62. Fernandez-Checa, J.C., Takikawa, H., Horie, T., Ookhtens, M., and Kaplowitz, N. (1992) *J. Biol. Chem.* **267**, 1667.
63. Chen, Z.S., Kawabe, T., Ono, M., Aoki, S., Sumizawa, T., Furukawa, T., Uchiumi, T., Wada, M., Kuwano, M., and Akiyama, S.I. (1999) *Mol. Pharmacol.* **56**, 1219.
64. Hagmann, W., Schubert, J., König, J., and Keppler, D. (2002) *Biol. Chem.* **383**, 1001.
65. Keppler, D., Jedlitschky, G., and Leier, I. (1998) *Methods Enzymol.* **292**, 607.
66. Jedlitschky, G., Hoffmann, U., and Kroemer, H.K. (2006) *Expert Opin. Drug Metab. Toxicol.* **2**, 351.
67. Kamisako, T., Leier, I., Cui, Y., König, J., Buchholz, U., Hummel-Eisenbeiss, J., and Keppler, D. (1999) *Hepatology* **30**, 485.
68. Nies, A.T., Cantz, T., Brom, M., Leier, I., and Keppler, D. (1998) *Hepatology* **28**, 1332.
69. Yamazaki, M., Akiyama, S., Ni'inuma, K., Nishigaki, R., and Sugiyama, Y. (1997) *Drug Metab. Dispos.* **25**, 1123.

70. Ishizuka, H., Konno, K., Naganuma, H., Sasahara, K., Kawahara, Y., Niinuma, K., Suzuki, H., and Sugiyama, Y. (1997) *J. Pharmacol. Exp. Ther.* **280**, 1304.
71. Cui, Y., König, J., and Keppler, D. (2001) *Mol. Pharmacol.* **60**, 934.
72. Matsushima, S., Maeda, K., Hayashi, H., Debori, Y., Schinkel, A.H., Schuetz, J.D., Kusuhara, H., and Sugiyama, Y. (2008) *Mol. Pharmacol,* **73**, 1474.
73. Huisman, M.T., Smit, J.W., Crommentuyn, K.M., Zelcer, N., Wiltshire, H.R., Beijnen, J.H., and Schinkel, A.H. (2002) *Aids* **16**, 2295.
74. Payen, L., Sparfel, L., Courtois, A., Vernhet, L., Guillouzo, A., and Fardel, O. (2002) *Cell Biol. Toxicol.* **18**, 221.
75. Masuda, M., I'lzuka, Y., Yamazaki, M., Nishigaki, R., Kato, Y., Ni'inuma, K., Suzuki, H., and Sugiyama, Y. (1997) *Cancer Res.* **57**, 3506.
76. Hulot, J.S., Villard, E., Maguy, A., Morel, V., Mir, L., Tostivint, I., William-Faltaos, D., Fernandez, C., Hatem, S., Deray, G., Komajda, M., Leblond V., and Lechat, P. (2005) *Pharmacognet. Genomics* **15**, 277.
77. Liedert, B., Materna, V., Schadendorf, D., Thomale, J., and Lage, H. (2003) *J. Invest. Dermatol.* **121**, 172
78. Materna, V., Liedert, B., Thomale, J., and Lage, H. (2005) *Int. J. Cancer* **115**, 393.
79. Mor-Cohen, R., Zivelin, A., Rosenberg, N., Shani, M., Muallem, S., and Seligsohn, U. (2001) *J. Biol. Chem.* **276**, 36923.
80. Paulusma, C.C., Kool, M., Bosma, P.J., Scheffer, G.L., ter Borg, F., Scheper, R.J., Tytgat, G.N., Borst, P., Baas, F., and Oude Elferink, R.P. (1997) *Hepatology* **25**, 1539.
81. Shoda, J., Suzuki, H., Sugiyama, Y., Hirouchi, M., Utsunomiya, H., Oda, K., Kawamoto, T., Matsuzaki, Y., and Tanaka, N. (2003) *Hepatol. Res.* **27**, 323.
82. Ito, S., leiri, I. Tanabe, M., Suzuki, A., Higuchi, S., and Otsubo, K. (2001) *Pharmacogenetics* **11**, 175.
83. Itoda, M., Saito, Y., Soyama, A., Saeki, M., Murayama, N., Ishida, S., Sai, K., Nagano, M., Suzuki, H., Sugiyama, Y., Ozawa, S., and Sawada, J.-I. (2002) *Drug Metab. Dispos.* **30**, 363.
84. Saito, S., lida, A., Sekine, A., Miura, Y., Ogawa, C., Kawauchi, S., Higuchi, S., and Nakamura, Y. (2002) *J. Hum. Genet.* **47**, 147.
85. Kartenbeck, J., Leuschner, U., Mayer, R., and Keppler, D. (1996) *Hepatology* **23**, 1061.

86. Dubin, I.N. and Johnson, F.B. (1954) *Medicine (Baltimore)* **33**, 155.
87. Sprinz, H. and Nelson, R.S. (1954) *Ann. Intern. Med.* **41**, 952.
88. Hashimoto, K., Uchiumi, T., Konno, T., Ebihara, T., Nakamura, T., Wada, M., Sakisaka, S., Maniwa, F., Amachi, T., Ueda, K., and Kuwano, M. (2002) *Hepatology* **36**, 1236.
89. Machida, I., Inagaki, Y., Suzuki, S., Hayashi, H., and Wakusawa, S. (2004) *Hepatol. Res.* **30**, 86.
90. Machida, I., Wakusawa, S., Sanae, F., Hayashi, H., Kusakabe, A., Ninomiya, H., Yano, M., and Yoshioka, K. (2005) *J. Gastroenterol.* **40**, 366.
91. Kajihara, S., Hisatomi, A., Mizuta, T., Hara, T., Ozaki, I., Wada, I., and Yamamoto, K. (1998) *Biochem. Biophys. Res. Commun.* **253**, 454.
92. Mor-Cohen, R., Zivelin, A., Rosenberg, N., Goldberg, I., and Seligsohn, U. (2005) *Hepatol. Res.* **31**, 104.
93. Toh, S., Wada, M., Uchiumi, T., Inokuchi, A., Makino, Y., Horie, Y., Adachi, Y., Sakisaka, S., and Kuwano, M. (1999) *Am. J. Hum. Genet.* **64**, 739.
94. Cebeauerova, D., Jirasek, T., Budisova, L., Mandys, V., Volf, V., Novotna, Z., Subhanova, I., Hrebicek, M., Elleder, M., and Jirsa, M. (2005) *Gastroenterology* **129**, 315.
95. Keitel, V., Kartenbeck, J., Nies, A.T., Spring, H., Brom, M., and Keppler, D. (2000) *Hepatology* **32**, 1317.
96. Keitel, V., Nies, A.T., Brom, M., Hummel-Eisenbeiss, J., Spring, H., and Keppler, D. (2003) *Am. J. Physiol. Gastrointest. Liver Physiol.* **284**, G165.
97. Chu, X.Y., Strauss, J.R., Mariano, M.A., Li, J., Newton, D.J., and Cai, X., Wang, R.W., Yabut, J., Hartley, D.P., Evans, D.C., and Evers, R. (2006) *J. Pharmacol. Exp. Ther.* **317**, 579.
98. Nezasa, K., Tian, X., Zamek-Gliszczynski, M.J., Patel, N.J., Raub, T.J., and Brouwer, K.L. (2006) *Drug Metab. Dispos.* **34**, 718.
99. Vlaming, M.L., Pala, Z., van Esch, A., Wagenaar, E., van Tellingen, O., de Waart, D.R., Oude Elferink, R.P., van de Wetering, K., and Schinkel, A.H. (2008) *Clin. Cancer Res.* **14**, 8152.
100. Kool, M., de Haas, M., Scheffer, G.L., Scheper, R.J., van Eijk, M.J., Juijn, J.A., Baas, F., and Borst, P. (1997) *Cancer Res.* **57**, 3537.
101. Belinsky, M.G., Bain, L.J., Balsara, B.B., Testa, J.R., and Kruh, G.D. (1998) *J. Natl. Cancer Inst.* **90**, 1735.
102. Kiuchi, Y., Suzuki, H., Hirohashi, T., Tyson, C.A., and Sugiyama, Y. (1998) *FEBS Lett.* **433**, 149.

103. König, J., Rost, D., Cui, Y., and Keppler, D. (1999) *Hepatology* **29**, 1156.
104. Scheffer, G.L., Kool, M., de Haas, M., de Vree, J.M., Pijnenborg, A.C., Bosman, D.K., Elferink, R.P., van der Valk, P., Borst, P., and Scheper, R.J. (2002) *Lab. Invest.* **82**, 193.
105. Scheffer, G.L., Kool, M., Heijn, M., de Haas, M., Pijnenborg, A.C., Wijnholds, J., van Helvoort, A., de Jong, M.C., Hooijberg, J.H., Mol, C.A., van der Linden, M., de Vree, J.M., van der Valk, P., Elferink, R.P., Borst, P., and Scheper, R.J. (2000) *Cancer Res.* **60**, 5269.
106. Soroka, C.J., Lee, J.M., Azzaroli, F., and Boyer, J.L. (2001) *Hepatology* **33**, 783.
107. Trauner, M., Arrese, M., Soroka, J., Ananthanarayanan, M., Koeppel, T.A., Schlosser, S.F., Suchy, F.J., Keppler, D., and Boyer, J.L. (1997) *Gastroenterology* **113**, 255.
108. Donner, M.G. and Keppler, D. (2001) *Hepatology* **34**, 351.
109. Cherrington, N.J., Hartley, D.P., Li, N., Johnson, D.R., and Klaassen, C.D. (2002) *J. Pharmacol. Exp. Ther.* **300**, 97.
110. Rost, D., Mahner, S., Sugiyama, Y., and Stremmel, W. (2002) *Am. J. Physiol. Gastrointest. Liver Physiol.* **282**, G720.
111. Chu, X.Y., Huskey, S.E., Braun, M.P., Sarkadi, B., Evans, D.C., and Evers, R. (2004) *J. Pharmacol. Exp. Ther.* **309**, 156.
112. Hiroshashi, T., Suzuki, H., and Sugiyama, Y. (1999) *J. Biol. Chem.* **274**, 15181.
113. Hirohashi, T., Suzuki, H., Takikawa, H., and Sugiyama, Y. (2000) *J. Biol. Chem.* **275**, 2905.
114. Kool, M., van der Linden, M., de Haas, M., Scheffer, G.L., de Vree, J.M., Smith, A.J., Jansen, G., Peters, G.J., Ponne, N., Scheper, R.J., Elferink, R.P., Baas, F., and Borst, P. (1999) *Proc. Natl. Acad. Sci. U.S.A.* **96,** 6914.
115. Zelcer, N., Saeki, T., Reid, G., Beijnen, J.H., and Borst, P. (2001) *J. Biol. Chem.* **276**, 46400.
116. Manautou, J.E., de Waart, D.R., Kunne, C., Zelcer, N., Goedken, M., Borst, P., and Elferink, R.O. (2005) *Hepatology* **42**, 1091.
117. Schuetz, J.D., Connelly, M.C., Sun, D., Paibir, S.G., Flynn, P.M., Srinivas, R.V., Kumar, A., and Fridland, A. (1999) *Nat. Med.* **5**, 1048.
118. Deeley, R.G., Westlake, C., and Cole, S.P. (2006) *Physiol. Rev.* **86**, 849.
119. Borst, P., de Wolf, C., and van de Wetering, K. (2007) *Pflugers Arch.* **453**, 661.

120. Russel, F.G., Koenderink, J.B., and Masereeuw, R. (2008) *Trends Pharmacol. Sci.* **29**, 200.
121. Lee, K., Belinsky, M.G., Bell, D.W., Testa, J.R., and Kruh, G.D. (1998) *Cancer Res.* **58**, 2741.
122. Zelcer, N., Reid, G., Wielinga, P., Kuil, A., van der Heijden, I., Schuetz, J.D., and Borst, P. (2003) *Biochem. J.* **371**, 361.
123. Lee, K., Klein-Szanto, A.J., and Kruh, G.D. (2000) *J. Natl. Cancer Inst.* **92**, 1934.
124. Rius, M., Nies, A.T., Hummel-Eisenbeiss, J., Jedlitschky, G., and Keppler, D. (2003) *Hepatology* **38**, 374.
125. Gradhand, U., Lang, T., Schaeffeler, E., Glaeser, H., Tegude, H., Klein, K., Fritz, P., Jedlitschky, G., Kroemer, H.K., Bachmakov, I., Anwald, B., Kerb, R., Zanger, U.M., Eichelbaum, M., Schwab, M., and Fromm, M.F. (2008) *Pharmacogenomics* **8**, 42.
126. van Aubel, R.A., Smeets, P.H., Peters, J.G., Bindels, R.J., and Russel, F.G. (2002) *J. Am. Soc. Neprol.* **13**, 595.
127. Nies, A.T., Spring, H., Thon, W.F., Keppler, D., and Jedlitschky, G. (2002) *J. Urol.* **167**, 2271.
128. Dazert, P., Meissner, K., Vogelgesang, S., Heydrich, B., Eckel, L., Bohm, M., Warzok, R., Kerb, R., Brinkmann, U., Schaeffeler, E., Schwab, M., Cascorbi, I., Jedlitschky, G., and Kroemer, H.K. (2003) *Am. J. Pathol.* **163**, 1567.
129. Nies, A.T., Jedlitschky, G., König, J., Herold-Mende, C., Steiner, H.H., Schmitt, H.P., and Keppler, D. (2004) *Neuroscience* **129**, 349.
130. Meyer Zu Schwabedissen, H.E., Grube, M., Heydrich, B., Linnemann, K., Fusch, C., Kroemer, H.K., and Jedlitschky, G. (2005) *Am. J. Pathol.* **166**, 39.
131. Jedlitschky, G., Burchell, B., and Keppler, D. (2000) *J. Biol. Chem.* **275**, 30069.
132. Klokouzas, A., Wu, C.P., van Veen, H.W., Barrand, M.A., and Hladky, S.B. (2003) *Eur. J. Biochem.* **270**, 3696.
133. Wu, C.P., Woodcock, H., Hladky, S.B., and Barrand, M.A. (2005) *Biochem. Pharmacol.* **69**, 1257.
134. Zhang, J., Alston, M.A., Huang, H., and Rabin, R.L. (2006) *Int. Immunol.* **18**, 485.
135. Jedlitschky, G., Tirschmann, K., Lubenow, L.E., Nieuwenhuis, H.K., Akkerman, J.W., Greinacher, A., and Kroemer, H.K. (2004) *Blood* **104**, 3603.

136. Oevermann, L., Scheitz, J., Starke, K., Köck, K., Kiefer, T., Dolken, G., Nießen, J., Greinacher, A., Siegmund, W., Zygmunt, M., Kroemer, H.K., Jedlitschky, G., and Ritter, C.A. (2009) *Int. J. Cancer* **124,** 2303.
137. Chen, Z.S., Lee, K., and Kruh, G.D. (2001) *J. Biol. Chem.* **276**, 33747.
138. Reid, G., Wielinga, P., Zelcer, N., De Haas, M., Van Deemter, L., Wijnholds, J., Balzarini, J., and Borst, P. (2003) *Mol. Pharmacol.* **63**, 1094.
139. Köck, K., Grube, M., Jedlitschky, G., Oevermann, L., Siegmund, W., Ritter, C.A., and Kroemer, H.K. (2007) *Clin. Pharmacokinet.* **46**, 449.
140. Ritter, C.A., Jedlitschky, G., Meyer zu Schwabedissen, H., Grube, M., Köck, K., and Kroemer, H.K. (2005) *Drug Metab. Rev.* **37**, 253.
141. Reid, G., Wielinga, P., Zelcer, N., van der Heijden, I., Kuil, A., de Haas, M., Wijnholds, J., and Borst, P. (2003) *Proc. Natl. Acad. Sci. U.S.A.* **100**, 9244.
142. Rius, M., Thon, W.F., Keppler, D., and Nies, A.T. (2005) *J. Urol.* **174**, 2409.
143. Lin, Z.P., Zhu, Y.L., Johnson, D.R., Rice, K.P., Nottoli, T., Hains, B.C., McGrath, J., Waxman, S.G., and Sartorelli, A.C. (2008) *Mol. Pharmacol.* **73**, 243.
144. Wijnholds, J., Mol, C.A., van Deemter, L., de Haas, M., Scheffer, G.L., Baas, F., Beijnen, J.H., Scheper, R.J., Hatse, S., De Clercq, E., Balzarini, J., and Borst, P. (2000) *Proc. Natl. Acad. Sci. U.S.A.* **97**, 7476.
145. Adachi, M., Sampath, J., Lan, L.B., Sun, D., Hargrove, P., Flatlety, R., Tatum, A., Edwards, M.Z., Wezeman, M., Matherly, L., Drake, R., and Schuetz, J. (2002) *J. Biol. Chem.* **277**, 38998.
146. Wielinga, P.R., Reid, G., Challa, E.E., van der Heijden, I., van Deemter, L., de Haas, M., Mol, C., Kuil, A.J., Groeneveld, E., Schuetz, J.D., Brouwer, C., De Abreu, R.A., Wijnholds, J., Beijnen, J.H., and Borst, P. (2002) *Mol. Pharmacol.* **62**, 1321.
147. Chen, Z.S., Lee, K., Walther, S., Raftogianis, R.B., Kuwano, M., Zeng, H., and Kruh, G.D. (2002) *Cancer Res.* **62**, 3144.
148. Wielinga, P., Hooijberg, J.H., Gunnarsdottir, S., Kathmann, I., Reid, G., Zelcer, N., van der Born, K., de Haas, M., van der Heijden, I., Kaspers, G., Wijnholds, J., Jansen, G., Peters, G., and Borst, P. (2005) *Cancer Res.* **65**, 4425.

149. Oguri, T., Achiwa, H., Sato, S., Bessho, Y., Takano, Y., Miyazaki, M., Muramatsu, H., Maeda, H., Niimi, T., and Ueda, R. (2006) *Mol. Cancer Ther.* **5**, 1800.
150. Savaraj, N., Wu, C., Wangpaichitr, M., Kuo, M.T., Lampidis, T., Robles, C., Furst, A.J., and Feun, L. (2003) *Int. J. Oncol.* **23**, 173.
151. Wakamatsu, T., Nakahashi, Y., Hachimine, D., Seki, T., and Okazaki, K. (2007) *Int. J. Oncol.* **31**, 1465.
152. Pratt, S., Shpard, R.L., Kandasamy, R.A., Johnston, P.A., Perry, W., 3rd., and Dantzig, A.H. (2005) *Mol. Cancer Ther.* **4**, 855.
153. Leggas, M., Adachi, M., Scheffer, G.L., Sun, D., Wielinga, P., Du, G., Mercer, K.E., Zhuang, Y., Panetta, J.C., Johnston, B., Scheper, R.J., Stewart, C.F., and Schuetz, J.D. (2004) *Mol. Cell. Biol.* **24**, 7612.
154. Tian, Q., Zhang, J., Tan, T.M., Chan, E., Duan, W., Chan, S.Y., Boelsterli, U.A., Ho, P.C., Yang, H., Bian, J.S., Huang, M., Zhu, Y.Z., Xiong, W., Li, X., and Zhou, S. (2005) *Pharm. Res.* **22**, 1837.
155. Norris, M.D., Smith, J., Tanabe, K., Tobin, P., Flemming, C., Scheffer, G.L., Wielinga, P., Cohn, S.L., London, W.B., Marshall, G.M., Allen, J.D., and Haber, M. (2005) *Mol. Cancer Ther.* **4**, 547.
156. Plasschaert, S.L., de Bont, E.S., Boezen, M., vander Kolk, D.M., Daenen, S.M., Faber, K.N., Kamps, W.A., de Vries, E.G., and Vellenga, E. (2005) *Clin. Cancer Res.* **11**, 8661.
157. Steinbach, D., Lengemann, J., Voigt, A., Hermann, J., Zintl, F., and Sauerbrey, A. (2003) *Clin. Cancer Res.* **9**, 1083.
158. Park, S., Shimizu, C., Shimoyama, T., Takeda, M., Ando, M., Kohno, T., Katsumata, N., Kang, Y.K., Nishio, K., and Fujiwara, Y. (2006) *Breast Cancer Res. Treat.* **99**, 9.
159. König, J., Hartel, M., Nies, A.T., Martignoni, M.E., Guo, J., Buchler, M.W., Friess, H., and Keppler, D. (2005) *Int. J. Cancer* **115**, 359.
160. Hagmann, W., Jesnowski, R., Faissner, R., Guo, C., and Lohr, J.M. (2008) *Pancreatology* **9**, 136.
161. Krishnamurthy, P., Schwab, M., Takenaka, K., Nachagari, D., Morgan, J., Leslie, M., Du, W., Boyd, K., Cheok, M., Nakauchi, H., Marzolini, C., Kim, R.B., Poonkuzhali, B., Schuetz, E., Evans, W., Relling, M., and Schuetz, J.D. (2008) *Cancer Res.* **68**, 4983.
162. Kool, M., van der Linden, M., de Haas, M., Baas, F., and Borst, P. (1999) *Cancer Res.* **59**, 175.

163. Cai, L., Struk, B., Adams, M.D., Ji, W., Haaf, T., Kang, H.L., Dho, S.H., Xu, X., Ringpfeil, F., Nancarrow, J., Zach, S., Schaen, L., Stumm, M., Niu, T., Chung, J., Lunze, K., Verrecchia, B., Goldsmith, L.A., Viljoen, D., Fuguera, L.E., Fuchs, W., Lebwohl, M., Uitto, J., Richards, R., Hohl, D., and Ramesar, R. (2000) *J. Mol. Med.* **78**, 36.
164. Ringpfeil, F., Lebwohl, M.G., Christiano, A.M., and Uitto, J. (2000) *Proc. Natl. Acad. Sci. U.S.A.* **96**, 6001.
165. Belinsky, M.G. and Kruh, G.D. (1999) *Br. J. Cancer* **80**, 1342.
166. Uitto, J. (2001) *J. Eur. Acad. Dermatol. Venereol.* **15**, 303.
167. Uitto, J., Pulkkinen, L., and Ringpfeil, F. (2001) *Trends Mol. Med.* **7**, 13.
168. Beck, K., Hayashi, K., Nishiguchi, B., Le Saux, O., Hayashi, M., and Boyd, C.D. (2003) *J. Histochem. Cytochem.* **51**, 887.
169. Homolya, L., Varadi, A., and Sarkardi, B. (2003) *Biofactors* **17**, 103.
170. Madon, J., Hagenbuch, B., Landmann, L., Meier, P.J., and Stiger, B. (2000) *Mol. Pharmacol.* **57**, 634.
171. Belinsky, M.G., Chen, Z.S., Shchaveleva, I., Zeng, H., and Kruh, G.D. (2002) *Cancer Res.* **62**, 6172.
172. Ilias, A., Urban, Z., Seidl, T.L., Le Saux, O., Sinko, E., Boyd, C.D., Sarkadi, B., and Varadi, A. (2002) *J. Biol. Chem.* **277**, 16860.
173. Bergen, A.A., Plomp, A.S., Hu, X., de Jong, P.T., and Gorgels, T.G. (2007) *Pflugers Arch.* **453**, 685.
174. Rene, J., de Vries, E.G., Nienhuis, E.F., Jansen, P.L., and Müller, M. (1999) *Br. J. Pharmacol.* **126**, 681.
175. Loe, D.W., Almquist, K.C., Deeley, R.G., and Cole, S.P. (1996) *J. Biol. Chem.* **271**, 9675.
176. Barnouin, K., Leier, I., Jedlitschky, G., Pourtier-Manzanedo, A., König, J., Lehmann, W.D., and Keppler, D. (1998) *Br. J. Cancer* **77**, 201.
177. Hooijberg, J.H., Broxterman, H.J., Kool, M., Assaraf, Y.G., Peters, G.J., Noordhuis, P., Scheper, R.J., Borst, P., Pinedo, H.M., and Jansen, G. (1999) *Cancer Res.* **59**, 2532.
178. Paumi, C.M., Ledford, B.G., Smitherman, P.K., Townsend, A.J., and Morrow, C.S. (2001) *J. Biol. Chem.* **276**, 7952.
179. Luo, F.R., Paranjpe, P.V., Guo, A., Rubin, E., and Sinko, P. (2002) *Drug Metab. Dispos.* **30**, 763.
180. Michoy, J.M., Van Bambeke, F., Mingeot-Leclercq, M.P., and Tulkens, P.M. (2004) *Antimicrob. Agents Chemother.* **48**, 2673.

181. Srinivas, R.V., Middlemas, D., Flynn, P., and Fridland, A. (1998) *Antimicrob. Agents Chemother.* **42**, 3157.
182. Olson, D.P., Scadden, D.T., D'Aquila, R.T., and De Paspuale, M.P. (2002) *AIDS* **16**, 1743.
183. Smitherman, P.K., Townsend, A.J., Kute, T.E., and Morrow, C.S. (2004) *J. Pharmacol. Exp. Ther.* **308**, 2004.
184. Evers, R., Kool, M., van Deemter, L., Janssen, H., Calafat, J., Oomen, L.C., Paulusma, C.C., Oude Elferink, R.P., Baas, F., Schinkel, A.H., and Borst, P. (1998) *J. Clin. Invest.* **101**, 1310.
185. Matsushima, S., Maeda, K., Kondo, C., Hirano, M., Sasaki, M., Suzuki, H., and Sugiyama, Y. (2005) *J. Pharmacol. Exp. Ther.* **314**, 1059.
186. van de Wetering, K., Zelcer, N., Kuil, A., Feddema, W., Hillebrand, M., Vlaming, M.L., Schinkel, A.H., Beijnen, J.H., and Borst, P. (2007) *Mol. Pharmacol.* **72**, 387.
187. Yamada, A., Maeda, K., Kamiyama, E., Sugiyama, D., Kondo, T., Shiroyanagi, Y., Nakazawa, H., Okano, T., Adachi, M., Schuetz, J.D., Adachi, Y., Hu, Z., Kusuhara, H., and Sugiyama, Y. (2007) *Drug Metab. Dispos.* **35**, 2166.
188. Zeng, H., Chen, Z.S., Belinsky, M.G., Rea, P.A., and Kruh, G.D. (2001) *Cancer Res.* **61**, 7225.
189. Zelcer, N., van de Wetering, K., Hillebrand, M., Sarton, E., Kuil, A., Wielinga, P.R., Tephly, T., Dahan, A., Beijnen, J.H., and Borst, P. (2005) *Proc. Natl. Acad. Sci. U.S.A.* **102**, 7274.
190. Xiong, H., Yoshinari, K., Brouwer, K.L., and Negishi, M. (2002) *Drug Metab. Dispos.* **30**, 918.
191. Wielinga, P.R., van der Heijden, I., Reid, G., Beijnen, J.H., Wijnholds, J., and Borst, P. (2003) *J. Biol. Chem.* **278**, 17664.
192. Ci, L., Kusuhara, H., Adachi, M., Schuetz, J.D., Takeuchi, K., and Sugiyama, Y. (2007) *Mol. Pharmacol.* **71**, 1591.
193. Uchida, Y., Kamiie, J., Ohtsuki, S., and Terasaki, T. (2007) *Pharm. Res.* **24**, 2281.
194. Hasegawa, M., Kusuhara, H., Adachi, M., Schuetz, J.D., Takeuchi, K., and Sugiyama, Y. (2007) *J. Am. Soc. Nephrol.* **18**, 37.

Chapter 9

ABC Transporter Mutations in Heritable Skin Diseases: The *ABCC6* and *ABCA12* Genes

Qiaoli Li, Qiujie Jiang and Jouni Uitto

1. Introduction

1.1 *The ABC family of transporters*

The human ABC gene family consists of 48 members divided into seven subgroups based on the general consensus of their structural characteristics.[1–3] The general features of ABC transporters include transmembrane domains (TMDs) consisting of transmembrane helical segments that are linked by intra- and extracellular peptide loops of varying sizes, as well as the presence of intracellular nucleotide binding folds (NBFs). Functional studies have suggested that ATP-driven dimerization of the NBFs, with concomitant restructuring of the dimer interface, could be linked to conformational changes in the TMDs.[4] Further understanding of the intramolecular structure/function relationships of ABC transporters has been provided by determination of high-resolution, three-dimensional structures of bacterial ABC export pumps.[5,6] These structures conform with earlier biochemical data and provide potentially suitable templates for homology modeling for human ABC transporter molecules.[7–10] In fact, models have been developed for several human proteins. For example, analysis of distribution of amino acid substitutions encountered in pseudoxanthoma elasticum (PXE) in the *ABCC6* gene (see below) has revealed that mutations of residues participating in domain-domain interactions and in contact between the two catalytic NBF domains are up to four-fold more frequent than the average mutational rate in the overall polypeptide chain, supporting the notion of structural basis of "informational transmission" for the NBF dimer interface.[10]

The ABC transporters are typically involved in translocation of substrates through polarized cells associated with physiological functions, such as extrusion of a variety of compounds from the liver or kidneys, absorption in the gut, or transport through the blood-brain barrier.[4] The ABC transporters were initially associated with multi-drug resistance due to their important role in cancer drug resistance as a result of their over expression in malignant cells.[3] However, it has been generally accepted that ABC transporters are also key players in physiological cellular defense systems. Many of the identified substrates are hydrophobic in nature, but the ABC transporters are also capable of transporting a variety of amphipathic anions and cations. While the individual ABC transporters have a characteristic repertoire of substrates, there is an overlapping substrate recognition pattern which allows the ABC transporter proteins to form a functional network capable of extruding a wide range of substrates.

The critical role of ABC transporters in normal physiology is attested to by demonstrations that mutations in several of the corresponding genes are associated with heritable diseases (Table 1).

In general, tissue-specific expression of these genes and the nature of the respective transport substrates determine the phenotypic spectrum of the disease. Several of these conditions have cutaneous manifestations, and in particular, two of these conditions, *viz.*, pseudoxanthoma elasticum (PXE) and Harlequin ichthyosis (HI) have primary manifestations in the skin.[11] These two disorders are due to mutations in the ABC transporter genes, *ABCC6* and *ABCA12*, respectively. This overview will concentrate on recent progress in understanding the molecular basis of

Table 1. Examples of the spectrum of human diseases due to mutations in ABC transporter genes.[a]

Gene	Associated diseases	Transporter function/substrates
ABCA1[b]	Tangier's disease; familial hypoalphalipoproteinemia; premature atherosclerosis	Cholesterol, phospholipids
ABCA3	Surfactant deficiency in newborns	Phospholipid metabolism
ABCA4[b]	Stargardt's disease; AR retinitis pigmentosa (some forms); AR cone-rod dystrophy	Photoreceptor retinoid/phospholipid transporter
ABCA12[b]	Lamellar ichthyosis type 2 (mild); Harlequin ichthyosis (severe)	Glucosylceramide, other epidermal lipids (?)
ABCB4/11	Progressive familial intrahepatic cholestasis-3/2	Bile-acid transport
ABCC2	Dubin-Johnson syndrome (hyperbilirubinemia II)	Bile-acid transport
ABCC6	Pseudoxanthoma elasticum	Organic anions (*in vitro*)
ABCC7	Cystic fibrosis	Chloride ion channel
ABCC8	Familial persistent hyperinsulinemic hypoglycemia of infancy; AD type 2 diabetes	Sulfonylurea receptor
ABCC9	Dilated cardiomyopathy with ventricular tachycardia	Regulatory subunit of cardiac K(ATP) channel
ABCD1	Adrenoleukodystrophy	Very long chain fatty acids
ABCG5/8	Sitosterolemia	Sterols

[a] Abbreviations: AD, autosomal dominant; AR, autosomal recessive.
[b] Mutations in these genes can result in a range of phenotypes depending on the types of mutations.

PXE and HI and will highlight the clinical implications of molecular genetics studies with regard to these disorders.

1.2 *The spectrum of heritable skin disorders*

Heritable skin, diseases represent a spectrum of clinical manifestations where the clinical findings at one end of the spectrum can be relatively minor, limited to the skin, hair and nails, while at the other end of the spectrum the cutaneous manifestations can be part of multi-organ pathology with considerable morbidity and mortality.[12] Due to their rarity in general, many of the genodermatoses pose a clinical challenge and the medical practitioner may not be aware of the salient clinical and genetic features of the condition. This challenge has been compounded by the complexity of classifications, often riddled with eponyms. Furthermore, in many of these conditions histopathology is not diagnostic, and the molecular basis remained for a long time unknown, thus precluding DNA-based diagnostic procedures. Over the past two decades, however, this situation has changed rapidly, and with the advent of molecular genetics in general, an increasingly large number of mutations have been linked to specific skin diseases. In fact, the molecular basis of some close to 300 distinct heritable skin diseases are now known and in many cases these genetic lesions readily explain the phenotypic manifestations of the disease. Examination of the mutation databases has revealed, however, both obvious candidate genes as well as a number of surprises. As discussed below, identification of mutations in the *ABCC6* gene in patients with PXE was entirely unexpected as this gene is primarily expressed in the liver and the kidneys, and at very low levels, if at all, in the tissues primarily affected in this disease. In contrast, identification of mutations in the *ABCA12* gene in patients with HI or with a form of lamellar ichthyosis (LI) could be anticipated due to the fact that the *ABCA12* transporter gene is expressed at relatively high level in human epidermis.

2. The Paradigms of Pseudoxanthoma Elasticum and Harlequin Ichthyosis

2.1 *Pseudoxanthoma elasticum*

2.1.1 *Clinical and genetic features of pseudoxanthoma elasticum*

PXE is a multi-system autosomal recessive disorder characterized by dystrophic mineralization of soft connective tissues in a number of organs, with primary manifestations in the skin, the eyes, and the arterial blood vessels.[13,14] The diagnosis is primarily made on the basis of cutaneous lesions, small yellowish papules on the predilection sites at flexural areas, i.e., the neck, axillae, groins and antecubital fossae. With time, these lesions progressively coalesce into larger plaques of inelastic leathery skin with yellowish hue. Histopathology of the affected skin reveals accumulation of elastin-containing material, and characteristically, these elastotic structures become progressively mineralized over the lifetime of an affected individual. The eye manifestations consist of angioid streaks in the retina, and bleeding of the choroid vessels can result in progressive loss of visual acuity leading occasionally to central blindness. The eye manifestations are complications of mineralization of the Bruch's membrane, an elastin-rich layer of the retina, which causes breakage of blood vessels and neovascularization with subsequent bleeding. The cardiovascular manifestations are derived from mineralization of the elastic layers in arterial blood vessels, and clinically present with gastrointestinal bleeding, intermittent claudication, hypertension, and occasionally, early myocardial infarction.

Since the primary pathology of PXE was initially noted to involve the elastic fibers in the affected tissues, the genes involved in the synthesis and assembly of the elastic fiber network were considered as the primary candidate genes for mutations. These gene/protein systems included elastin (*ELN*) on chromosome 7q, elastin-associated microfibrillar proteins, such as fibrillin 1 and fibrillin 2 (*FBN1* and *FBN2*) on chromosomes 15 and 5, and lysyl oxidase (*LOX*) also on chromosome 15. However, genetic linkage analyses of inheritance of the disease excluded these chromosomal regions.[15,16] Subsequent attempts to map the "PXE gene", utilizing positional cloning approaches, provided evidence for linkage to a critical interval in the short arm of chromosome 16.[17,18] Sequencing of the candidate genes in this region resulted in identification of the *ABCC6* gene as the one harboring mutations causing PXE.[19–22] The critical role of the *ABCC6* gene in the pathogenesis of PXE was confirmed by development of transgenic mice through targeted ablation of the corresponding mouse gene, *Abcc6*.[23,24] These *Abcc6*$^{-/-}$ mice recapitulate cardinal histopathologic and ultrastructural features of human PXE, including progressive soft tissue mineralization.

2.1.2 *The ABCC6 gene/protein system*

PXE is caused by mutations in the *ABCC6* gene, which encodes a transmembrane transporter protein ABCC6 (also known as multidrug resistance-associated protein 6 — MRP6). The *ABCC6* gene spans ~75 kb of the human genome on human chromosomal region 16p13.1 and consists of 31 exons (Fig. 1A). This chromosomal region also contains two closely related, but nonfunctional, 5′ pseudogenes, *ABCC6*-ψ1 and *ABCC6*-ψ2, which correspond to exons 1–9 and 1–4 of *ABCC6*, respectively.[25] A close sequence similarity between the two pseudogenes and the coding gene has complicated mutation detection in *ABCC6*, although allele-specific primers have been designed for each exon facilitating analysis without interference from the pseudogene sequences.[26]

The *ABCC6* gene is transcribed into a ~5 kb mRNA and translated into a ~165 kD protein of 1503 amino acids (Fig. 1B). As is characteristic for all ABC transporters, the ABCC6 protein is predicted to contain two nucleotide binding folds, ~115 and 174 amino acids, respectively. Protein database programs, coupled with experimental data and established ABCC1 membrane topology, suggest a consensus that ABCC6 has three hydrophobic TMDs consisting of 5, 6 and 6 transmembrane helices. Based on sequence homology with other members of the C family of ABC transporter proteins, particularly the prototypic ABCC1, the ABCC6 protein has been postulated to function as an efflux transporter. The *ABCC6* gene is primarily expressed in the liver, to a lesser extent in the proximal tubules of kidneys, and at a lower level, if at all, in a number of tissues, including those directly affected in PXE (the skin, the eyes, and the cardiovascular system).[27,28] *In vitro* studies have shown that ABCC6 can function as a transmembrane transporter of anionic, glutathione-conjugated molecules,[29] but its *in vivo* substrate specificity is currently unknown.

2.1.3 *ABCC6 mutation database*

Well over 300 distinct mutations have been identified in the *ABCC6* gene representing over 1000 mutant alleles in families with PXE.[14,26,30] The types of mutations include missense and nonsense mutations in 28 of the 31 exons of the gene, intronic mutations causing mis-splicing, and small deletions and insertions as well as large deletions spanning different parts of the entire coding region and sometimes including flanking genes (Fig. 1A). Two mutations are recurrent, and, specifically, the nonsense mutation p.R1141X in exon 24 is particularly frequent in Caucasian individuals, accounting for approximately 30% of all PXE mutations. A recurrent *AluI*-mediated deletion of exons 23–29 (DEL23–29; p.A999-S1403del) of the *ABCC6* gene has been found in at least one allele in up to 20% of the US and 12% of the European patients. Additional recurrent nonsense mutations (p.Q378X in exon 9, p.R518X in exon 12, and p.R1164X in exon 24), as well as clustering of the missense mutations to exons 24 and 28, corresponding to the NBFs, has permitted the development of streamlined mutation detection strategies to facilitate identification of mutations in the *ABCC6* gene (Fig. 1B). Consequently, mutation analysis is now readily available for individuals with a family history of PXE and can be used for confirmation of the clinical diagnosis, carrier detection, and presymptomatic identification of affected individuals at risk for PXE.[14]

2.1.4 *Pathomechanistic consequences of ABCC6 mutations*

While the ABCC6 protein has been postulated to be an efflux pump, a notion supported by *in vitro* data, the physiological role of this protein and its substrate specificity *in vivo* remain largely unknown. Furthermore, the precise pathomechanistic links between the *ABCC6* mutations and the pathological mineralization

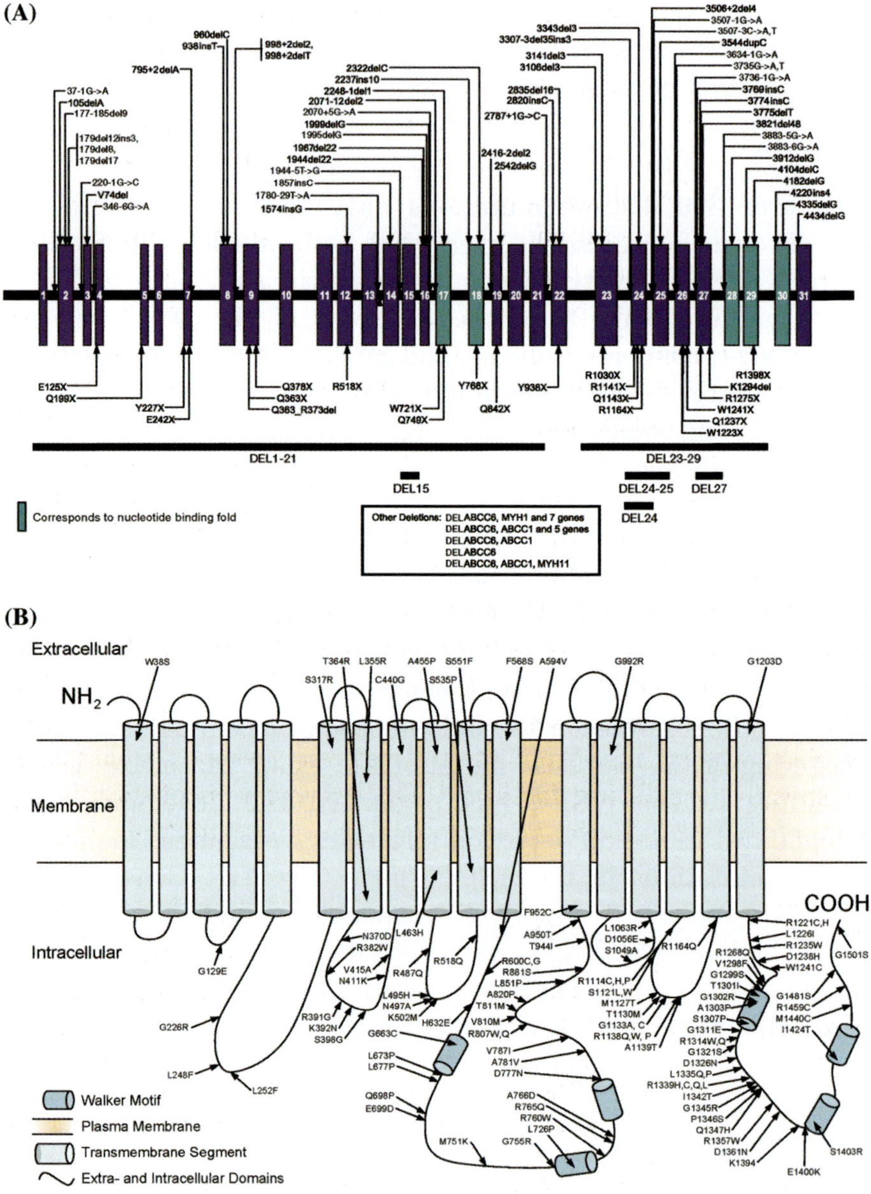

Fig. 1. Schematic representation of the *ABCC6* gene **(A)** and the predicted structure of the corresponding protein, ABCC6 **(B)**, together with known mutations indicated by the arrows.

of peripheral connective tissues in PXE remain unclear. It has been proposed, however, that PXE is a metabolic disorder and the absence of functional ABCC6 activity, primarily in the liver, results in deficiency of circulating factor(s) that are physiologically required to prevent aberrant mineralization under normal calcium and phosphate homeostatic conditions.[31,32] In support of this, so-called "metabolic hypothesis", are several clinical and experimental observations in patients with PXE as well as in the *Abcc6*$^{-/-}$ mice. First, the clinical findings in patients with PXE are rarely present at birth, and the average age of diagnosis is in early teens. The ensuing progression of the disease demonstrates slow, yet continued, accumulation of mineral deposits in soft connective tissues of the affected organs. In *Abcc6*$^{-/-}$ mice, no evidence of aberrant mineralization is noted until ~5 weeks after birth, thus paralleling the onset of the disease in humans. Secondly, sera from patients with PXE as well as from *Abcc6*$^{-/-}$ mice lack the capacity to prevent calcium/phosphate precipitation in an *in vitro* assay system utilizing human smooth muscle cell cultures.[33] Serum from patients with PXE, when added to cell culture medium, has also been shown to alter the assembly of elastic fibers in fibroblast cultures, suggesting the presence of specific circulating factors.[34] The critical role of circulating factors in the development of tissue mineralization is also

---

Fig. 1. (*Continued*) **(A)** The gene consists of 31 exons drawn to scale with the corresponding introns. Note that exons 17–18 and 28–30 (green) correspond to the nucleotide binding folds 1 and 2, respectively. The positions of nonsense mutations are indicated below the gene structure, while small insertions and deletions resulting in frame-shift and premature termination codon are depicted above the gene. The positions of large intragenic deletions are indicated below the gene, and a number of mutations resulting in deletion of the entire *ABCC6* and contiguous genes are shown in the box. **(B)** The schematic representation of the ABCC6 protein depicts three transmembrane domains consisting of 5, 6 and 6 transmembrane helical segments, respectively. The segments representing three Walker motifs within the nucleotide binding folds are also indicated (blue). The positions of the mutations resulting in amino acid substitutions are indicated. Modified from Ref. 30, with permission.

supported by grafting studies in which muzzle skin from *Abcc6*$^{-/-}$ mice and their wild-type littermates were grafted onto the back of wild-type or *Abcc6*$^{-/-}$ mice.[35] The muzzle skin of mice contains vibrissae surrounded by connective tissue capsules which become reproducibly mineralized in *Abcc6*$^{-/-}$ mice at ~5 weeks of age, thus serving as an early and quantifiable biomarker of the mineralization process in this mouse model of PXE.[23,33] When muzzle skin from wild-type mice was grafted onto the back of *Abcc6*$^{-/-}$ mice at four weeks of age, mineralization ensued in subsequent two months. In contrast, no mineralization was observed in the muzzle skin from *Abcc6*$^{-/-}$ mice transplanted onto the back of the corresponding wild-type mice. In this model, the vascular capillaries in the graft regress while new vascular ingrowth occurs from the recipient wound bed to replace the regressing vessels. The survival of skin grafts is, therefore, dependent of reestablishment of circulation in the graft from the recipient animal, and the grafts are supplied by blood of the recipient mice. As skin grafts from the *Abcc6*$^{-/-}$ mice did not develop mineralization when placed onto the wild-type mice but the skin from wild-type mouse showed mineralization after grafting onto the *Abcc6*$^{-/-}$ mice, these findings suggest that circulating factors in the recipients' blood play a critical role in determining the degree of mineralization of the graft.[35] Nevertheless, the characteristics of the circulating factor(s) remain unknown.

Intriguing insight into the potential pathomechanisms of PXE has been recently provided by observations that defects in another gene, *GGCX*, can result in a PXE-like phenotype.[36–38] This gene encodes an enzyme, γ-glutamyl carboxylase, which catalyzes γ-carboxylation of glutamine residues in a number of proteins. This reaction requires vitamin K in its reduced form (KH_2) as a cofactor. The target proteins subject to γ-carboxylation by the GGCX protein are the vitamin K-dependent coagulation factors synthesized and activated by γ-glutamyl carboxylation in the liver.[14] In the absence of the GGCX activity, the coagulation factors remain in inactive form, and bleeding disorder ensues. The development of PXE-like cutaneous features, with characteristic connective tissue mineralization in the dermis, has been suggested to involve the matrix gla protein (MGP), which also requires activation by γ-glutamyl carboxylation catalyzed by GGCX.[39] Under normal physiologic conditions, fully active MGP is a powerful anti-mineralization factor, as illustrated by the fact that targeted ablation of the corresponding gene in mice results in extensive vascular mineralization.[40] Thus, in its under-carboxylated, inactive form, MGP may not be able to prevent the mineralization process. It has been proposed, therefore, that ABCC6 may serve as a transporter of reduced vitamin K, thus resulting in deficiency of MGP γ-glutamyl carboxylation in peripheral cells, such as fibroblasts, and mineralization of the adjacent extracellular matrix of connective tissue ensues.[14,41] While this hypothesis is being tested in a number of experimental systems, the postulate of the potential role of vitamin K or its derivatives, if proven correct, raises the possibility that administration of an appropriate form of this vitamin may counteract the manifestations of PXE.

2.2 *Ichthyosis*

2.2.1 *Clinical and genetic heterogeneity of ichthyoses*

The skin manifestations in PXE are primarily of cosmetic concern. In contrast, at the other end of the spectrum of skin diseases caused by mutations in ABC transporter genes, HI is a devastating congenital skin condition that often leads to premature demise of newborns during the early postnatal period.

Ichthyosis (in reference to fish scale) comprises a group of both acquired and heritable keratinization disorders with dry, hyperkeratotic and scaly skin.[42] While the phenotypic spectrum of ichthyosiform dermatoses is extremely broad with either limited or extensive involvement of the skin, among the inherited forms, three clinically and genetically distinct subtypes have been identified. Ichthyosis vulgaris, an autosomal dominant and relatively common form of dry and scaly skin is due to mutations in the filaggrin gene.[43–45] The X-linked ichthyosis is an X-linked recessive condition due to deficient steroid sulfatase activity resulting in fine scales associated with corneal opacities.[46] LI in itself is a heterogeneous group of autosomal recessive disorders with large, plaque-like, brown scales over most of the body, associated with ectropion and alopecia. The classic form of LI is due to mutations in the *TGM1* gene encoding transglutaminase catalyzing cross-linking of proteins in the stratum corneum, i.e., the uppermost layer of epidermis.[47,48] Finally, a related, yet distinct, disorder of keratinization is non-bullous congenital ichthyosiform erythroderma (NBCIE), an autosomal recessive condition, characterized by red and scaly skin without blisters.

HI is a rare, extremely severe form of ichthyosis, most closely associated with the LI group of these disorders.[49] The neonates with HI are encased in a thick skin which severely restricts their movement. As the skin dries out, it forms hard, diamond-shaped plaques, and the normal facial features are affected with distortion of the lips, eyelids and ears. In general, the newborns with HI die within a few days of birth because of respiratory distress, bacterial infections and feeding difficulties. A few of these affected individuals survive, however, and the skin eventually resembles severe non-bullous congenital ichthyosiform erythroderma or lamellar ichthyosis.

The unifying histopathologic feature in all forms of ichthyosis is hyperkeratosis, reflecting alterations in the terminal differentiation of stratum corneum, the outermost layer of the skin, which serves as a functional barrier crucial for physiological homeostasis. Specifically, the epidermis serves as a protective barrier that excludes the invasion of harmful organisms, such as viruses, bacteria and fungi, as well as other antigenic particles.[43] A correctly functioning barrier is also critical for the maintenance of physiological homeostasis of hydration and electrolyte balance. The formation of the outer physical barrier results from differentiation of keratinocytes within the stratified epithelium, the top layer of skin facing the environment. The principal cells within the epidermis, the keratinocytes, traverse from the lower compartment, the basal layer, adjacent to the dermal-epidermal basement membrane, to the upper layers. There they undergo differentiation, characterized by altered expression of proteins according to the cells' location within the epidermis and the anatomical site of the skin. As the keratinocytes traverse the epidermis, they begin to accumulate increasing amounts of keratinocyte-specific proteins that are involved in terminal differentiation. Once the cells reach the stratum corneum, they denucleate and shed their DNA, and the terminally differentiated corneocytes pack tightly into a resilient and impermeable layer, which, in addition to proteins, consists of intercellular lipids. In fact, synthesis of the appropriate lipids and their extrusion to the extracellular space by keratinocytes is critical for proper assembly of the stratum corneum.

During the final stages of epidermal differentiation, a highly ordered structure, the cornified envelope (CE), forms below the plasma membrane of terminally differentiated keratinocytes. The CE, essential for normal barrier function, is composed of several proteins, including filaggrin and keratin intermediate filaments, and these proteins are cross-linked extensively by transglutaminases. Thus, terminal differentiation is a highly coordinated process in which keratinocytes alter the gene expression profile, leading to production of an impermeable barrier at the outer surface of the skin. Consequently, mutations in the genes involved in the formation of the stratum corneum, such as filaggrin in ichthyosis vulgaris, and transglutaminases in lamellar ichthyosis, explain the abnormalities noted in the epidermis.

2.2.2 *ABCA12 mutations underlie HI and some forms of LI*

The molecular basis of HI has remained unknown until recently, but it is now well established that mutations in the *ABCA12* gene underlie this disorder.[50,51] Initial approaches utilizing single-nucleotide polymorphism-based chip technology and homozygosity mapping of families with individuals affected with HI placed the candidate locus on chromosomal region 2q35, and microsatellite markers narrowed the interval to consist of six genes.[51] Several previous observations pointed to *ABCA12* as a candidate gene within the critical region. First, a characteristic ultrastructural feature of the epidermis in HI is abnormality in the localization of epidermal lipids, together with abnormal ultrastructural epidermal lamellar granules.[52] *ABCA12* has been suggested to encode a transmembrane transporter protein which from sequence homology to several other ABC family members, was thought to be involved in the transport of lipids.[53] Secondly, the *ABCA12* gene was previously shown to harbor missense mutations in a milder form of ichthyosis, lamellar ichthyosis type 2 (LI2), with some resemblance to the phenotype in patients with HI who survive beyond the immediate postnatal period.[54] As of today, a total of 53 distinct mutations have been identified in the *ABCA12* gene (Fig. 2).

Subsequently, expression of ABCA12 was demonstrated in lamellar granules, and in normal epidermal keratinocytes there is an upregulation of the expression of ABCA12 in association with physiological keratinization of human epidermis.[55] Mutations in the *ABCA12* gene result in congested lipid secretion in the skin of patients with HI, and it has been suggested that ABCA12 transports glucosylceramide, the major lipid of the stratum corneum of the epidermis. Finally, lamellar granule-mediated lipid secretion was resumed in the cultured keratinocytes of patients upon transfer of the wild-type *ABCA12* gene.[50] Thus, it is clear that mutations in the *ABCA12* transporter gene underlie HI.

The critical role of ABCA12 in skin abnormalities in HI was confirmed by observations on mice lacking *Abca12* expression. In some of these studies, *Abca12*$^{-/-}$ mice were developed by targeted ablation of the corresponding mouse gene, while another study identified HI mice as part of the recessive mouse N-ethyl-N-nitrosourea mutagenesis screen.[56–58] Both animal models displayed many of the hallmarks of HI, including hyperkeratosis, loss of barrier function, and defects in lipid homeostasis. Further analysis demonstrated lamellar granule abnormalities and defective ceramide distribution in the epidermis of these mice, illustrating that Abca12 plays a pivotal role in maintaining lipid balance in the skin.[58] Examination of keratinocytes cultured from mice lacking *Abca12* expression demonstrated that these cells had severely impaired lipid efflux leading to intracellular accumulation of neutral lipids. One of the studies also identified Abca12 as a mediator of Abca1-regulated cellular cholesterol efflux, a finding that may have significant implications for other diseases of lipid metabolism and homeostasis.[57]

2.2.3 *Characteristics of the ABCA12 gene/protein system*

The A subfamily of ABC transporters consists of 12 members that share a high degree of sequence conservation.[53] Mutations in some of these genes have been described to cause inherited diseases related to lipid transport, such as surfactant deficiency and HI (Table 1). The ABCA12 protein is encoded by the *ABCA12* gene residing on the 12q35 region of human genome and consisting of 53 exons (Fig. 2A). The gene is transcribed to a ~9.1 kb mRNA transcript which encodes a protein of 2,595 amino acids (Fig. 2B). While the precise nucleotide sequence of the *ABCA12* gene and of the corresponding cDNA are readily available from the databases, molecular modeling of the corresponding protein has resulted in somewhat different predicted structures for the protein structure.[50,54] While the presence of two NBFs is well established, the number of TMDs and the transmembrane helices has not been established with certainty. In fact, predictions of the protein structure by ENSEMBL (http://www.ensembl.org/index.html) and HMMTOP (http://www.enzim.hu/hmmtop) programs suggested the presence of 13 transmembrane helices. However, incorporation of experimental data into the predictions suggests the structure shown in Fig. 2B; i.e., an N-terminal intracellular tail — one transmembrane helix — large extracellular loop 1 — TMD1 consisting of five or seven transmembrane helices — NBF1 — one transmembrane helix — large extracellular loop 2 — TMD2 consisting of five or seven

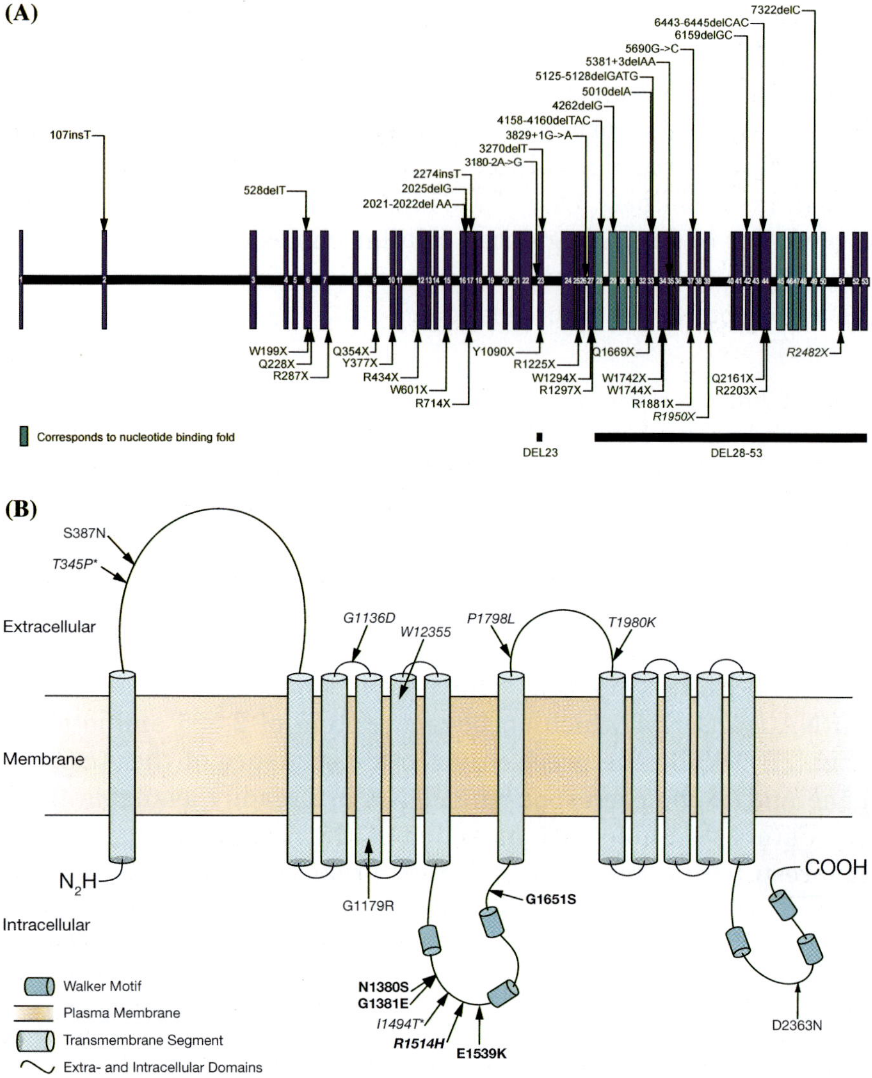

Fig. 2. Schematic representation of the *ABCA12* gene **(A)** and the predicted topography of the corresponding protein ABCA12 **(B)**, together with known mutations in *ABCA12*.

(A) The gene consists of 53 exons drawn to scale with the corresponding introns. The positions of nonsense mutations are indicated below the gene, while mutations representing small insertions or deletions and resulting in frame-shift and premature termination codon, are depicted above the gene structure; these transmembrane helices — NBF2 — intracellular C-terminus. It should be noted that protein predictions suggest that presence of a 45-amino acid signal sequence at the N-terminus, and its removal would eliminate the first transmembrane helix (see Fig. 2B). In general, however, the secondary structures of the proteins of the A family of ABC transporters are not as well characterized as the members of the C family.

3. Clinical Implications of Mutation Analysis in PXE and HI

As indicated above, identification of mutations in the *ABCC6* gene has allowed establishment of DNA-based clinical testing to confirm the diagnosis of PXE and their use for presymptomatic analysis of individuals at risk for the disease. By extension, these tests can be applied to prenatal genetic diagnosis in pregnancies

Fig. 2. (*Continued*) mutations are associated with Harlequin ichthyosis phenotypes but the two mutations shown in Italics have also been reported in NBCIE. Positions of two larger intragenic deletions are indicated below the structure. **(B)** The ABCA12 protein is predicted to contain four transmembrane domains, with 1, 5, 1 and 5 transmembrane segments, respectively. Please note that this model is based on protein transmembrane topology predictions by different programs, together with experimental data. According to the particular programs used, in fact the second and fourth transmembrane domain may contain either 5 or 7 transmembrane segments. It should be noted that a signal peptide prediction program SignalP 3.0 Server suggests the presence of a 45 amino acid N-terminal signal peptide with a peptidase cleavage site between amino acids 45 and 46. This would remove the first transmembrane segment depicted in the Figure. In contrast, other programs (SOSUI signal, Signal-CF, and SIG-Pred) did not recognize the presence of a signal sequence. Thus, the precise number and the exact topography of the transmembrane segments are not known. This figure also demonstrates the presence of mutations in Harlequin ichthyosis (plain text), in lamellar ichthyosis type 2 (bold), and in non-bullous congenital ichthyosiform erythroderma (Italics), with those associated with multiple skin malignant neoplasia indicated by an asterisk. Note that the precise positions of these missense mutations along the protein may change if different topography models are utilized.

at risk for recurrence. This can be performed by chorionic villus sampling as early as the 10th week of gestation, thus providing the parents and healthcare providers critical information on the fetal genotype during the first trimester. In case of PXE, however, there has been very little demand for prenatal testing. In contrast, DNA-based prenatal diagnosis of HI has been established and can be offered to the families at risk for recurrence of this autosomal recessive disorder in a family with previous history of an affected child.[59] Prior to identification of the *ABCA12* gene as the one harboring mutations in HI, prenatal testing was available only through fetal skin biopsy, an invasive procedure that is usually performed at 18–20 weeks gestation, i.e., towards the end of the second trimester. The early DNA-based testing is rapidly replacing the fetal skin biopsy as a procedure of "standard of care" for testing fetuses at risk for recurrence of this disease in a family with previous history of HI.

As an extension of DNA-based prenatal testing, the feasibility of preimplantation genetic diagnosis for HI and related keratinization disorders can be established. In this procedure, DNA-based diagnosis is performed before the pregnancy is initiated in the context of *in vitro* fertilization.[60] Specifically, blastomer analysis of an 8-cell embryo can be performed and only the embryos that lack the deleterious mutations in the *ABCA12* gene are implanted into the uterus for initiation of the pregnancy. Finally, identification of the mutations and detailed understanding of the pathomechanisms will allow development of molecular therapies for these devastating, currently intractable, disorders.

Acknowledgments

The authors thank Dr. András Váradi, Institute of Enzymology of the Hungarian Academy of Sciences, for insightful comments. Carol Kelly and Nanita Barchi assisted in preparation of the manuscript. The authors' original studies were supported by the U.S. Department of Health and Human Services, the National Institutes of Health, and the Dermatology Foundation.

References

1. Dean, M., Rzhetsky, A., and Allikmets, R. (2001) *Genome Res.* **11**, 1156.
2. Tusnády, G.E., Sarkadi, B., Simon, I., and Váradi, A. (2006) *FEBS Lett.* **580**, 1017.
3. Lage, H. (2008) *Cell Mol. Life Sci.* **65**, 2145.
4. Szakács, G., Váradi, A., Özvegy-Laczka, C., and Sarkadi, B. (2008) *Drug Discov. Today* **13**, 379.
5. Dawson, R.J. and Locher, K.P. (2006) *Nature* **443**, 180.
6. Dawson, R.J. and Locher, K.P. (2007) *FEBS Lett.* **581**, 935.
7. Zolnerciks, J.K., Wooding, C., and Linton, K.J. (2007) *FASEB J.* **21**, 3937.
8. Li, Y.F., Polgar, O., Okada, M., Esser, L., Bates, S.E., and Xia, D. (2007) *J. Mol. Graph. Medel.* **25**, 837.
9. Ravna, A.W. and Sager, G. (2008) *Bioorg. Med. Chem. Lett.* **18**, 3481.
10. Fülöp, K., Barna, L., Symmons, O., Závodszky, P., and Váradi, A. (2009) *Res. Commun.* **379**, 706.
11. Uitto, J. (2005) *Trends Mol. Med.* **11**, 341.
12. Pulkkinen, L., Ringpfeil, F., and Uitto, J. (2002) *J. Am. Acad. Dermatol.* **47**, 91.
13. Uitto, J., Li, Q., and Jiang, Q. (2010) *J. Invest. Dermatol.* **130**, 661.
14. Li, Q., Jiang, Q., Pfendner, E., Váradi, A., and Uitto, J. (2009) *Exp. Dermatol.* **18**, 1.
15. Christiano, A.M., Lebwohl, M.G., Boyd, C.D., and Uitto, J. (1992) *J. Invest. Dermatol.* **99**, 660.
16. Raybould, M.C., Birley, A.J., Moss, C., Hultén, M., and McKeown, C.M. (1994) *Clin. Genet.* **45**, 48.
17. Le Saux, O., Urban, Z., Göring, H.H., Csiszar, K., Pope, F.M., Richards, A., Pasquali-Ronchetti, I., Terry, S., Bercovitch, L., Lebwohl, M.G., Breuning, M., van den Berg, P., Kornet, L., Ott, J., de Jong, P.T., Bergen, A.A., and Boyd, C.D. (1999) *Genomics* **62**, 1.
18. Cai, L., Struk, B., Adams, M.D., Ji, W., Haaf, T., Kang, H.L., Dho, S.H., Xu, X., Ringpfeil, F., Nancarrow, J., Zäch, S., Schaen, L., Stumm, M., Niu, T., Chung, J., Lunze, K., Verrecchia, B., Goldsmith, L.A., Viljoen, D., Figuera, L.E., Fuchs, W., Lebwohl, M., Uitto, J., Richards, G., Hohl, D., and Ramesar, R. (2000) *J. Mol. Med.* **78**, 36.

19. Ringpfeil, F., Lebwohl, M.G., Christiano, A.M., and Uitto, J. (2000) *Proc. Natl. Acad. Sci. U.S.A.* **97**, 6001.
20. Le Saux, O., Urban, Z., Tschuch, C., Csiszar, K., Bacchelli, B., Quaglino, D., Pasquali-Ronchetti, I., Pope, F.M., Richards, A., Terry, S., Bercovitch, L., de Paepe, A., and Boyd, C.D. (2000) *Nat. Genet.* **25**, 223.
21. Bergen, A.A., Plomp, A.S., Schuurman, E.J., Terry, S., Breuning, M., Dauwerse, H., Swart, J., Kool, M., van Soest, S., Baas, F., ten Brink, J.B., and de Jong, P.T. (2000) *Nat. Genet.* **25**, 228.
22. Struk, B., Cai, L., Zäch, S., Ji, W., Chung, J., Lumsden, A., Stumm, M., Huber, M., Schaen, L., Kim, C.A., Goldsmith, L.A., Viljoen, D., Figuera, L.E., Fuchs, W., Munier, F., Ramesar, R., Hohl, D., Richards, R., Neldner, K.H., and Lindpaintner, K. (2000) *J. Mol. Med.* **78**, 282.
23. Klement, J.F., Matsuzaki, Y., Jiang, Q.J., Terlizzi, J., Choi, H.Y., Fujimoto, N., Li, K., Pulkkinen, L., Birk, D.E., Sundberg, J.P., and Uitto, J. (2005) *Mol. Cell. Biol.* **25**, 8299.
24. Gorgels, T.G., Hu, X., Scheffer, G.L., van der Wal, A.C., Toonstra, J., de Jong, P.T., van Kuppevelt, T.H., Levelt, C.N., de Wolf, A., Loves, W.J., Scheper, R.J., Peek, R., and Bergen, A.A. (2005) *Hum. Mol. Genet.* **14**, 1763.
25. Pulkkinen, L., Nakano, A., Ringpfeil, F., and Uitto, J. (2001) *Hum. Genet.* **109**, 356.
26. Pfendner, E., Uitto, J., Gerard, G.F., and Terry, S.F. (2008) *Expert Opin. Med. Diagn.* **2**, 1.
27. Belinsky, M.G. and Kruh, G.D. (1999) *Br. J. Cancer* **80**, 1342.
28. Scheffer, G.L., Hu, X., Pijnenborg, A.C., Wijnholds, J., Bergen, A.A., and Scheper, R.J. (2002) *Lab. Invest.* **82**, 515.
29. Illiás, A., Ubran, Z., Seidl, T.L., Le Saux, O., Sinkó, E., Boyd, C.D., Sarkadi, B., and Váradi, A. (2002) *J. Biol. Chem.* **277**, 16860.
30. Pfendner, E.G., Vanakker, O., Terry, S.F., Vourthis, S., McAndrew, P.E., McClain, M.R., Fratta, S., Marais, A.S., Hariri, S., Coucke, P.J., Ramsay, M., Viljoen, D., Terry, P.F., De Paepe, A., Uitto, J., and Bercovitch, L.G. (2007) *J. Med. Genet.* **44**, 621.
31. Uitto, J., Pulkkinen, L., and Ringpfeil, F. (2001) *Trends Mol. Med.* **7**, 13.
32. Jiang, Q. and Uitto, J. (2006) *J. Invest. Dermatol.* **126**, 1440.
33. Jiang, Q., Li, Q., and Uitto, J. (2007) *J. Invest. Dermatol.* **127**, 1392.

34. Le Saux, O., Bunda, S., VanWart, C.M., Douet, V., Got, L., Martin, L., and Hinek, A. (2007) *J. Invest. Dermatol.* **126**, 1497.
35. Jiang, Q., Endo, M., Dibra, F., Wang, K., and Uitto, J. (2009) *J. Invest. Dermatol.* **129**, 348.
36. Vanakker, O.M., Martin, L., Gheduzzi, D., Leroy, B.P., Loeys, B.L., Guerci, V.I., Matthys, D., Terry, S.F., Coucke, P.J., Pasquali-Ronchetti, I., and de Paepe, A. (2007) *J. Invest. Dermatol.* **127**, 581.
37. Li, Q., Grang, D.K., Armstrong, N.L., Whelan, A.J., Hurley, M.Y., Rishavy, M.A., Hallgren, K.W., Berkner, K.L., Schurger, L.J., Jiang, Q., and Uitto, J. (2009) *J. Invest. Dermatol.* **129**, 553.
38. Li, Q., Schurgers, J., Smith, A.C.M., Tsokos, M., Uitto, and Cowen, E.W. (2009) *Am. J. Pathol.* **174**, 534.
39. Schurgers, L.J., Cranenburg, E.C., and Vermeer, C. (2008) *Thromb. Haemost.* **100**, 593.
40. Luo, G., Ducy, P., McKee, M.D., Pinero, G.J., Loyer, E., Behringer, R.R. and Karsenty, G. (1997) *Nature* **386**, 78.
41. Borst, P., van de Wetering, K., and Schlingemann, R. (2008) *Cell Cycle* **7**, 1575.
42. Schmuth, M., Gruber, R., Elias, P.M., and Williams, M.L. (2007) *Adv. Dermatol.* **23**, 231.
43. McGrath, J.A. and Uitto, J. (2008) *Trends Mol. Med.* **14**, 20.
44. Brown, S.J. and Irvine, A.D. (2008) *Semin. Cutan. Med. Surg.* **27**, 128.
45. Brown, S.J. and McLean, W.H.I. (2008) *J. Invest. Dermatol.* **129**, 543.
46. Elias, P.M., Crumrine, D., Rassner, U., Hachem, J.P., Menon, G.K., Man, W., Choy, M.H., Leypoldt, L., Reingold, K.R., and Williams, M.L. (2004) *J. Invest. Dermatol.* **122**, 314.
47. Akiyama, M. and Shimizu, H. (2008) *Exp. Derm.* **17**, 373.
48. Richard, G. (2004) *Am. J. Med. Genet. Part C (Semin. Med. Genet.)* **131c**, 32.
49. Akiyama, M. (2006) *J. Dermatol. Sci.* **42**, 83.
50. Akiyama, M., Sugiyama-Nakagiri, Y., Sakai, K., McMillan, J.R., Goto, M., Arita, K., Tsuji-Abe, Y., Tabata, N., Matsuoka, K., Sasaki, R., Sawamura, D., and Shimizu, H. (2005) *J. Clin. Invest.* **115**, 1708.
51. Kelsell, D.P., Norgett, E.E., Unsworth, H., The, M.T., Cullup, T., Mein, C.A., Dopping-Hepenstal, P.J., Dale, B.A., Tadini, G., Fleckman, P., Stephens, K.G., Sybert, V.P., Mallory, S.B., North, B.V., Witt, D.R., Sprecher, E., Taylor, A.E., Ilchyshyn, A., Kennedy, C.T., Goodyear, H., Moss, C., Paige, D., Harper, J.I., Young, B.D.,

Leigh, I.M., Eady, R.A., and O'Toole, E.A. (2005) *Am. J. Hum. Genet.* **76**, 794.

52. Akiyama, M. (2006) *Arch. Dermatol.* **142**, 914.
53. Kaminski, W.E., Piehler, A., and Wenzel, J.J. (2006) *Biochim. Biophys. Acta* **1762**, 510.
54. Lefévre, C., Audebert, S., Jobard, F., Bouadjar, B., Lakhdar, H., Boughdene-Stambouli, O., Blanchet-Bardon, C., Heilig, R., Foglio, M., Weissenbach, J., Lathrop, M., Prud'homme, J.F., and Fischer, J. (2003) *Hum. Mol. Genet.* **12**, 2369.
55. Sakai, S., Akiyama, M., Sugiyama-Nakagiri, Y., McMillan, J.R., Sawamura, D., and Shimizu, H. (2007) *Exp. Dermatol.* **16**, 920.
56. Yanagi, T., Akiyama, M., Nishihara, H., Sakai, K., Nishie, W., Tanaka, S., and Shimizu, H. (2008) *Hum. Mol. Genet.* **17**, 3075.
57. Smyth, I., Hacking, D.F., Hilton, A.A., Mukhamedova, N., Meikle, P.J., Ellis, S., Slattery, K., Collinge, J.E., de Graaf, A., Bahlo, M., Sviridov, D., Kile, B.T., and Hilton, D.J. (2008) *PLoS Genet.* **4**, e1000192.
58. Zuo, Y., Zhuang, D.Z., Han, R., Issac, G., Manning, J.J., McKee, M., Welti, R., Brissette, J.L., Fitzgerald, M.L., and Freeman, M.W. (2008) *J. Biol. Chem.* **283**, 36624.
59. Yanagi, T., Akiyama, M., Sakai, K., Nagasaki, A., Ozawa, N., Kosaki, R., Sago, H., and Shimizu, H. (2008) *J. Am. Acad. Dermatol.* **58**, 653.
60. Spits, C. and Sermon, K. (2009) *Prenat. Diagn.* **29**, 50.

Chapter 10

CFTR: Understanding the Cause and Influencing the Outcome of a Major Genetic Disease*

Tamás Hegedüs, Martina Gentzsch and John R. Riordan

1. CFTR Function

1.1 *CFTR is an anion channel*

Before its cloning and allocation to the then unnamed ABC transporter family, there was circumstantial evidence that the autosomal recessive gene mutated in patients with cystic fibrosis might code for a protein involved in epithelial ion permeability.[1] The sequence of CFTR cDNA was at least partially consistent with this expectation since the conceptual translation product appeared to be a poly-topic membrane protein.[2] However, the presence of two consensus NBD sequences together with the membrane spanning sequences suggested that it was a transporter rather than a channel. This prompted considerable debate as transporters and channels were viewed as being entirely distinct structurally as well as functionally.[3] Interestingly this view has softened as examples of members of families of membrane proteins serving either function have become apparent.[4,5] The impact of the substitution of charged

* This work is supported by the NIH and the Cystic Fibrosis Foundation.

residues within predicted transmembrane helices on channel activity in CFTR expressing cells provided indirect evidence that CFTR was itself a chloride channel[6] while direct evidence was provided by purification and reconstitution of the protein.[7] The anion conductive properties of CFTR have been extensively characterized and described in excellent reviews.[8] In addition to its selectivity for halides, of which chloride is the most physiologically relevant, CFTR also plays an important role in bicarbonate permeability either directly or indirectly in cooperation with chloride/bicarbonate exchangers.[9,10] This function may contribute to the regulation of pH in the fluid layer on epithelial surfaces such as those in the airways[11] and may be crucial to mucin secretion that is aberrant in cystic fibrosis.[12,13]

The ion permeation pathway or pore through the molecule has not yet been defined. However, mutagenesis of residues including cysteine scanning and modification has revealed the likely participation of amino acids along TM6 with positively charged residues at either end perhaps attracting negative ions.[14,15] Residues in other helices including TM1, TM5, TM8 and TM12 also influence pore properties.[16,17] Homology models of CFTR provide a possible basis for an overall arrangement of all twelve TMs[18,19] and hence the ion translocation pathway[20] but these models remain to be tested rigorously.

1.2 *Regulation of CFTR*

Control of the still to be defined ion channel gate is exerted at two levels: phosphorylation at multiple sites by PKA and the binding of ATP at the two composite sites at the interface between the NBDs. The overriding control by phosphorylation occurs primarily through the release of restraints on channel activity caused by the unphosphorylated R domain.[21–24] The structure of the R domain is unknown and indeed it may be largely unstructured, with order-disorder transitions occurring in the local regions within the domain proximal to different PKA phosphorylation sites.[25,26] An NMR study, employing the isolated

R and NBD1 domains, found that phosphorylation of these R domain sites reduced the tendency of adjacent sequences to form helices and to interact with NBD1.[25] R domain segments also may interact with other portions of CFTR including NBD2.[27] However the mechanism by which the phosphorylated R domain permits gating of the channel is far from clear. PKA phosphorylation favors the association of CFTR's two NBDs as assessed by chemical cross-linking.[28] This closed conformation of the two NBDs may in turn promote the actively gating state of the ion pore formed by the membrane-integrated domains. Alternatively or in addition, phosphorylated R domain may directly interact with the cytoplasmic portions of the long transmembrane helices and thereby disrupt the channel closed state. Very recent studies of the effects of substitutions of residues influencing the stability of some of these helices showed altered sensitivity of channel activation to both PKA and ATP.[29]

As in all ABC transporters, two ATP binding sites form at the interface between the two NBDs but these sites are highly asymmetric in CFTR as the first site (NBS1), composed of the Walker motifs of NBD1 and the signature motif of NBD2, is non-hydrolytic whereas the second site (NBS2) hydrolyzes ATP.[30] Thus rather than being immediately hydrolyzed when the composite NBS1 forms, intact ATP remains occluded there.[31] The lack of hydrolytic activity appears to be due to non-canonical residues at key locations on both sides of the composite NBS1, i.e., substitutions E573S and H605S in the NBD1 Walker B and H loop motifs, respectively and G1348H and G1350H in the NBD2 signature motif. This is depicted schematically in the cartoon in Fig. 1. The relationship between ATP binding and hydrolysis at these asymmetric sites and channel opening and closing is of fundamental interest but is not yet clearly resolved. In one model, opening of a phosphorylated channel requires ATP binding-dependent closure of the interface between the two NBDs followed by closing of the channel when the interface is opened by hydrolysis of one of the two bound ATPs.[32,33] This model is based primarily on the observation that substitution of

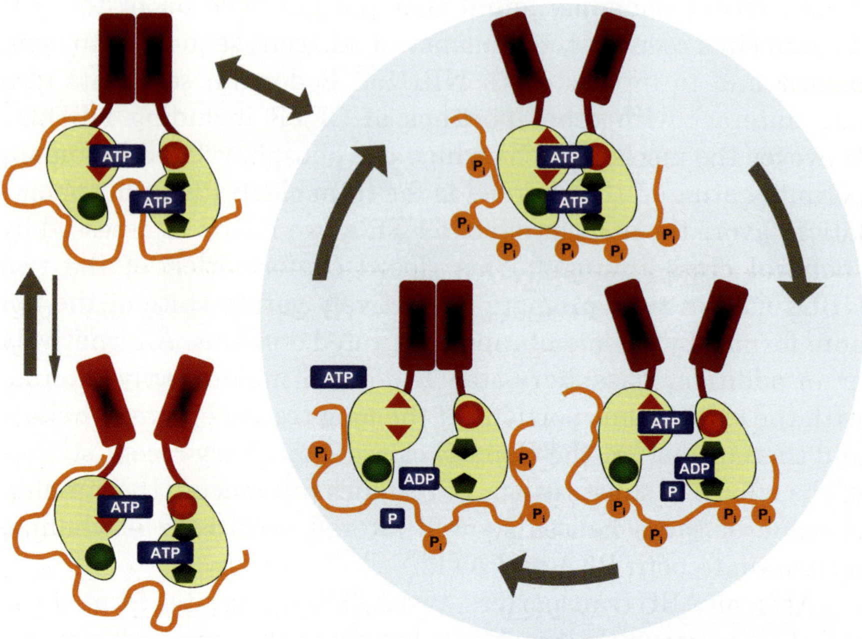

Fig. 1. CFTR as a phosphorylation-controlled hydrolyzable-ligand gated channel. The cartoon depicts CFTR gating occurring by a reversible shift in equilibrium between pre-existing conformational states (left upper and lower panels). Shown in the circle to the right is the coupling of the gating process to ATP binding and dissociation at NBS1 (red) and binding and hydrolysis at NBS2 (green) after R domain phosphorylation by PKA. Red triangles denote major ATP binding site involving Walker motifs in NBD1, green pentagons, the corresponding site in NBD2; red and green balls the signature motifs in NBD2 and NBD1, respectively. The orange ribbon is an entirely artistic depiction of the R domain, emphasizing its disordered character. Additional description in the text.

key catalytic residues in NBD2 (e.g., E1371Q or H1402A) greatly prolongs the channel open state.[32] This locking open has led to the reasonable conclusion that ATP hydrolysis is necessary for channel closing and, by extrapolation, that ATP binding is required for and in fact "drives" opening.[33] However, such obligatory direct coupling between the ATP hydrolysis and channel gating cycles is incompatible with a number of other observations. Non-hydrolytic native CFTR channels in the sweat gland were first described by Quinton and Reddy.[34] Ramjeesingh *et al.*[35] reported the lack of tight coupling between the hydrolysis and gating cycles of purified and reconstituted CFTR. Robust gating is observed in planar lipid bilayers under a variety of non-hydrolytic conditions. These include the wild-type protein in the absence of magnesium ions or in the presence of non-hydrolyzable ATP analogs,[36] as well as C-terminally truncated constructs that completely lack the hydrolytic NBD2.[37]

Thus, the gating mechanism *per se* is not strictly dependent on ATP hydrolysis even though the presence of the bound nucleotide at both sites may contribute to the maintenance of the open state. Channel opening, i.e., disturbance of the closed state may occur stochastically due to thermal fluctuations prior to stabilization by the ATP bound configuration of the NBDs. In this model, illustrated in Fig. 1, there is a low but finite probability of channel opening in the absence of ATP bridging of the interface between the NBDs even though there is binding of the nucleotide to the individual NBDs (left images). However, the open probability is greatly increased when, in the phosphorylated channel, each ATP is bound at both sides of the two composite sites (upper image in the circle). As detailed above these sites differ from each other with respect to their ability to hydrolyze ATP, with the lower (green) being hydrolytic and the upper (red), non-hydrolytic. Under channel gating conditions this model posits that there is dissociation from both sites, ATP from NBS1 and its hydrolysis products from NBS2, albeit at quite different rates, perhaps permitting channel closing. By analogy with a mechanistic scheme of ABC transporter operation in which opening of the interface between NBDs changes the orientation of the MSDs and accomplishes transport, it has been tempting to postulate that channel closure may be similarly linked to separation of the NBDs. However, as yet there is no evidence of a direct 1:1 coupling between ATP mediated closure of the NBD dimer and channel opening. In fact, comparison of hydrolysis and gating rates suggests that the latter may be approximately 20-fold more rapid than the former.[38] While channel activity is undoubtedly influenced by ATP binding and hydrolysis, the ATP bound, closed NBD configuration may promote an actively gating state, whereas the channel is quiet when the NBD dimer is open. In this scenario, the tightly opposed NBDs, probably via interactions with the coupling helices in the MSD cytoplasmic loops, destabilize the channel closed state. Consistent with this possibility, restricting the dynamics of these interactions by chemical cross-linking leaves the channel in a stable closed state.[39] It is possible also that the influence of ATP binding on the channel is not primarily due to a change in the relationship between the two NBDs. Instead intra-NBD conformational changes could influence channel state.

2. Structural Properties of CFTR

2.1 *Full-length CFTR*

Similar to other ABC proteins, CFTR contains two large transmembrane domains (MSD1 and 2) and two nucleotide binding domains (NBD1 and 2). Each nucleotide binding domain is connected to the C-terminus of the corresponding transmembrane domain. As with some other members of the ABCC subfamily, the two NBDs are quite different as emphasized above.[30,40] The two halves of the molecule are joined by the regulatory (R) domain, which contains a large number of phosphorylation sites and is unique to CFTR.[41] Most of the extracellular loops between membrane spanning helices are very short and may not be exposed to the extracellular space.[42] The two N-glycosylation sites (N894 and N900) are located in the longest, fourth extracellular loop.[43] Homology models and preliminary confirmatory experimental evidence indicate that the transmembrane helices extend far into the cytoplasm with relatively short sequences between them mediating the connection between the nucleotide binding domains and membrane embedded parts.[18] CFTR is a large transmembrane protein that makes the determination of its 3D structure at the atomic level extremely challenging. These challenges are exacerbated by the low natural abundance of the protein, difficulties in obtaining soluble non-aggregated preparations from heterologous over-expression systems, and its highly dynamic properties and disordered regions, including the large (~200 amino acids), relatively unstructured R domain.

Expression and purification efforts made it possible to generate 2D crystals for electron crystallography revealing two crystal forms for CFTR at ~20 Å resolution.[44] The low-resolution structures showed distinct similarity to structures of P-glycoprotein.[45,46] Both of the CFTR crystal forms exhibit a high density and a low density region when viewed from the side, parallel to the crystal plane, corresponding to the transmembrane domains and the cytoplasmic domains, respectively.[44] In some crystals, a large gap appears on the side below the putative transmembrane region, which was suggested to provide an exit channel for ions similar to other ion channels.[47,48] Higher resolution images are being obtained by cryo-EM of well ordered crystals at multiple tilt angles with the beginning of detection of TM helices (M. Rosenberg and R. Ford, unpublished). The position of the highly dynamic R domain is not visible as a defined density by common crystallographic methods, but information has been obtained from labeling with Ni-NTA nanogold binding to an exogenously inserted hexa-histidine sequence.[49,50] Since the labeling revealed the localization of only a small segment of the R domain, the location of the whole region remains to be more completely defined. A recent cryo-EM analysis has provided additional data indicating less defined density in the same region, consistent with the expected property of a disordered region.[51] The same study also has more clearly localized the disposition of the NBDs, especially that of NBD2 in the overall CFTR protein density.

In the 2D crystals CFTR molecules are monomeric, while in single-particle experiments dimers are observed.[49,50] The members of the dimer are tilted with respect to each other in a manner that may not be reflective of the native configuration in the membrane under physiological conditions. There is a variety of biochemical and electrophysiological data that attempts to determine the native quaternary structure of CFTR[52–56] without general agreement of whether it is monomeric, dimeric or of a higher order. In cells both the N- and C-termini bind adaptor proteins and this may contribute to both spatial and temporal organization. These interactions probably enable the clustering and immobilization of populations of CFTR molecules in the cell surface membrane.[57–59] However, a recent study has provided evidence based on single molecule fluorescent imaging that the functional unit of the channel is a single molecule[60] and this seems most consistent with current understanding of primary structure-function relationships.

2.2 *Nucleotide binding domains*

The nucleotide binding domains of CFTR exhibit the common ABC NBD structure with a β-sheet subdomain similar to the core domain of F1 ATPases and an α-helical subdomain characteristic of ABC proteins. CFTR NBD2 is very similar to NBDs from other ABC transporters based on sequence comparison, while NBD1 has several unique features.[61] NBD1 harbors a flexible region between the first two β-strands, called the regulatory insertion (residues 404–435), as it contains a PKA phosphorylation site. The C-terminal portion of the NBD1 crystal structure was termed the regulatory extension (RE) at the C-terminus[61] but it can more accurately be considered as the N-terminal portion the R domain.[62] The linker region between the β- and α-subdomains in NBD1 is shorter than in NBDs of other ABC transporters, lacking a short β-sheet and an α-helix that they contain. It has been difficult to generate crystals not only of NBD1, but also of NBD2. The latter domain required attachment of the regulatory domain of MalK and multiple mutations for successful crystallization and the two NBD proteins did not associate with each other even though they both exhibit strong self-association.[63]

CFTR NBD1 has received special attention, since the Phe508 residue is located on the surface of its α-subdomain, at the end of a short helix following the Q-loop.[61,64] Determination of the crystal structure of the domain with this major disease causing mutation was a milestone in the field. Since it was expected that the deletion of Phe508 might disrupt domain folding, initially it seemed surprising that structures of NBD1 constructs in the presence or absence of Phe508 are highly similar. The presence of suppressor and solubilizing mutations[65] used to aid expression of soluble and crystallizable preparations also appear not to have a significant effect on the overall structure. The most obvious difference caused by ΔF508 is the shortening of the surface loop between two helices, where F508 is located. Both the shortening of this region and the missing aromatic side-chain influences, not only the folding kinetics of NBD1 but also the interaction between the NBD1 and transmembrane domains, eventually leading to the degradation of the protein.[18,64] A computational study indicated increased dynamics of the F508 loop[66] and this has been confirmed by hydrogen-deuterium exchange experiments.[67] In another loop, the S7-H6 loop (formed by amino acids 572–583), separated from the 508 position, which exhibits different conformations in different crystals, residues have been identified computationally to play a role in stabilizing or destabilizing certain folding intermediates in the presence or absence of Phe508.[68] It has become increasingly apparent that the rescue of the ΔF508 CFTR function requires correcting domain folding and the restoration of domain-domain interactions. Moreover, it has been reported that Phe508 deletion affects the post-translational folding of NBD2 and most probably its association with NBD1.[69] This association may play an important role in CFTR maturation in spite of the fact that NBD2 is not absolutely essential for channel function of the wild-type protein. NBD1 and NBD2 associate in a head-to-tail conformation similar to that found in other ABC transporters as demonstrated by Mense *et al.*[28] To dissect the details of the interactions between the domains and the functional mechanism, an atomic-level structure is needed for NBD1 and NBD2 together. A modeled dimer structure has been generated using an NBD1 crystal structure and an NBD2 homology model overlaid on the structural alignment of several crystal structures of NBD dimers.[70] A recent study used a different approach involving docking of NBD2 onto NBD1 to avoid clashes.[71] However, the relationship of these models to reality remains unclear as experimental structural determination has not yet been possible with either the intact CFTR protein or the combined, individually expressed NBD1 and NBD2 that have exhibited no tendency to associate with each other.[61] The latter observation points strongly to the requirement of the rest of the CFTR protein for the appropriate association of the two NBDs.

2.3 *Membrane domains*

Structural information on the transmembrane domains is very sparse because of the difficulties of working with large hydrophobic polypeptides. In spite of this, a series of interesting studies have been done by Deber and colleagues, who investigated the effect of CF causing mutations located in the transmembrane region.[72–75] Helix-loop-helix constructs have been expressed and changes in their dynamics and structure due to mutations have been studied by SDS-PAGE, FRET, and NMR.[72–75] Although these methods have not provided high resolution 3D information on even these isolated peptide TM segments, the results indicate a highly delicate balance of helix conformation and dynamics, important for both function and maturation of CFTR. Clarke and coworkers[76] found that ΔF508 disrupted interactions between the N- and C-terminal transmembrane domains, indicating that transmembrane helices are severely affected by this distant mutation. So far high resolution structural information on the transmembrane domains relies exclusively on homology modeling. Since CFTR belongs to the ABCC exporter subfamily despite functioning as a channel, high resolution structures of exporters have been used as template.[18,77] Similar to the crystallized bacterial ABC exporters, Sav1866, MsbA, and P-glycoprotein, CFTR in these models also exhibits crossing-over of each of the transmembrane domains to interact with the opposite nucleotide binding domain. Indeed, this was confirmed by chemical cross-linking between specifically

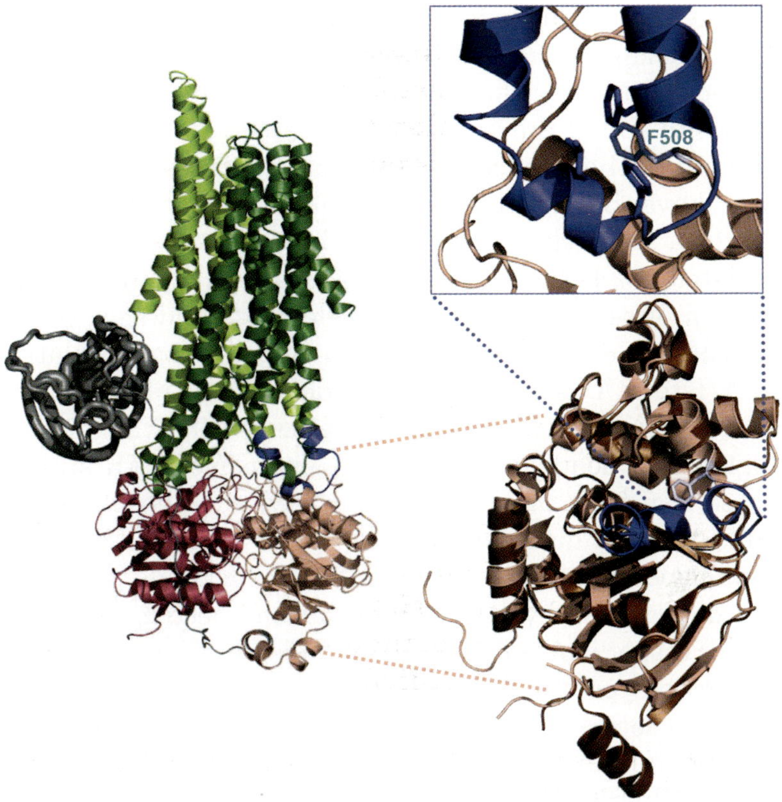

Fig. 2. Structural hallmarks of CFTR. CFTR has a crossing-over configuration of domains similar to ABC exporters. Its unique feature is the presence of the intrinsically disordered R domain (gray, left). The coupling helices in the loops between the long TM helices connect the NBDs and TMDs. The coupling helix in CL4 interacting with NBD1 is highlighted in blue and also shown is a cut-away view from the plane of the bilayer (right, bottom). Hydrophobic amino acids in this coupling helix together with F508 from NBD1 form a hydrophobic cluster that may stabilize the interaction between NBD1 and TMD2.

inserted cysteine residues.[18] As indicated in Fig. 2, F508 participates directly in one of the "domain-swapping" interfaces. The similarity of at least this aspect of CFTR tertiary structure to that of active exporters implies some level of mechanistic similarity between channel and exporter function. That is conformational changes in the NBDs during the catalytic cycle may be transduced to the MSDs through the coupling helices in the cytoplasmic loops between membrane spanning helices at the opposite end of the protein.[39]

Although the structure-function relationships of the R domain have been extensively studied, they remain poorly understood. An early investigation, based on the CD spectra of the isolated polypeptide,[78] showed that the R domain had a low proportion of secondary structural elements and that the small helical content decreased even further upon phosphorylation by PKA. A later CD study could not detect any phosphorylation dependent changes in the R domain, which was disordered throughout its sequence.[79] A recent extensive NMR study also indicated that both the unphosphorylated and phosphorylated R domain is largely disordered.[25] This observation is based on the limited dispersion (~8–8.7 ppm) of the backbone amide resonances determined by ^{1}H-^{15}N correlation spectra, compared to the larger dispersion (~7–10 ppm) of folded proteins. Intrinsically disordered polypeptides, including the R domain, do not have a well defined secondary structure, but they are characterized by a highly dynamic, heterogeneous structural ensemble with rapid conversion from one conformation to another.[80,81] Many of these conformations exhibit local and limited residual structures, which are able to undergo disorder-to-order transitions stabilized by a binding partner.[82] In some cases protein-protein interactions have been reported to be promoted by phosphorylation of residues in disordered regions.[83] However, in the case of the R domain the alpha helical propensity of residual elements and interactions within this region are decreased upon phosphorylation and the binding of the R domain to NBD1 is abrogated by phosphorylation, as revealed in the NMR experiments.[25] Since these experiments used isolated recombinant R and NBD1 polypeptides the important findings do not exclude interactions of R domain elements with other parts of CFTR, including NBD2[27] and possibly other cytoplasmic regions. A widely accepted hypothesis in the field is that the unphosphorylated R domain intercalates between NBD1 and NBD2 inhibiting their functional association (see below). This hypothesis is based on some of the NBD1 crystal structures, in which the N-terminal end of the R domain covers the potential dimer interface. However, these crystal structures do not reflect the steric constraints operating in the full-length protein, where the flexible R domain part may not be able to flip into the dimer interaction surface. Indeed, amino acid 605, which is covered by the N-terminal part of the R domain in some of the NBD1 crystal structures, can be accessed by chemical cross-linkers in the full length protein.[28] Furthermore, phosphorylation by PKA stimulates the channel activity of an active CFTR that is entirely devoid of NBD2,[29,37] emphasizing that the R domain acts at sites other than the interface between the NBDs.

In summary, high resolution structural information on full length CFTR so far relies largely on homology models based on Sav1866, MsbA and P-glycoprotein structures.[18,77,84,85] NBD2 and the complex of the two nucleotide binding domains can be modeled by homology with the structures of NBDs of other ABC proteins. The low sequence similarity in the transmembrane regions requires careful adjustments in the sequence alignment and also the incorporation of as many experimental constraints as possible in order to generate informative models. However, the available homology models provide an important step towards understanding the function of CFTR and the mechanisms underlying disease-causing mutations in addition to serving as starting points for mutagenesis studies, rational drug design and molecular dynamic simulations.

3. CFTR Synthesis and Assembly

3.1 *Biosynthetic processing of CFTR*

A great deal of attention has been drawn to the biosynthetic processing of CFTR since this process is impaired by disease-causing mutations. In general, the pathways followed and mechanisms involved are common to those shared by other eukaryotic membrane glycoproteins including other ABC transporters. Indeed mutations affecting the folding and assembly of several of these have been identified in a range of different human diseases. This is not surprising since a missense mutation or other small perturbation at any position in these long polypeptides has a much higher probability of influencing folding than a specific functional site. This would be especially true if there were a strict interdependence between domains in the assembly of these complex multidomain proteins. In addition, CFTR may be an extreme example of susceptibility to conformational instability because of its content of the R domain, which is large and unstructured. When the heterologous expression of wild-type and ΔF508 CFTR was first compared, it was observed not only that the mutant failed to be converted from an immature ER associated form to a mature form competent for ER export, but also that even the wild-type underwent that conversion very inefficiently.[86] Only approximately 30% of co-translationally core-glycosylated full length polypeptide chains met the requirements of the COP-II based ER export machinery, providing the first indication that CFTR has difficulty in achieving a stable folded state.[87] The fate of the remaining ~70% of newly synthesized chains became apparent when CFTR was the first membrane protein found to be a substrate for the 26S proteasome.[88,89] Polyubiquitination, which signals extraction from the ER membrane and delivery to the proteasome can occur before translation is complete.[87] During translation the conformational state is both nurtured by chaperones and scrutinized by other components of the ERAD system in a multistep balancing act. The consequences for the trafficking of the protein through the various membrane compartments of the secretory and endocytic pathways are described in the following section. Here we concentrate on events within the protein itself.

The main evidence for a conformational difference between the immature and mature forms of CFTR comes from differential sensitivity to limited proteolytic digestion, with the former being generally more susceptible to cleavage than the latter.[90] However, the relatively small number of accessible cleavage sites even in the nascent polypeptide indicates that considerable folding into compact structures occur early in at least several of the domains.[91] Nevertheless, additional observations have indicated that associations among domains are absolutely crucial to stabilization of the full-length protein. As the polypeptide chain grows from the N-terminus, formation of the first three major domains to the end of the R domain is insufficient to generate a structure that satisfies requirements for ER export.[37,92] On the other hand, constructs extending to include the fourth domain (MSD2) are exported, acquire complex oligosaccharide chains in the Golgi and form chloride channels, albeit with low open probability.[37] This finding initially came as somewhat of a surprise because it indicated that association of the two NBDs, a hallmark of ABC protein structure and function, apparently is not required for formation of a stable unit recognized as competent to proceed through the secretory pathway. One might speculate that the two NBDs may not closely associate until late in the assembly and trafficking of the molecule to its functional site in the plasma membrane so that unproductive ATP hydrolysis is avoided. In any case, there is now strong evidence from several laboratories that this core four domain assembly (MSD1/NBD1/R/MSD2) satisfies cellular quality control systems. Interestingly, this ability to achieve a stable mature state in the absence of the second nucleotide binding domain may not be shared by other ABC proteins as P-glycoprotein apparently does not mature when NBD2 is deleted.

Some insight into the interactions among the first four domains that may account for their combined stability comes from the contacts that are revealed in the structural models as well as the NMR studies of Baker *et al.*,[25] showing associations of multiple segments of the R domain with NBD1. The transmembrane helices in MSD1 and MSD2 are extensively intercalated and the so-called coupling helices in CL1 of MSD1 and CL3 and CL4 of MSD2 can apparently maintain contact with NBD1 in the absence of NBD2. It may seem surprising that the unfilled contacts that NBD2 normally makes with CL1 and CL2 loops from MSD1 and with CL3 from MSD2 do not result in aberrant associations or instability. However, perhaps R domain sequences may be able to occupy these cytoplasmic loop sites or they may remain covered by molecular chaperones. What is clear is that CFTR domains are highly dependent on each other for achievement of their final functional state. This became apparent in the studies of Du *et al.*,[69] Younger *et al.*[94] and Cui *et al.*,[37] but is most convincingly demonstrated by the recent exhaustive experiments of Du and Lukacs.[92] They showed that neither any of the individual, nor indeed combinations of two or three domains were able to pass cellular quality control systems. Thus, four domains from the N-terminus to beyond MSD2 were required. Emphasizing the crucial role of the NBD/CL interactions, the absence of the PHE508 residue, which participates directly in the NBD1/CL4 interface, prevents maturation of the four domain, ΔNBD2 construct.

From the general ABC protein perspective, it is interesting to note that the hallmark association between the two NBDs, essential to function, is apparently not required to form a stable structure that can traffic and moreover can exhibit some channel function. Despite this, NBD2 requires the four upstream N-terminal domain unit to achieve its final compact conformation as indicated by the fact that the NBD is very much more sensitive to limited proteolysis in ΔF508 CFTR[37,69] or when it is expressed as an individual domain (unpublished observations). In fact, it is striking that the ΔF508 mutation in NBD1 results in failure to achieve final compact states by all of the domains, emphasizing their inter-dependence. This rationalizes the fact that other disease-associated mutations in all domains, including those at domain interfaces also impair its overall folding and assembly of CFTR.[92]

The domain assembly process of course occurs sequentially during biosynthesis and consequently N-terminal mutations might be expected to have the greatest impact. In fact, many

NBD1 mutations have been identified in cystic fibrosis patients, but the fact that mutations all along the sequence can cause mis-assembly confirms that domain interdependence is a two way process. In the case of the major NBD1 mutation, ΔF508, second site suppressor mutations within the domain effectively restore folding, assembly and function of the entire protein.[95,96] As ΔF508 suppressor mutations have not yet been found elsewhere in the protein, this would seem to indicate that the intra-domain suppressors may act primarily by influencing NBD1 itself. However, the overall high resolution 3D structures of wild-type and ΔF508 NBD1s are not demonstrably different.[97] Several of the surface loops may be mobile as they are not highly ordered and small changes in sensitivity to limited proteolysis are observed.[37,69] A recent extensive analysis of several NBD1 crystal structures and hydrogen-deuterium exchange experiments described in detail the dynamic behavior of these surface sites. The results indicated that only the dynamics of the 507–511 loop was consistently different in the presence or absence of F508.[67] The suppressor mutations and other NBD1 substitutions that improve the solubility of the isolated domain presumably do influence associations with other domains. Experiments, in which "halves" of CFTR are co-expressed, with the discontinuity either before or after the R domain, reveal that the NBD1 suppressor mutations are unable to restore maturation of the ΔF508 protein whereas in the intact full length protein they are.[98] This suggests that domain assembly may be ordered not only spatially but also temporally since the halves are formed simultaneously when co-expressed but sequentially in the intact CFTR molecule.

Since CFTR shares many structural features with other ABC exporters, including the so-called domain swapping between their N- and C-terminal halves not apparent in ABC importers, it seems likely that they may share at least some aspect of their folding and assembly pathways. Consistent with this possibility, deletion of aromatic residues at positions in NBD1 corresponding to that of F508 in CFTR also causes the failure of the conformational maturation of both P-glycoprotein[99] and

MRP1.[100] It may be of interest to scrutinize the locations of disease-causing missense mutations in other human ABC transporters to discern whether they may be most likely to disrupt folding or transport function.

3.2 *Role of molecular chaperones in CFTR biogenesis*

Given the complex biosynthetic assembly of CFTR, it is not surprising that multiple chaperones are involved. The nascent CFTR polypeptide chain interacts extensively with chaperone and cochaperone complexes on both sides of the ER membrane and some of these associations are prolonged in the case of the ΔF508 CFTR polypeptide chain, which is unable to be exported.[94,101] These chaperones and cochaperones, integrated with protein processing enzymes and components of the ERAD system can both promote and terminate progression through the biosynthetic and assembly pathway. The Hsc-Hsp 40/70 set of cytoplasmic chaperones and cochaperones appear to have an impact at more than one point along the growing chain. Although up and down regulation of the abundance of these chaperones can influence the amount of immature CFTR in cells, such manipulations do not bring about a significant increase in the conversion to the mature form.[102] The lifetime and fate of newly synthesized CFTR is most strongly influenced by Hsp90 and its cochaperones.[103] When the ability of Hsp90 to bind its clients is inhibited by an ansamycin drug, CFTR nascent chains immediately succumb to proteasomal degradation. This interaction with Hsp90 seems to be an absolute requirement to enable either wild-type or mutant CFTR to avoid ERAD. This is consistent with the fact that Hsp90 clients in general tend to be proteins which, especially during biosynthesis, are unstable[104] or unassembled parts of protein complexes.[105] Interestingly, other ABC transporters which are not inherently unstable such as P-glycoprotein and MRP1 do not appear to be acutely dependent on Hsp90.[103] This may point to CFTR's unique unstructured R domain as a contributor to the requirement for chaperoning by Hsp90.

Altering the level of expression of Hsp90 cochaperones has been found to have a strong influence on the stability of immature CFTR. One of these cochaperones, Aha1, normally destabilizes client binding to Hsp90 by stimulating its ATPase activity.[106] Significantly, treatment of cells with Aha1 siRNA was found to cause accumulation of immature CFTR and enabled some maturation of ΔF508. Changing expression levels of other Hsp90 cochaperones also influenced immature CFTR stability, but did not enable maturation of ΔF508.

While the Hsp70 and 90 chaperone systems and cochaperones that link them, such as Hop, play a direct role in CFTR processing, additional cytoplasmic chaperones have also been implicated. These include small heat shock proteins[107] and cysteine string proteins, recently shown to increase immature CFTR association with the Hsp70 dependent ubiquitin ligase, CHIP to promote degradation.[108] A broad proteomic screen of CFTR co-immunoprecipitating proteins also identified GRP75 and Hsp105, but the functional significance of this has not been explored further.[106] Since most CFTR processing occurs at the ER with even a large amount of the wild-type subjected to ERAD, the influence of chaperones of the ER membrane and lumen might be expected to be important. For example, calnexin, optimally situated at the cytoplasmic face of the ER membrane is able to bind the untrimmed oligosaccharide chains of CFTR in the first extracytoplasmic loop of MSD2.[109] Since this association between calnexin and CFTR is prolonged in the case of ΔF508 and other mutants that are not exported from the ER, the possibility was considered that this binding might be responsible for the retention of these mutants at the ER. However, this is not the case as ΔF508 cannot be exported even in the complete absence of association with calnexin.[43] More recently it has been proposed that calnexin binding contributes to the stabilization of MSD2, enabling it to integrate appropriately with the more N-terminal parts of the protein.[110] However, this requirement would appear to be a quantitative rather than absolute since active CFTR channels can be synthesized without interaction with calnexin.[43]

It has been suggested that the luminal counterpart of calnexin, calreticulin may also bind to CFTR and possibly contribute to transport out of the ER.[111] Additional ER luminal chaperones detected in CFTR immunoprecipitates include GRP78, cyclophilin B, reticulocalbin and calumenin.[106] The latter two EF-hand, Ca^{2+}-binding proteins of the CREC family have been reported to interact with proteins of the translocon through which newly synthesized proteins cross the ER membrane, but their possible influence on CFTR translocation has not been investigated.

Thus, the CFTR nascent chain encounters multiple chaperones during the course of its biosynthesis. In addition to aiding the folding of individual domains, chaperone binding is likely required to shield more N-terminal surfaces destined to interact with C-terminal domains as they are formed. However, if these inter-domain interfaces fail to form appropriately, the chaperone bound polypeptide is presented to a series of different ubiquitinylating enzymes in order to be tagged for proteasomal degradation.

4. Intracellular Trafficking of CFTR

The CFTR membrane glycoprotein is synthesized on ER-associated ribosomes, core-glycosylated at the ER, transported to the Golgi for oligosaccharide trimming and extension and finally delivered to the apical surface. In some cell lines and under certain circumstances, CFTR proteins may be able to bypass Golgi compartments[112–114] and reach the cell surface using an alternative route that might involve endosomal pathways.

A negatively charged short di-acidic motif (DXD), in a surface loop just N-terminal of the Walker B motif in NBD1, facilitates ER export of CFTR,[115] while exposure of arginine-framed tripeptides (RXR), also in NBD1, contribute to its retention at the ER.[116,117] On the other hand it has become obvious that N-linked glycans of CFTR do not contribute to retention of the protein at the ER but may be important for stability at the apical membrane.[43] Interestingly, in this membrane the C-terminus of CFTR interacts with multiple PDZ proteins that mediate association with the actin cytoskeleton and tethers CFTR to scaffolding proteins in macromolecular complexes.[118] The association with the actin cytoskeleton is also mediated by filamin that binds to a sequence near the N-terminus of CFTR.[119]

From the plasma membrane CFTR is rapidly internalized;[120,121] however, the majority of wild-type CFTR, is then recycled to the cell surface.[122,123] Endocytic CFTR resides in early endosomes and recycling endosomes and a small amount is also transferred to late endosomes, multivesicular bodies and lysosomes for degradation.[122,124] The endocytic routing of CFTR was found to be regulated by multiple proteins, including components of vesicle coats, adaptor complexes, SNARE proteins, Rab GTPases and PDZ-proteins.[125,126] Internalization of CFTR from the plasma membrane, which occurs via clathrin-coated vesicles (Fig. 3), is mediated by a tyrosine-based signal (YXXΦ, Φ = hydrophobic amino acid) and a dileucine motif in the C-terminus of CFTR, involving binding to the AP-2 complex.[127–130] Several small GTPases of the Rab family have been shown to regulate endocytic trafficking of CFTR. Thus, delivery of CFTR to early endosomes is dependent on Rab5 while recycling of wild-type CFTR from endosomes to the plasma membrane is promoted by Rab11 and Rme1.[124,131] Another Rab GTPase, Rab7, was shown to enhance the degradation of CFTR by facilitating targeting to lysosomes.[124] Interestingly, endocytic recycling of CFTR and routing to lysosomes and trafficking appears also to be regulated by interactions with the PDZ proteins, i.e., CAL, NHERF1 and SNX27.[132–134] Other proteins reported to affect the endocytic trafficking of CFTR and its expression levels at the plasma membrane are multiple syntaxins[53,135–138] and the SNARE proteins VAMP8 and Vti1b.[139] Finally, components of the actin-based cytoskeleton appear to play a significant role in the endocytic routing of CFTR. Myosin VI was shown to participate in CFTR internalization,[140] whereas myosin Vb facilitates recycling from

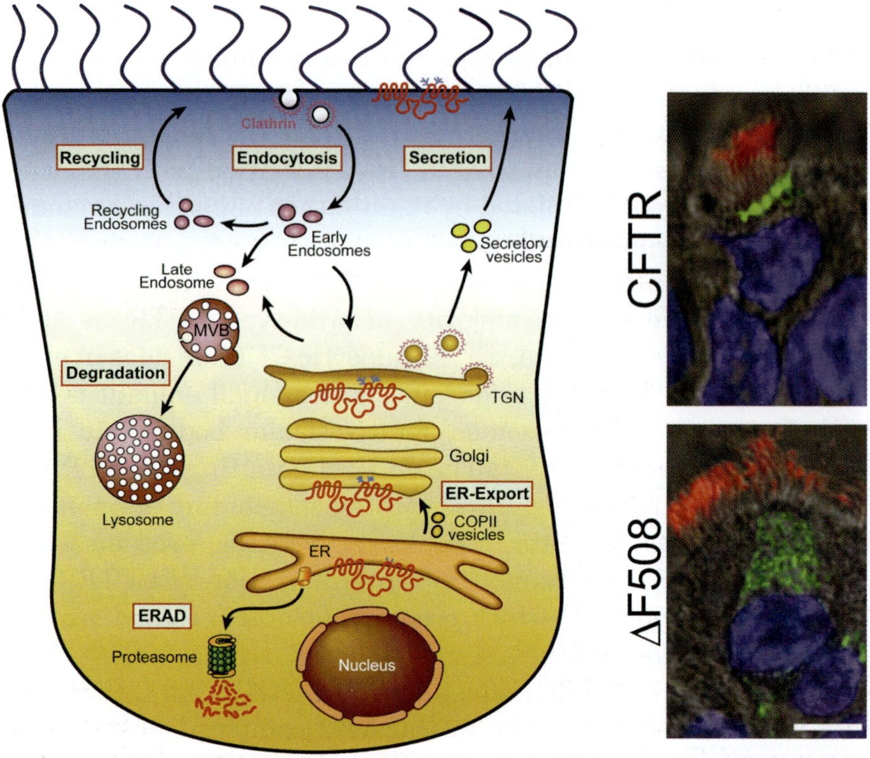

(A) (B)

Fig. 3. Intracellular trafficking routes of CFTR. **(A)** CFTR is core-glycosylated at the ER. ΔF508 is recognized as misfolded by ER quality control and targeted for proteasomal degradation. Wild-type CFTR traffics through the Golgi and core sugars are elongated. CFTR reaches the apical membrane and is internalized by clathrin-mediated endocytosis. Endocytic CFTR recycles back to the membrane or is directed to late endosomes, multivesicular bodies (MVB) and lysosomes for degradation. **(B)** In WT expressing differentiated primary human airway epithelial cells CFTR localized to the apical membrane, while in ΔF508 expressing cells the protein is present only intracellularly as shown by confocal microscopy after fluorescent labeling of CFTR (bar = 10 μM).

endosomes back to the plasma membrane.[141] Strikingly, inhibition of actin-polymerization significantly decreased the cell surface density of CFTR,[142] suggesting that the sub-apical cytoskeleton contributes to CFTR retention at that membrane.

ΔF508 CFTR is retained by quality control systems at the ER in its immature, core-glycosylated form and the mutant protein is subsequently degraded by ER-associated degradation (ERAD) with the consequence that no protein is transported to the cell surface.[87,143] The inability to transport human ΔF508 CFTR to the apical membrane has been confirmed in different expression systems and tissues including native airways, intestinal epithelia and sweat duct cells.[144–146] Interestingly, the processing of CFTR appears to be species-dependent as pig and mouse ΔF508 CFTR are partially processed like the wild-type protein.[147]

There is great interest in elucidating protein interactions that regulate intracellular trafficking pathways of CFTR, because interference with key determinants may facilitate the identification of pharmacological targets for rescue of cell surface expression of ΔF508. As outlined in the previous section, in the ER, multiple molecular chaperones have been identified that assist the folding of the nascent CFTR polypeptide from both sides of the ER membrane as it is synthesized.[101,106] Immature forms of both wild-type and ΔF508 CFTR interact with the cytoplasmic chaperone, Hsp70[148] and the ER lectin-chaperone, calnexin.[109] The Hsp90 complex seems to be a particularly important player in monitoring the folding process of CFTR, and up- or down-regulation of the expression of components of this machinery was shown to promote ΔF508 maturation.[106] The membrane protein Derlin-1 and the cytoplasmic chaperone Hsp70 are both associated with distinct ubiquitylating proteins that recognize misassembled CFTR as a substrate.[94,149] Polyubiquitylated CFTR is finally degraded by ERAD in a process that involves the 26S proteasome.

A potential therapeutic approach that addresses the trafficking defect of ΔF508 CFTR is to identify a means of promoting the transit of ΔF508 CFTR to the plasma membrane. Treatment with low temperature or chemical chaperones such as glycerol *in vitro* has previously been shown to partially overcome the processing block of the mutant protein. More recently several new small molecule corrector compounds have also been identified that allow a proportion of ΔF508 molecules to mature. Such rescued ΔF508 protein trafficks to the cell surface and has some cAMP-regulated chloride channel function. However, when ΔF508 and other processing variants of CFTR have reached the plasma membrane, either by temperature shift, drug rescue or alteration of arginine-framed tripeptide ER-retention signals, they are thermally unstable and disappear from this location much more rapidly than the wild type protein.[117,123,124,150,151] For temperature-rescued ΔF508 CFTR, it has been shown that its half-life at the cell surface is at least 5 times shorter than that of wild-type CFTR.[122,123,152] Apparently ΔF508 CFTR that has proceeded to the plasma membrane and endocytic compartments is subjected to an additional mechanism of quality control. Details of the recognition machinery that distinguishes wild-type from mutant forms of CFTR and targets them towards divergent pathways are not yet entirely understood. However, it was demonstrated that misfolded mutant CFTR protein is subjected to ubiquitylation and recognition by a ubiquitin-dependent endosomal sorting complex that includes Hrs, STAM-2, TSG101, Vps25 and Vps32.[122] Thus, marking misfolded CFTR with ubiquitin does not function solely as a degradation signal at the ER, but may also be a signal that directs the mutant CFTR protein to lysosomal degradation.

Another covalent modification of CFTR that appears to affect its intracellular trafficking is the phosphorylation of the R-domain.[153] Stimulation of protein phosphorylation by cAMP-dependent protein kinase A and by protein kinase C appears to result in an increase in the number of CFTR channels at the plasma membrane through augmentation of transfer to the cell surface and decreased internalization.[120,121,153–156]

5. Overcoming Cystic Fibrosis Causing Mutations

Since cystic fibrosis is caused by recessive mutations in a single gene, as soon as the *cftr* gene was discovered, gene replacement therapy was considered the most direct molecular therapeutic approach to treatment. While this expectation has not yet been realized, remarkable progress has been made with other approaches based on the defects caused by specific mutations. ΔF508, the most common, and several other CFTR mutations result in folding defective proteins that are cleared by cellular quality control systems. Other mutations impair channel gating (e.g., G551D) and there are several that result in premature termination of translation (e.g., W1282X). The overall disturbance of epithelial ion transport resulting from the lack of CFTR activity can be lessened by manipulation of another ion channel involved in this process, the epithelial sodium channel (ENaC), which mediates sodium reabsorption.[157] Activation of chloride channels other than CFTR offer yet another potential means to compensate for CFTR's absence.[158,159] It has not yet been determined whether any of these are upregulated in CF patents but if they are the effects apparently are not sufficiently compensatory to ameliorate disease.

Gene replacement therapy would be applicable to all patients, regardless of the type of mutation and stabilization of the ΔF508 protein with a small molecule would be effective in the majority of patients since ~90% have this mutation on at least one allele. However additional mutation-specific molecular therapeutic approaches are also being pursued. For example the missense substitution G551D is present in a few percent of patients and an effective small molecule that restores its channel activity has been developed and progressed through very encouraging clinical trials. The nearly complete loss of channel activity caused by this single residue substitution in the NBD1 signature sequence is partially overcome by the compound VTX-770.[160] Several premature termination of translation mutants are known for CFTR, some of which are quite common in certain ethnic populations and progress is being made in the utilization of aminoglycoside antibiotics and other compounds to promote read-through.[161,162] Although still at an early experimental stage, manipulation of splicing factors in cultured cells is being explored to overcome splice site mutations.[163]

5.1 *Gene replacement therapy*

The primary targets for CF gene therapy are the terminally differentiated cells of the small airways. It is more difficult to stably deliver genes into non-dividing cells and damage of these cells, caused by the toxicity of the delivery vector, is also a limiting factor.[164,165] The efficiency of delivery is the other important issue. The efficacy of restoration by gene delivery is affected by both physical and immunological barriers, which are more impenetrable in CF patients compared to healthy individuals because of the impacted mucus and the presence of various immune-cells in a chronic inflammatory state.[166,167] These and other factors have limited the effectiveness of trials with non-viral and adenovirus vectors, but more promising preliminary results have been obtained with adeno-associated and lenti-viral based vectors.

Phospholipid/cholesterol mixtures have been employed as non-viral delivery vehicles. The lipids are mixed with DNA to form nuclease resistant, positively charged liposomes that can fuse with the cell membrane and enter the cell.[168] Unlike viruses, liposome carriers do not pose limitations on the size of the DNA. However, several types of cationic liposome systems have cytotoxic effects and result in increased pro-inflammatory cytokines with infiltration of neutrophils and macrophages. Cationic polymers, other than lipids such as polyethyleneimine and poly-L-lysine (poly-K) have also been employed.[169] To minimize immune responses and to stabilize these polymers at physiological salt concentrations, the influence of modification of the side chains of the lysines with PEG has also been explored.[170] In a clinical trial, 8 of 12 patients treated with PEGylated poly-K-CFTR cDNA in their lower airways exhibited restored nasal potential for six days without any associated detrimental effects.

The most extensively used vectors have been adenovirus (Ad) serotypes 2 and 5. Although Ad can infect a wide variety of cell types, including terminally differentiated airway epithelial cells, their receptors are located on the basolateral surface of these cells, restricting delivery from the accessible apical side.[171]

Moreover, Ad caused a significant immunological response and cytotoxicity, and the presence of the recombinant vector could not be detected beyond 30 days.[172,173] A more advantageous system for CFTR gene delivery is provided by Adeno-associated viruses (AAV).[164] AAV is not pathogenic and these viruses are able to infect a wide variety of cells, even those that are differentiated and non-dividing. AAVs do not perturb the host's genome and the viral genetic material exists mainly as episomes. Typically, AAV delivery of recombinant genes results in persistent expression for the lifespan of the transfected cell.[174] However, this advantage may be limited by the elevated turnover of bronchial epithelial cells in CF patients. The lack of permissive receptors on these cells also limits the efficiency of delivery.[174] Nevertheless, in Phase I trials, where AAV was introduced by nasal and endobronchial routes, dose-dependent expression was reported for approximately 30 days.[175,176] This process was safe and well tolerated but was without detectable improvement in the patients' pulmonary function. Repeated administration of the vector has not been possible because of the increased level of neutralizing AAV antibodies.[177] However, new AAV serotypes are being developed that allow for re-administration with decreased immunogenicity and site-specific integration into the host genome.[178]

In vitro and *in vivo* experiments with lentiviral vectors to deliver genes into mice hold promise for avoidance of many of the problems associated with other methods. Lentiviruses are members of the *Retroviridae* family that can carry a large amount of genetic information. Early studies indicated that lentiviral vectors elicit a much reduced immune response compared to adenoviral vectors.[179] Vectors can be highly concentrated and can be pseudotyped with envelope glycoproteins from viruses that recognize receptors on the apical membrane surface of airway epithelia. For example, experiments with influenza-HA pseudotyped lentiviral vectors have demonstrated effective infection of mouse tracheal epithelial cells from both the basolateral and apical side.[180] Notably, there was effective delivery to the ciliated epithelial cells in which CFTR is normally expressed. Administration of high titer (>10^9 IU/ml) HA-pseudotyped lentiviral vectors by nasal administrations led to a higher level expression in the trachea and lungs than in the nose.[181] The low level of transduction of nasal cells may be caused by the lower temperature and the presence of a different host defense system compared to the lower airways. There was evidence of at least some lacZ expression in 70% of the small and large airways in the lung after transduction with a lacZ-containing vector. Vector transgene expression lasted for over six months and the level of persistent vector expression was increased when inadvertently, expression in hematopoietic cells was inhibited by including four tandem copies of a target sequence for a hematopoietic-cell specific micro RNA sequence.[181]

It may be anticipated that, despite the extremely demanding requirements of practicable gene replacement therapy for CF, this objective will ultimately be reached by further advances in overcoming or circumventing the numerous limiting factors.

5.2 *Small molecules as CFTR correctors and potentiators*

High throughput screening has identified a growing number of small-molecule corrector compounds that result in surface expression of a portion of ΔF508 CFTR.[182–184] These include pirazols (e.g., VRT 532), quinazoline derivatives (e.g., VRT 325), aminoarylthiazoles (e.g., Corr-2b), quinazolinylamino-pyrimidinones (e.g., Corr-3a), methylbithiazole analogs (e.g., Corr-4a) and sildenafil analogs (e.g., KM11060).[182–185] VRT-532 was originally identified as a potentiator compound[183] but was later shown to have some corrector activity.[186] VRT-325 does not show strict specificity for ΔF508 CFTR because it also promotes maturation of other misfolded membrane proteins, i.e., the structurally related P-glycoprotein mutant G268V[186,187] and the hERG cardiac potassium channel mutant G601S.[183] The small molecule correctors, Corr 2b, VRT-532 and Corr-4a seem to be more specific for the rescue of ΔF508 CFTR maturation.[93,182,186,188] Most corrector compounds cause only a modest increase in the maturation of CFTR processing mutants so that the yield of mature protein is usually not more than 5–10% of that of wild-type CFTR. However, the combination of different corrector compounds, i.e., VRT-325, together with Corr-2b or Corr-4a, appears to have an additive effect on ΔF508 rescue.[189] The mechanism by which these correctors rescue ΔF508 CFTR from ER-retention is not completely clear. Corrector compounds may either promote ΔF508 CFTR folding or protect misfolded CFTR from ERAD by allowing it to bypass quality control systems. Recent studies suggest that certain corrector compounds (e.g., VRT-325 and Corr-4a) may bind to mutant CFTR directly[93] and may facilitate domain folding and assembly at the ER.[188,190–192]

More recently, a new approach to rescue ΔF508 CFTR heterologously expressed in several mammalian cell types by inhibition of histone deacetylase has been described.[193] In addition to increasing the transcription of CFTR, these inhibitors probably act by modulating transcription of CFTR-interacting proteins such as cytosolic chaperones and other components that affect CFTR folding, stability and trafficking.

Because the ΔF508 deletion results in a thermally unstable protein[117,150] and both the functional and biochemical half-lives of the rescued mutant protein are greatly shortened, corrector compounds must stabilize the protein as well as restoring folding and function. It is important to note that folding yield and subsequent stability of CFTR mutants are not always strictly coupled.[69] Unfortunately, the corrector compounds analyzed so far (e.g., VRT-325 and Corr-4a) caused only a partial extension of the lifetime of the rescued ΔF508 protein.[152,182,183] While endocytic turnover of temperature-rescued ΔF508 CFTR has been characterized to some degree,[122,124] much less information is available on the endocytic fate of ΔF508 CFTR rescued by small-molecule correctors. However, the recycling of the mutant protein seems to be impaired even in the presence of the corrector compounds tested so far.[152,194]

Currently there is a great enthusiasm over the measurable improvements in CF patient parameters in early clinical trials with the VX-770 potentiator compound, which activates G551D CFTR.[160] In addition, the VX-809 corrector caused a small reduction in the sweat chloride concentration of homozygous ΔF508 patients, according to unpublished reports. Additional screening programs[184] and other efforts to develop yet more effective compounds are continuing.[195]

References

1. Quinton, P.M. (1999) *Physiol. Rev.* **79**, S3.
2. Riordan, J.R., Rommens, J.M., Kerem, B., Alon, N., Rozmahel, R., Grzelczak, Z., Zielenski, J., Lok, S., Plavsic, N., Chou, J.L., *et al.* (1989) *Science* **245,** 1066.
3. Hyde, S.C., Emsley, P., Hartshorn, M.J., Mimmack, M.M., Gileadi, U., Pearce, S.R., Gallagher, M.P., Gill, D.R., Hubbard, R.E., and Higgins, C.F. (1990) *Nature* **346**, 362.
4. Accardi, A. and Miller, C. (2004) *Nature* **427,** 803.
5. Levin, E.J., Quick, M., and Zhou, M. (2009) *Nature* **462**, 757.
6. Anderson, M.P., Gregory, R.J., Thompson, S., Souza, D.W., Paul, S., Mulligan, R.C., Smith, A.E., and Welsh, M.J. (1991) *Science* **253**, 202.
7. Bear, C.E., Li, C.H., Kartner, N., Bridges, R.J., Jensen, T.J., Ramjeesingh, M., and Riordan, J.R. (1992) *Cell* **68**, 809.
8. Linsdell, P., Evagelidis, A., and Hanrahan, J.W. (2000) *Biophys. J.* **78,** 2973.
9. Seidler, U., Blumenstein, I., Kretz, A., Viellard-Baron, D., Rossmann, H., Colledge, W.H., Evans, M., Ratcliff, R., and Gregor, M. (1997) *J. Physiol. (Lond.)* **505**, 411.
10. Choi, J.Y., Muallem, D., Kiselyov, K., Lee, M.G., Thomas, P.J., and Muallem, S. (2001) *Nature* **410**, 94.
11. Paradiso, A.M., Coakley, R.D., and Boucher, R.C. (2003) *J. Physiol. (Lond.)* **548**, 203.
12. Kuver, R., Wong, T., Klinkspoor, J.H., and Lee, S.P. (2006) *Am. J. Physiol. Gastrointest. Liver Physiol.* **291**, G1148.
13. Garcia, M.A., Yang, N., and Quinton, P.M. (2009) *J. Clin. Invest.* **119**, 2613.

14. Liu, X., Smith, S.S., and Dawson, D.C. (2003) *J. Exp. Zoolog. Part A Comp. Exp. Biol.* **300**, 69.
15. Beck, E.J., Yang, Y., Yaemsiri, S., and Raghuram, V. (2008) *J. Biol. Chem.* **283**, 4957.
16. McDonough, S., Davidson, N., Lester, H.A., and McCarty, N.A. (1994) *Neuron* **13**, 623.
17. Fatehi, M. and Linsdell, P. (2009) *J. Membr. Biol.* **228**, 151.
18. Serohijos, A.W., Hegedus, T., Aleksandrov, A.A., He, L., Cui, L., Dokholyan, N.V., and Riordan, J.R. (2008) *Proc. Natl. Acad. Sci. U.S.A.* **105**, 3256.
19. Mornon, J.P., Lehn, P., and Callebaut, I. (2009) *Cell. Mol. Life Sci.* **66**, 3469.
20. Alexander, C., Ivetac, A., Liu, X., Norimatsu, Y., Serrano, J.R., Landstrom, A., Sansom, M., and Dawson, D.C. (2009) *Biochemistry* **48**, 10078.
21. Cheng, S.H., Rich, D.P., Marshall, J., Gregory, R.J., Welsh, M.J., and Smith, A.E. (1991) *Cell* **66**, 1027.
22. Gadsby, D.C. and Nairn, A.C. (1999) *Adv. Second Messenger Phosphoprotein Res.* **33**, 79.
23. Chang, X.B., Tabcharani, J.A., Hou, Y.X., Jensen, T.J., Kartner, N., Alon, N., Hanrahan, J.W., and Riordan, J.R. (1993) *J. Biol. Chem.* **268**, 11304.
24. Seibert, F.S., Chang, X.B., Aleksandrov, A.A., Clarke, D.M., Hanrahan, J.W., and Riordan, J.R. (1999) *Biochim. Biophys. Acta* **1461**, 275.
25. Baker, J.M., Hudson, R.P., Kanelis, V., Choy, W.Y., Thibodeau, P.H., Thomas, P.J., and Forman-Kay, J.D. (2007) *Nat. Struct. Mol. Biol.* **14**, 738.
26. Hegedus, T., Serohijos, A.W., Dokholyan, N.V., He, L., and Riordan, J.R. (2008) *J. Mol. Biol.* **378**, 1052.
27. Chappe, V., Irvine, T., Liao, J., Evagelidis, A., and Hanrahan, J.W. (2005) *EMBO J.* **24**, 2730.
28. Mense, M., Vergani, P., White, D.M., Altberg, G., Nairn, A.C., and Gadsby, D.C. (2006) *EMBO J.* **25,** 4728.
29. Wang, W., Wu, J., Bernard, K., Li, G., Wang, G., Bevensee, M.O., and Kirk, K.L. (2010) *Proc. Natl. Acad. Sci. U.S.A.* **107**, 3888.
30. Aleksandrov, L., Aleksandrov, A.A., Chang, X.B., and Riordan, J.R. (2002) *J. Biol. Chem.* **277**, 15419.

31. Aleksandrov, L.A., Aleksandrov, A.A., and Riordan, J.R. (2008) *Biochem. J.* **416**, 129.
32. Vergani, P., Lockless, S.W., Nairn, A.C., and Gadsby, D.C. (2005) *Nature* **433**, 876.
33. Muallem, D. and Vergani, P. (2009) *Philos. Trans. R. Soc. Lond. B Biol. Sci.* **364**, 247.
34. Quinton, P.M. and Reddy, M.M. (1992) *Nature* **360**, 79.
35. Ramjeesingh, M., Li, C., Garami, E., Huan, L.J., Galley, K., Wang, Y., and Bear, C.E. (1999) *Biochemistry* **38**, 1463.
36. Aleksandrov, A.A., Chang, X., Aleksandrov, L., and Riordan, J.R. (2000) *J. Physiol. (Lond.)* **528**(2), 259.
37. Cui, L., Aleksandrov, L., Chang, X.B., Hou, Y.X., He, L., Hegedus, T., Gentzsch, M., Aleksandrov, A., Balch, W.E., and Riordan, J.R. (2007) *J. Mol. Biol.* **365**, 981.
38. Aleksandrov, A.A., Aleksandrov, L., and Riordan, J.R. (2002) *FEBS Lett.* **518**, 183.
39. He, L., Aleksandrov, A.A., Serohijos, A.W., Hegedus, T., Aleksandrov, L.A., Cui, L., Dokholyan, N.V., and Riordan, J.R. (2008) *J. Biol. Chem.* **283**, 26383.
40. Szabo, K., Szakacs, G., Hegedus, T., and Sarkadi, B. (1999) *J. Biol. Chem.* **274**, 12209.
41. Dulhanty, A.M. and Riordan, J.R. (1994) *FEBS Lett.* **343**, 109.
42. Hammerle, M.M., Aleksandrov, A.A., and Riordan, J.R. (2001) *J. Biol. Chem.* **276**, 14848.
43. Chang, X.B., Mengos, A., Hou, Y.X., Cui, L., Jensen, T.J., Aleksandrov, A., Riordan, J.R., and Gentzsch, M. (2008) *J. Cell Sci.* **121**, 2814.
44. Rosenberg, M.F., Kamis, A.B., Aleksandrov, L.A., Ford, R.C., and Riordan, J.R. (2004) *J. Biol. Chem.* **279**, 39051.
45. Rosenberg, M.F., Callaghan, R., Modok, S., Higgins, C.F., and Ford, R.C. (2005) *J. Biol. Chem.* **280**, 2857.
46. Rosenberg, M.F., Kamis, A.B., Callaghan, R., Higgins, C.F., and Ford, R.C. (2003) *J. Biol. Chem.* **278**, 8294.
47. Sato, C., Ueno, Y., Asai, K., Takahashi, K., Sato, M., Engel, A., and Fujiyoshi, Y. (2001) *Nature* **409**, 1047.
48. Long, S.B., Campbell, E.B., and Mackinnon, R. (2005) *Science* **309**, 897.
49. Awayn, N.H., Rosenberg, M.F., Kamis, A.B., Aleksandrov, L.A., Riordan, J.R., and Ford, R.C. (2005) *Biochem. Soc. Trans.* **33**, 996.

50. Mio, K., Ogura, T., Mio, M., Shimizu, H., Hwang, T.C., Sato, C., and Sohma, Y. (2008) *J. Biol. Chem.* **283**, 30300.
51. Zhang, L., Aleksandrov, L.A., Zhao, Z., Birtley, J.R., Riordan, J.R., and Ford, R.C. (2009) *J. Struct. Biol.* **167**, 242.
52. Marshall, J., Fang, S., Ostedgaard, L.S., O'Riordan, C.R., Ferrara, D., Amara, J.F., Hoppe, H.T., Scheule, R.K., Welsh, M.J., Smith, A.E., *et al.* (1994) *J. Biol. Chem.* **269**, 2987.
53. Li, C., Roy, K., Dandridge, K., and Naren, A.P. (2004) *J. Biol. Chem.* **279**, 24673.
54. Eskandari, S., Wright, E.M., Kreman, M., Starace, D.M., and Zampighi, G.A. (1998) *Proc. Natl. Acad. Sci. U.S.A.* **95**, 11235.
55. Schillers, H., Shahin, V., Albermann, L., Schafer, C., and Oberleithner, H. (2004) *Cell. Physiol. Biochem.* **14**, 1.
56. Chen, J.H., Chang, X.B., Aleksandrov, A.A., and Riordan, J.R. (2002) *J. Membr. Biol.* **188**, 55.
57. Bates, I.R., Hebert, B., Luo, Y., Liao, J., Bachir, A.I., Kolin, D.L., Wiseman, P.W., and Hanrahan, J.W. (2006) *Biophys. J.* **91**, 1046.
58. Haggie, P.M., Kim, J.K., Lukacs, G.L., and Verkman, A.S. (2006) *Mol. Biol. Cell* **17**, 4937.
59. Jin, S., Haggie, P.M., and Verkman, A.S. (2007) *Biophys. J.* **93**, 1079.
60. Haggie, P.M. and Verkman, A.S. (2008) *J. Biol. Chem.* **283**, 23510.
61. Lewis, H.A., Buchanan, S.G., Burley, S.K., Conners, K., Dickey, M., Dorwart, M., Fowler, R., Gao, X., Guggino, W.B., Hendrickson, W.A., Hunt, J.F., Kearins, M.C., Lorimer, D., Maloney, P.C., Post, K.W., Rajashankar, K.R., Rutter, M.E., Sauder, J.M., Shriver, S., Thibodeau, P.H., Thomas, P.J., Zhang, M., Zhao, X., and Emtage, S. (2004) *EMBO J.* **23**, 282.
62. Csanady, L., Chan, K.W., Nairn, A.C., and Gadsby, D.C. (2005) *J. Gen. Physiol.* **125**, 43.
63. Atwell, S., Brouillette, C.G., Conners, K., Emtage, S., Gheyi, T., Guggino, W.B., Hendle, J., Hunt, J.F., Lewis, H.A., Lu, F., Protasevich, II, Rodgers, L.A., Romero, R., Wasserman, S.R., Weber, P.C., Wetmore, D., Zhang, F.F., and Zhao, X. (2010) *Protein Eng. Des. Sel.* **23**, 375.
64. Thibodeau, P.H., Brautigam, C.A., Machius, M., and Thomas, P.J. (2005) *Nat. Struct. Mol. Biol.* **12**, 10.
65. Pissarra, L.S., Farinha, C.M., Xu, Z., Schmidt, A., Thibodeau, P.H., Cai, Z., Thomas, P.J., Sheppard, D.N., and Amaral, M.D. (2008) *Chem. Biol.* **15**, 62.

66. Wieczorek, G. and Zielenkiewicz, P. (2008) *J. Cyst. Fibros.* **7**, 295.
67. Lewis, H.A., Wang, C., Zhao, X., Hamuro, Y., Conners, K., Kearins, M.C., Lu, F., Sauder, J.M., Molnar, K.S., Coales, S.J., Maloney, P.C., Guggino, W.B., Wetmore, D.R., Weber, P.C., and Hunt, J.F. (2010) *J. Mol. Biol.* **396**, 406.
68. Serohijos, A.W., Hegedus, T., Riordan, J.R., and Dokholyan, N.V. (2008) *PLoS Comput. Biol.* **4**, e1000008.
69. Du, K., Sharma, M., and Lukacs, G.L. (2005) *Nat. Struct. Mol. Biol.* **12**, 17.
70. Callebaut, I., Eudes, R., Mornon, J.P., and Lehn, P. (2004) *Cell. Mol. Life Sci.* **61**, 230.
71. Huang, S.Y., Bolser, D., Liu, H.Y., Hwang, T.C., and Zou, X. (2008) *J. Mol. Graph. Model.* **27**, 822.
72. Partridge, A.W., Melnyk, R.A., and Deber, C.M. (2002) *Biochemistry* **41**, 3647.
73. Choi, M.Y., Cardarelli, L., Therien, A.G., and Deber, C.M. (2004) *Biochemistry* **43**, 8077.
74. Wehbi, H., Gasmi-Seabrook, G., Choi, M.Y., and Deber, C.M. (2008) *Biochim. Biophys. Acta* **1778**, 79.
75. Choi, M.Y., Partridge, A.W., Daniels, C., Du, K., Lukacs, G.L., and Deber, C.M. (2005) *J. Biol. Chem.* **280**, 4968.
76. Chen, E.Y., Bartlett, M.C., Loo, T.W., and Clarke, D.M. (2004) *J. Biol. Chem.* **279**, 39620.
77. Mornon, J.P., Lehn, P., and Callebaut, I. (2008) *Cell. Mol. Life Sci.* **65**, 2594.
78. Dulhanty, A.M. and Riordan, J.R. (1994) *Biochemistry* **33**, 4072.
79. Ostedgaard, L.S., Baldursson, O., Vermeer, D.W., Welsh, M.J., and Robertson, A.D. (2000) *Proc. Natl. Acad. Sci. U.S.A.* **97**, 5657.
80. Dunker, A.K., Silman, I., Uversky, V.N., and Sussman, J.L. (2008) *Curr. Opin. Struct. Biol.* **18**, 756.
81. Tompa, P. and Fuxreiter, M. (2008) *Trends Biochem. Sci.* **33**, 2.
82. Oldfield, C.J., Cheng, Y., Cortese, M.S., Romero, P., Uversky, V.N., and Dunker, A.K. (2005) *Biochemistry* **44**, 12454.
83. Solt, I., Magyar, C., Simon, I., Tompa, P., and Fuxreiter, M. (2006) *Proteins* **64**, 749.
84. Ward, A., Reyes, C.L., Yu, J., Roth, C.B., and Chang, G. (2007) *Proc. Natl. Acad. Sci. U.S.A.* **104**, 19005.

85. Aller, S.G., Yu, J., Ward, A., Weng, Y., Chittaboina, S., Zhuo, R., Harrell, P.M., Trinh, Y.T., Zhang, Q., Urbatsch, I.L., and Chang, G. (2009) *Science* **323**, 1718.
86. Cheng, S.H., Gregory, R.J., Marshall, J., Paul, S., Souza, D.W., White, G.A., O'Riordan, C.R., and Smith, A.E. (1990) *Cell* **63**, 827.
87. Kopito, R.R. (1999) *Physiol. Rev.* **79**, S167.
88. Ward, C.L., Omura, S., and Kopito, R.R. (1995) *Cell* **83**, 121.
89. Jensen, T.J., Loo, M.A., Pind, S., Williams, D.B., Goldberg, A.L., and Riordan, J.R. (1995) *Cell* **83**, 129.
90. Zhang, F., Kartner, N., and Lukacs, G.L. (1998) *Nat. Struct. Biol.* **5**, 180.
91. Kleizen, B., van Vlijmen, T., de Jonge, H.R., and Braakman, I. (2005) *Mol. Cell* **20**, 277.
92. Du, K. and Lukacs, G.L. (2009) *Mol. Biol. Cell* **20**, 1903.
93. Wang, Y., Loo, T.W., Bartlett, M.C., and Clarke, D.M. (2007) *J. Biol. Chem.* **282**, 33247.
94. Younger, J.M., Chen, L., Ren, H.Y., Rosser, M.F., Turnbull, E.L., Fan, C.Y., Patterson, C., and Cyr, D.M. (2006) *Cell* **126**, 571.
95. Teem, J.L., Berger, H.A., Ostedgaard, L.S., Rich, D.P., Tsui, L.C., and Welsh, M.J. (1993) *Cell* **73**, 335.
96. DeCarvalho, A.C., Gansheroff, L.J., and Teem, J.L. (2002) *J. Biol. Chem.* **277**, 35896.
97. Lewis, H.A., Zhao, X., Wang, C., Sauder, J.M., Rooney, I., Noland, B.W., Lorimer, D., Kearins, M.C., Conners, K., Condon, B., Maloney, P.C., Guggino, W.B., Hunt, J.F., and Emtage, S. (2005) *J. Biol. Chem.* **280**, 1346.
98. He, L., Aleksandrov, L.A., Jensen, T.J., Nesbitt, K.L., and Riordan, J.R. (2010) *FASEB J.* **24**, 3103.
99. Loo, T.W., Bartlett, M.C., and Clarke, D.M. (2002) *J. Biol. Chem.* **277**, 27585.
100. Buyse, F., Vandenbranden, M., and Ruysschaert, J.M. (2004) *FEBS Lett.* **578**, 145.
101. Amaral, M.D. (2004) *J. Mol. Neurosci.* **23**, 41.
102. Jiang, C., Fang, S.L., Xiao, Y.F., O'Connor, S.P., Nadler, S.G., Lee, D.W., Jefferson, D.M., Kaplan, J.M., Smith, A.E., and Cheng, S.H. (1998) *Am. J. Physiol.* **275**, C171.
103. Loo, M.A., Jensen, T., Cui, J.L., Hou, Y., Chang, X.B., and Riordan, J.R. (1998) *EMBO J.* **17**, 6879.

104. Neckers, L., Mollapour, M., and Tsutsumi, S. (2009) *Trends Biochem. Sci.* **34**, 223.
105. Neckers, L., Tsutsumi, S., and Mollapour, M. (2009) *Nat. Struct. Mol. Biol.* **16**, 235.
106. Wang, X., Venable, J., LaPointe, P., Hutt, D.M., Koulov, A.V., Coppinger, J., Gurkan, C., Kellner, W., Matteson, J., Plutner, H., Riordan, J.R., Kelly, J.W., Yates, J.R. 3rd, and Balch, W.E. (2006) *Cell* **127**, 803.
107. Ahner, A., Nakatsukasa, K., Zhang, H., Frizzell, R.A., and Brodsky, J.L. (2007) *Mol. Biol. Cell* **18**, 806.
108. Schmidt, B.Z., Watts, R.J., Aridor, M., and Frizzell, R.A. (2009) *J. Biol. Chem.* **284**, 4168.
109. Pind, S., Riordan, J.R., and Williams, D.B. (1994) *J. Biol. Chem.* **269**, 12784.
110. Rosser, M.F., Grove, D.E., Chen, L., and Cyr, D.M. (2008) *Mol. Biol. Cell* **19**, 4570.
111. Harada, K., Okiyoneda, T., Hashimoto, Y., Ueno, K., Nakamura, K., Yamahira, K., Sugahara, T., Shuto, T., Wada, I., Suico, M.A., and Kai, H. (2006) *J. Biol. Chem.* **281**, 12841.
112. Luo, Y., McDonald, K., and Hanrahan, J.W. (2009) *Biochem. J.* **419**, 211.
113. Bannykh, S.I., Bannykh, G.I., Fish, K.N., Moyer, B.D., Riordan, J.R., and Balch, W.E. (2000) *Traffic* **1**, 852.
114. Yoo, J.S., Moyer, B.D., Bannykh, S., Yoo, H.M., Riordan, J.R., and Balch, W.E. (2002) *J. Biol. Chem.* **277**, 11401.
115. Wang, X., Matteson, J., An, Y., Moyer, B., Yoo, J.S., Bannykh, S., Wilson, I.A., Riordan, J.R., and Balch, W.E. (2004) *J. Cell Biol.* **167**, 65.
116. Chang, X.B., Cui, L., Hou, Y., Jensen, T.J., Aleksandrov, A.A., Mengos, A., and Riordan, J.R. (1999) *Mol. Cell* **4**, 137.
117. Hegedus, T., Aleksandrov, A., Cui, L., Gentzsch, M., Chang, X.B., and Riordan, J.R. (2006) *Biochim. Biophys. Acta* **1758**, 565.
118. Guggino, W.B. and Stanton, B.A. (2006) *Nat. Rev. Mol. Cell Biol.* **7**, 426.
119. Thelin, W.R., Kesimer, M., Tarran, R., Kreda, S.M., Grubb, B.R., Sheehan, J.K., Stutts, M.J., and Milgram, S.L. (2005) *J. Biol. Chem.* **280**, 41512.
120. Lukacs, G.L., Segal, G., Kartner, N., Grinstein, S., and Zhang, F. (1997) *Biochem. J.* **328**, 353.

121. Prince, L.S., Workman, R.B. Jr., and Marchase, R.B. (1994) *Proc. Natl. Acad. Sci. U.S.A.* **91**, 5192.
122. Sharma, M., Pampinella, F., Nemes, C., Benharouga, M., So, J., Du, K., Bache, K.G., Papsin, B., Zerangue, N., Stenmark, H., and Lukacs, G.L. (2004) *J. Cell Biol.* **164**, 923.
123. Lukacs, G.L., Chang, X.B., Bear, C., Kartner, N., Mohamed, A., Riordan, J.R., and Grinstein, S. (1993) *J. Biol. Chem.* **268**, 21592.
124. Gentzsch, M., Chang, X.B., Cui, L., Wu, Y., Ozols, V.V., Choudhury, A., Pagano, R.E., and Riordan, J.R. (2004) *Mol. Biol. Cell* **15**, 2684.
125. Okiyoneda, T., and Lukacs, G.L. (2007) *Biochim. Biophys. Acta* **1773**, 476.
126. Ameen, N., Silvis, M., and Bradbury, N.A. (2007) *J. Cyst. Fibros.* **6**, 1.
127. Hu, W., Howard, M., and Lukacs, G.L. (2001) *Biochem. J.* **354**, 561.
128. Prince, L.S., Peter, K., Hatton, S.R., Zaliauskiene, L., Cotlin, L.F., Clancy, J.P., Marchase, R.B., and Collawn, J.F. (1999) *J. Biol. Chem.* **274**, 3602.
129. Weixel, K.M. and Bradbury, N.A. (2000) *J. Biol. Chem.* **275**, 3655.
130. Peter, K., Varga, K., Bebok, Z., McNicholas-Bevensee, C.M., Schwiebert, L., Sorscher, E.J., Schwiebert, E.M., and Collawn, J.F. (2002) *J. Biol. Chem.* **277**, 49952.
131. Picciano, J.A., Ameen, N., Grant, B., and Bradbury, N.A. (2003) *Am. J. Physiol., Cell Physiol.* **285**, C1009.
132. Swiatecka-Urban, A., Duhaime, M., Coutermarsh, B., Karlson, K.H., Collawn, J., Milewski, M., Cutting, G.R., Guggino, W.B., Langford, G., and Stanton, B.A. (2002) *J. Biol. Chem.* **277**, 40099.
133. Cheng, J., Wang, H., and Guggino, W.B. (2004) *J. Biol. Chem.* **279**, 1892.
134. McDermott, M.I., Thelin, W.R., Lyons, P., Gentzsch, M., Hong, W., Stutts, M.J., and Milgram, S.L. (2008) *Mol. Biol. Cell* **19**, 256.
135. Cormet-Boyaka, E., Di, A., Chang, S.Y., Naren, A.P., Tousson, A., Nelson, D.J., and Kirk, K.L. (2002) *Proc. Natl. Acad. Sci. U.S.A.* **99**, 12477.
136. Naren, A.P., Nelson, D.J., Xie, W., Jovov, B., Pevsner, J., Bennett, M.K., Benos, D.J., Quick, M.W., and Kirk, K.L. (1997) *Nature* **390**, 302.
137. Bilan, F., Thoreau, V., Nacfer, M., Derand, R., Norez, C., Cantereau, A., Garcia, M., Becq, F., and Kitzis, A. (2004) *J. Cell. Sci.* **117**, 1923.

138. Peters, K.W., Qi, J., Watkins, S.C., and Frizzell, R.A. (1999) *Am. J. Physiol.* **277**, C174.
139. Bilan, F., Nacfer, M., Fresquet, F., Norez, C., Melin, P., Martin-Berge, A., Costa de Beauregard, M.A., Becq, F., Kitzis, A., and Thoreau, V. (2008) *Exp. Cell Res.* **314**, 2199.
140. Swiatecka-Urban, A., Boyd, C., Coutermarsh, B., Karlson, K.H., Barnaby, R., Aschenbrenner, L., Langford, G.M., Hasson, T., and Stanton, B.A. (2004) *J. Biol. Chem.* **279**, 38025.
141. Swiatecka-Urban, A., Talebian, L., Kanno, E., Moreau-Marquis, S., Coutermarsh, B., Hansen, K., Karlson, K.H., Barnaby, R., Cheney, R.E., Langford, G.M., Fukuda, M., and Stanton, B.A. (2007) *J. Biol. Chem.* **282**, 23725.
142. Ganeshan, R., Nowotarski, K., Di, A., Nelson, D.J., and Kirk, K.L. (2007) *Biochim. Biophys. Acta* **1773**, 192.
143. Riordan, J.R. (1999) *Am. J. Hum. Genet.* **64**, 1499.
144. Kartner, N., Augustinas, O., Jensen, T.J., Naismith, A.L., and Riordan, J.R. (1992) *Nat. Genet.* **1**, 321.
145. Mall, M., Kreda, S.M., Mengos, A., Jensen, T.J., Hirtz, S., Seydewitz, H.H., Yankaskas, J., Kunzelmann, K., Riordan, J.R., and Boucher, R.C. (2004) *Gastroenterology* **126**, 32.
146. Kreda, S.M., Mall, M., Mengos, A., Rochelle, L., Yankaskas, J., Riordan, J.R., and Boucher, R.C. (2005) *Mol. Biol. Cell* **16**, 2154.
147. Ostedgaard, L.S., Rogers, C.S., Dong, Q., Randak, C.O., Vermeer, D.W., Rokhlina, T., Karp, P.H., and Welsh, M.J. (2007) *Proc. Natl. Acad. Sci. U.S.A.* **104**, 15370.
148. Yang, Y., Janich, S., Cohn, J.A., and Wilson, J.M. (1993) *Proc. Natl. Acad. Sci. U.S.A.* **90**, 9480.
149. Turnbull, E.L., Rosser, M.F., and Cyr, D.M. (2007) *BMC Biochem.* **8**(1), S11.
150. Sharma, M., Benharouga, M., Hu, W., and Lukacs, G.L. (2001) *J. Biol. Chem.* **276**, 8942.
151. Swiatecka-Urban, A., Brown, A., Moreau-Marquis, S., Renuka, J., Coutermarsh, B., Barnaby, R., Karlson, K.H., Flotte, T.R., Fukuda, M., Langford, G.M., and Stanton, B.A. (2005) *J. Biol. Chem.* **280**, 36762.
152. Varga, K., Goldstein, R.F., Jurkuvenaite, A., Chen, L., Matalon, S., Sorscher, E.J., Bebok, Z., and Collawn, J.F. (2008) *Biochem. J.* **410**, 555.

153. Lewarchik, C.M., Peters, K.W., Qi, J., and Frizzell, R.A. (2008) *J. Biol. Chem.* **283**, 28401.
154. Bertrand, C.A. and Frizzell, R.A. (2003) *Am. J. Physiol. Cell Physiol.* **285**, C1.
155. Chappe, F., Loewen, M.E., Hanrahan, J.W., and Chappe, V. (2008) *J. Pharmacol. Exp. Ther.* **327**, 226.
156. Howard, M., Jiang, X., Stolz, D.B., Hill, W.G., Johnson, J.A., Watkins, S.C., Frizzell, R.A., Bruton, C.M., Robbins, P.D., and Weisz, O.A. (2000) *Am. J. Physiol., Cell Physiol.* **279**, C375.
157. Boucher, R.C. (2007) *Trends Mol. Med.* **13**, 231.
158. Galietta, L.J. (2009) *Biophys. J.* **97**, 3047.
159. Duran, C., Thompson, C.H., Xiao, Q., and Hartzell, H.C. (2010) *Annu. Rev. Physiol.* **72**, 95.
160. Van Goor, F., Hadida, S., Grootenhuis, P.D., Burton, B., Cao, D., Neuberger, T., Turnbull, A., Singh, A., Joubran, J., Hazlewood, A., Zhou, J., McCartney, J., Arumugam, V., Decker, C., Yang, J., Young, C., Olson, E.R., Wine, J.J., Frizzell, R.A., Ashlock, M., and Negulescu, P. (2009) *Proc. Natl. Acad. Sci. U.S.A.* **106**, 18825.
161. Kerem, E. (2006) *Paediatr. Respir. Rev.* **7**(1), S166.
162. Kerem, E., Hirawat, S., Armoni, S., Yaakov, Y., Shoseyov, D., Cohen, M., Nissim-Rafinia, M., Blau, H., Rivlin, J., Aviram, M., Elfring, G.L., Northcutt, V.J., Miller, L.L., Kerem, B., and Wilschanski, M. (2008) *Lancet* **372**, 719.
163. Nissim-Rafinia, M. and Kerem, B. (2006) *Prog. Mol. Subcell. Biol.* **44**, 233.
164. Mueller, C. and Flotte, T.R. (2008) *Clin. Rev. Allergy Immunol.* **35**, 164.
165. Flotte, T.R. (2007) *J. Cell. Physiol.* **213**, 301.
166. Terheggen-Lagro, S.W., Rijkers, G.T., and van der Ent, C.K. (2005) *J. Cyst. Fibros.* **4**(2), 15.
167. Conese, M., Copreni, E., Di Gioia, S., De Rinaldis, P., and Fumarulo, R. (2003) *J. Cyst. Fibros.* **2**, 129.
168. Boussif, O., Lezoualc'h, F., Zanta, M.A., Mergny, M.D., Scherman, D., Demeneix, B., and Behr, J.P. (1995) *Proc. Natl. Acad. Sci. U.S.A.* **92**, 7297.
169. Ziady, A.G., Kelley, T.J., Milliken, E., Ferkol, T., and Davis, P.B. (2002) *Mol. Ther.* **5**, 413.

170. Ziady, A.G., Gedeon, C.R., Miller, T., Quan, W., Payne, J.M., Hyatt, S.L., Fink, T.L., Muhammad, O., Oette, S., Kowalczyk, T., Pasumarthy, M.K., Moen, R.C., Cooper, M.J., and Davis, P.B. (2003) *Mol. Ther.* **8**, 936.
171. Lee, J.H., Zabner, J., and Welsh, M.J. (1999) *Hum. Gene Ther.* **10**, 603.
172. Brody, S.L., Metzger, M., Danel, C., Rosenfeld, M.A., and Crystal, R.G. (1994) *Hum. Gene Ther.* **5**, 821.
173. Crystal, R.G., McElvaney, N.G., Rosenfeld, M.A., Chu, C.S., Mastrangeli, A., Hay, J.G., Brody, S.L., Jaffe, H.A., Eissa, N.T., and Danel, C. (1994) *Nat. Genet.* **8**, 42.
174. Nakai, H., Yant, S.R., Storm, T.A., Fuess, S., Meuse, L., and Kay, M.A. (2001) *J. Virol.* **75**, 6969.
175. Flotte, T., Carter, B., Conrad, C., Guggino, W., Reynolds, T., Rosenstein, B., Taylor, G., Walden, S., and Wetzel, R. (1996) *Hum. Gene Ther.* **7**, 1145.
176. Flotte, T.R., Zeitlin, P.L., Reynolds, T.C., Heald, A.E., Pedersen, P., Beck, S., Conrad, C.K., Brass-Ernst, L., Humphries, M., Sullivan, K., Wetzel, R., Taylor, G., Carter, B.J., and Guggino, W.B. (2003) *Hum. Gene Ther.* **14**, 1079.
177. Moss, R.B., Rodman, D., Spencer, L.T., Aitken, M.L., Zeitlin, P.L., Waltz, D., Milla, C., Brody, A.S., Clancy, J.P., Ramsey, B., Hamblett, N., and Heald, A.E. (2004) *Chest* **125**, 509.
178. Li, W., Zhang, L., Johnson, J.S., Zhijian, W., Grieger, J.C., Ping-Jie, X., Drouin, L.M., Agbandje-McKenna, M., Pickles, R.J., and Samulski, R.J. (2009) *Mol. Ther.* **17**, 2067.
179. Sinn, P.L., Arias, A.C., Brogden, K.A., and McCray, P.B. Jr. (2008) *J. Virol.* **82**, 10684.
180. McKay, T., Patel, M., Pickles, R.J., Johnson, L.G., and Olsen, J.C. (2006) *Gene Ther.* **13**, 715.
181. Patel, M., Giddings, A., and Olsen, J. (2009) American Society of Gene Therapy: 12th Annual Meeting **17**, S30.
182. Pedemonte, N., Lukacs, G.L., Du, K., Caci, E., Zegarra-Moran, O., Galietta, L.J., and Verkman, A.S. (2005) *J. Clin. Invest.* **115**, 2564.
183. Van Goor, F., Straley, K.S., Cao, D., Gonzalez, J., Hadida, S., Hazlewood, A., Joubran, J., Knapp, T., Makings, L.R., Miller, M., Neuberger, T., Olson, E., Panchenko, V., Rader, J., Singh, A., Stack, J.H., Tung, R., Grootenhuis, P.D., and Negulescu, P. (2006) *Am. J. Physiol. Lung Cell Mol. Physiol.* **290**, L1117.

184. Robert, R., Carlile, G.W., Pavel, C., Liu, N., Anjos, S.M., Liao, J., Luo, Y., Zhang, D., Thomas, D.Y., and Hanrahan, J.W. (2008) *Mol. Pharmacol.* **73**, 478.
185. Yoo, C.L., Yu, G.J., Yang, B., Robins, L.I., Verkman, A.S., and Kurth, M.J. (2008) *Bioorg. Med. Chem. Lett.* **18**, 2610.
186. Wang, Y., Bartlett, M.C., Loo, T.W., and Clarke, D.M. (2006) *Mol. Pharmacol.* **70**, 297.
187. Loo, T.W., Bartlett, M.C., and Clarke, D.M. (2005) *Mol. Pharm.* **2**, 407.
188. Wang, Y., Loo, T.W., Bartlett, M.C., and Clarke, D.M. (2007) *Mol. Pharmacol.* **71**, 751.
189. Wang, Y., Loo, T.W., Bartlett, M.C., and Clarke, D.M. (2007) *Biochem. J.* **406**, 257.
190. Loo, T.W., Bartlett, M.C., Wang, Y., and Clarke, D.M. (2006) *Biochem. J.* **395**, 537.
191. Loo, T.W., Bartlett, M.C., and Clarke, D.M. (2008) *Biochem. J.* **413**, 29.
192. Grove, D.E. and Cyr, D.M. (2007) *Pediatr. Pulmonol.* **30**, 225.
193. Hutt, D.M., Herman, D., Rodrigues, A.P., Noel, S., Pilewski, J.M., Matteson, J., Hoch, B., Kellner, W., Kelly, J.W., Schmidt, A., Thomas, P.J., Matsumura, Y., Skach, W.R., Gentzsch, M., Riordan, J.R., Sorscher, E.J., Okiyoneda, T., Yates, J.R. 3rd, Lukacs, G.L., Frizzell, R.A., Manning, G., Gottesfeld, J.M., and Balch, W.E. (2010) *Nat. Chem. Biol.* **6**, 25.
194. Cholon, D.M., O'Neal, W.K., Randell, S.H., Riordan, J.R., and Gentzsch, M. (2010) *Am. J. Physiol. Lung Cell Mol. Physiol.* **298**, L304.
195. Mills, A.D., Yoo, C., Butler, J.D., Yang, B., Verkman, A.S., and Kurth, M.J. (2010) *Bioorg. Med. Chem. Lett.* **20**, 87.

Chapter 11

ABCC8: Physiology and Pathology*

David Ortiz, Yumiko Nakamura, Hiroshi Hashiguchi, Joseph Bryan and Lydia Aguilar-Bryan

1. Introduction

K_{ATP} channels are formed by a K^+ conducting, inwardly rectifying pore surrounded by four high or low affinity sulfonylurea receptors, SUR1 or SUR2, respectively, members of the ATP binding cassette (ABC) superfamily that contribute to the regulation of channel activity.[1–5] SURs belong to the "C" subfamily of ABC proteins whose members include the multidrug resistance-associated proteins (MRPs), poly-specific efflux pumps, and the cystic fibrosis transmembrane conductance regulator (CFTR), an ATP-gated chloride channel whose dysfunction causes cystic fibrosis, the most common recessive disease in Caucasians. SURs currently have no known transport activity, but are unique among ABC proteins in using ATP binding and/or hydrolysis to regulate the gating of an associated ion conducting pore.

K_{ATP} channels serve as metabolic sensors and modulate membrane electrical activity in response to changes in cell metabolism. The channels are involved in multiple physiological processes

* This work was supported by the Thrasher Research Fund (LA-B), ADA 1–10-BS-21 and NIH Grant DK84242 (JB)

that include a critical role in glucose stimulated insulin secretion from β-cells localized in the pancreatic islets of Langerhans and a protective role in cardiac myocytes during ischemia. Additionally, K_{ATP} channels are found in other tissues including skeletal muscle, vascular and non-vascular smooth muscle and brain.[1,6–9] This review focuses on the neuroendocrine channel, $(SUR1/K_{IR}6.2)_4$, mainly in the pancreatic islet, its pathological states, implications for treatment, and some of its biochemical and pharmacological properties.

2. Channel Topology and Assembly

SURs have the ABC "core": two transmembrane domains with six helices (TMD 1 and 2) each linked with a nucleotide-binding domain (NBD 1 and 2) all on a single polypeptide (Fig. 1). Along with several other members of the ABCC subfamily, SURs contain a third amino terminal transmembrane domain (TMD0) composed of five helices tethered to TMD1 by a cytosolic linker chain termed "L0".[10,11] TMD0 and L0 are the principal domains interacting with the K_{IR}.[12,13]

$K_{IR}6.2$ is a member of the potassium inward rectifier superfamily, named as such since they conduct K^+ ions into cells more efficiently than out. With a physiologic K^+ gradient, cytosolic $[K^+]_c$ greater than extracellular $[K^+]_o$, $K_{IR}6.2$ conducts an outward current. $K_{IR}6.2$ consists of two transmembrane helices, M1 and M2. M2 lines the permeation pathway in the K_{IR} tetramer and M1 is outward facing and accessible for interaction with the SUR subunit. Additionally, $K_{IR}6.1$ and $K_{IR}6.2$ are the only members of the K_{IR} family which require assembly with another protein to be active. A third helix, termed the "slide" helix, lies on the cytoplasmic face of the membrane at a right angle to M1 and M2 and plays a role in regulation of channel gating.[14]

The $(SUR1/K_{IR}6.2)_4$ channel is an obligate hetero-octamer in that each subunit contains an –RKR– ER-retention signal that only allows fully formed channels to exit the ER and traffic to the plasma membrane.[15] Channels are believed to begin as

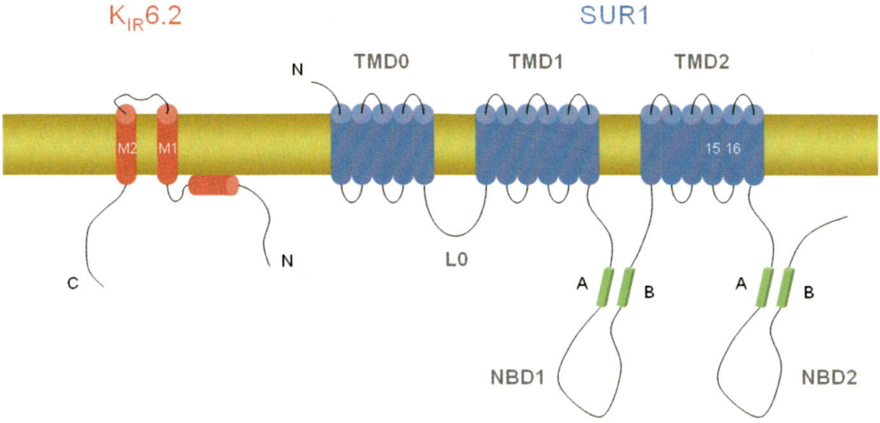

Fig. 1. SUR1 contains the classic core ABC domains: two nucleotide binding domains (NBD 1 and 2) each contributing Walker A and B motifs to two composite nucleotide binding pockets, as well as two transmembrane domains (TMD 1 and 2) with six helices each and a third amino-terminal domain, TMD0, with five helices connected by the cytoplasmic linker, L0. K_{IR}6.2 consists of two transmembrane helices, a third, "slide" helix that is involved in the regulation of channel gating and cytoplasmic amino and carboxyl termini that are involved in ATP binding. The binding pocket for sulfonylureas is believed to contain two subsites: site B includes L0 and KIR6.2 amino terminus and site A includes the cytosolic loop between helices 15 and 16.

SUR1-K_{IR}6.2 dimers, which form early to prevent degradation of K_{IR}, and then assemble to form hetero-octamers.[16] Deletion of the C-terminus of SUR1 prevents exit from the ER.[17]

3. Dual Regulation of K_{ATP} by Nucleotides

K_{ATP} channels are subject to complex and finely tuned regulation by nucleotides, phosphoinositides, long-chain acyl-CoA esters, Zn^{2+}, and other factors. While most of the known regulators of K_{ATP} channels either inhibit or stimulate the channel, nucleotides exhibit a dual regulation. Studies in which K_{IR} pores, lacking their ER retention signals, are expressed in the absence of SUR1 indicate that nucleotides inhibit the channel by binding directly to an adenine-selective, nucleotide-diphosphate binding pocket in $K_{IR}6.1$ and $K_{IR}6.2$ composed of residues from amino and carboxyl-termini of adjacent K_{IR} subunits.[12,18–22] Nucleotide inhibition is antagonized by both phosphoinositides and long-chain acyl-CoA esters interacting directly with the K_{IR}.[23–32] Co-expression of SUR1 has three main effects on the channel causing: (1) a MgATP-dependent *stimulation* of the channel, (2) an increase in the affinity of $K_{IR}6.2$ for inhibitory ATP, and (3) an increase in the maximum open probability (see Fig. 2). The first effect, the MgATP dependent stimulation via SUR1, implies the binding and/or hydrolysis of MgATP at the SUR1 NBDs induces a conformational change that stimulates the pore, however many aspects of this mechanism are unclear.

SUR1 contains two nucleotide binding pockets located at the NBD dimer interface, hence each NBD contributes residues to each binding pocket. In this review, NBD1 refers to the nucleotide binding pocket to which NBD1 contributes the Walker A and B motifs. Similar to other members of the ABCC subfamily, SUR1

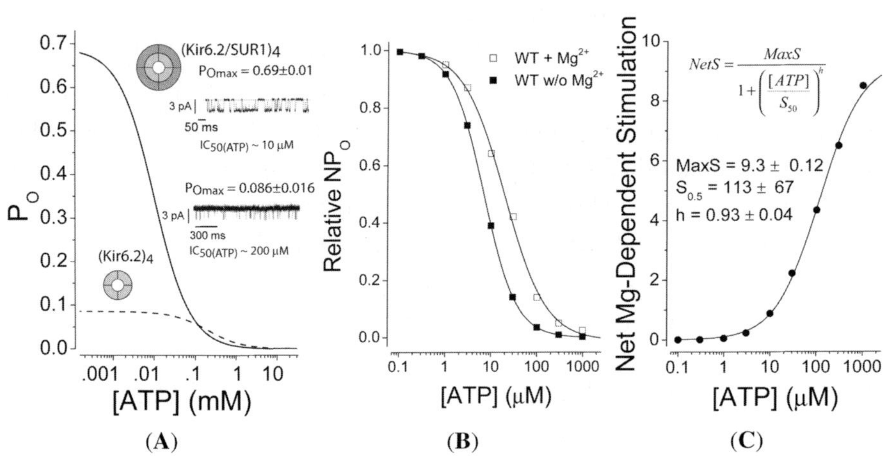

Fig. 2. Three actions of SUR1 on the Kir6.2 pore — **(A)** SUR1 increases the P_{Omax} and sensitivity of the pore to ATP^{4-}. The traces, values and curves are based on published data.[12,19] **(B)** Inhibition of K_{ATP} channels by ATP in the presence and absence of Mg^{2+}. **(C)** Net Mg-dependent, SUR1-dependent stimulation (Relative activity with Mg^{2+}/Relative activity without Mg^{2+} — 1) of the Kir6.2 pore derived from data in (B). The line is a fit to a modified Hill equation.

has asymmetric NBDs. NBD2 has a non-canonical signature sequence, FSQGQ instead of LSGGQ, and the Walker B motif of NBD1 has an aspartate next to the highly conserved D instead of a glutamate. This substitution is important because glutamate at this position is implicated in the catalytic activity of ABC proteins.[33] In agreement with this, photolabeling studies indicate that NBD1 of SUR1 is capable of binding ATP in the presence or absence of Mg^{2+}, but has little, or perhaps no, intrinsic ATPase activity, whereas NBD2 appears to bind and hydrolyze MgATP.[34–36] Both isolated NBDs fused to the maltose binding protein, either expressed alone or together, exhibit low ATPase activities.[37] However, whether both NBDs in the full-length receptor are able to hydrolyze ATP remains to be established.

How the catalytic cycle of SUR1 correlates with the kinetics of channel gating is not understood. There are initial reports on the K_m, 100 μM[37] and 400 μM[38] for SUR1 and a fusion of SUR1 with $K_{IR}6.2$, respectively, and V_{max}, 9 nmol P_i/min/mg[37] and 110 nmol P_i/min/mg,[38] for the ATPase activity of purified SUR1 in mixed detergent/lipid micelles. The ATP concentration at which 50% of K_{ATP} channel activity is inhibited (IC_{50}) is in the low (5–10) μM range[19] in the absence of Mg^{2+} required for ATP hydrolysis by SUR1, while higher concentrations, ~500 μM, are required for stimulation by SUR1.[39] Cytosolic factors that might affect the ATPase activity of SUR1 under physiologic conditions remain to be determined. The current working hypothesis is that the energy of ATP binding and hydrolysis cycles SUR1 through a series of distinct enzymatic intermediates (i.e., pre-hydrolytic, post-hydrolytic, etc.) and that one or more post-hydrolytic conformations act, via an undefined allosteric mechanism, to stabilize an open state of the pore. Studies on the cardiac K_{ATP} channel, SUR2A/$K_{IR}6.2$, argue that a MgADP bound post-hydrolytic intermediate of SUR2A stimulates K_{ATP} channel activity,[40–42] while pre-hydrolytic states do not. MgADP stimulates the activity of isolated channels,[43,44] presumably by product inhibition and stabilization of a post-hydrolytic conformation, thus K_{ATP} channels are considered to be sensors of the ATP/ADP ratio.

Mutations in either SUR1 or Kir6.2 that disrupt the balance between nucleotide-dependent stimulation and inhibition can result in tonic inhibition or hyper-activation of K_{ATP} channels and thus lead to over or under secretion of insulin, congenital hyperinsulinism (CHI) and neonatal diabetes mellitus (NDM), respectively. Mutations in SUR1, which cause a MgATP-dependent hyperstimulation of the channel, are a major cause of NDM. However, how these mutations affect the ATPase activity of SUR1 in order to hyperstimulate the channel is not understood. Assuming that a MgADP-bound post-hydrolytic intermediate(s) is stimulatory presents several possibilities. Little is known about SUR1 ATPase kinetics, but assuming catalysis is the rate limiting step, then a SUR1 NDM mutation could increase the catalysis rate and cause SUR1 to spend a greater proportion of time in a stimulatory post-hydrolytic conformation, thus hyperstimulating the pore. Alternatively, the dissociation rate of the ADP product could be decreased, slowing the overall turnover rate to produce an accumulation of SUR1 in the post-hydrolytic conformation and thus increase stimulation. Recent studies of two different SUR1-ND mutants using isolated NBD-MBP fusion constructs reach conclusions that support both possibilities.[45,46] Similarly, it is not clear how K_{ATP} openers (i.e., diazoxide) or inhibitors that bind directly to SUR1 affect its ATPase activity.

4. Pharmacology

SURs are the target for sulfonylureas, oral hypoglycemic agents that inhibit K_{ATP} channels and stimulate insulin release. Both NDM and Type 2 diabetes mellitus (T2DM) are treated with sulfonylureas, e.g., glibenclamide, while hyperinsulinism is commonly treated with K_{ATP} channel openers such as diazoxide, which also binds to SUR1. The mechanism by which the binding of drug on SUR1 affects gating of the K_{IR} pore remains unclear. One appealing hypothesis is that glibenclamide binds to and stabilizes one or more pre-hydrolytic conformations of SUR1 thus reducing channel activity; conversely channel openers like diazoxide bind to and stabilize the stimulatory, ADP-bound conformation.

The binding of sulfonylureas to SUR1 has been studied extensively. [^{3}H]-glibenclamide binds with high, subnanomolar affinity to both apo-SUR1 and SUR1 associated with ATP^{4-}, whereas MgATP has a negative allosteric effect on binding.[50–52] As cited above, NBD1 binds ATP^{4-}, K_D values in the low μM range,[53] similar binding with NBD2 requires Mg^{2+}. Based on results with other ABC proteins this suggests that the NBDs of SUR1 will "dimerize" poorly in ATP^{4-} and that it is the "inward facing" SUR1 conformations that bind [^{3}H]-glibenclamide with high affinity. MgATP can bind to both NBDs and cause the NBDs to dimerize and the transmembrane domains (TMDs) to rearrange from inward to outward facing orientations. Subsequent hydrolysis at NBD2 would produce the post-hydrolytic K_{ATP}-stimulatory conformation. [^{3}H]-glibenclamide binding studies are consistent with the hypothesis that the post-hydrolytic MgADP bound conformation and perhaps the pre-hydrolytic MgATP bound state, both "outward facing", have significantly reduced affinities for glibenclamide.[52] Structurally, this suggests that the rearrangement in the TMDs that results from NBD dimerization alters the binding pocket for glibenclamide to lower its affinity (Fig. 3). Whether hydrolysis is necessary to switch to the lower affinity conformation remains to be established. This switching from a high- to a low-affinity binding pocket is reminiscent of the change in affinity for transport substrates in other ABC proteins upon ATP binding and hydrolysis.

The sulfonylurea-binding pocket is a bipartite structure with two subsites: "A" and "B". Site A contains the cytosolic loop between helices 15 and 16, and site B contains L0 and the amino terminus of $K_{IR}6.2$.[3,54,55] K_{ATP} openers (e.g., diazoxide) require MgATP and hydrolysis to bind to SUR1.[56] Binding of openers does not occur in the presence of non-hydrolyzable analogs, and MgATP-dependent binding is abolished in the presence of saturating MgAMP-PNP, suggesting that openers bind well to the post- but not pre-hydrolytic states of SUR1. Two distinct binding pockets for openers have been demonstrated. Two residues in TM17 have

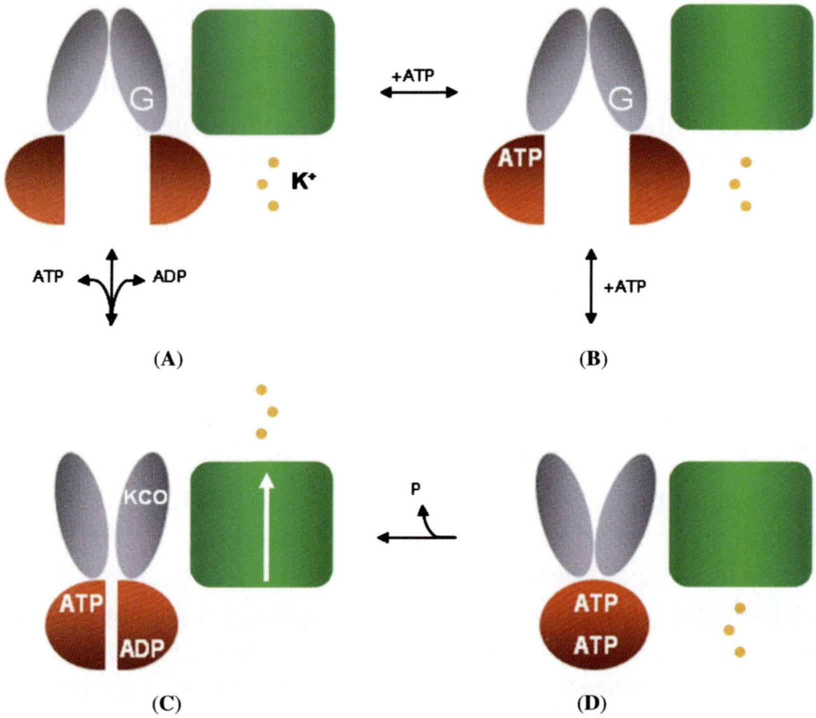

Fig. 3. Simplified scheme to correlate the SUR1 enzymatic intermediates with K_{ATP} channel activity and binding of drugs. (K_{IR} pore in green, K^+ ions in orange, SUR1 TMDs in blue and NBDs in red. Cartoons depict changes in the ABC core based on crystal structures of ABC exporters (i.e., Msba, PGP, SAV1866) in both "inward" and "outward" facing conformations.[47–49] Studies performed on the cardiac K_{ATP} channel argue that a post-hydrolytic MgADP bound intermediate (D) attains a conformation that stimulates the K_{IR} pore. Consistent with this argument, K_{ATP} channel openers (KCOs) such as pinacidil require MgATP to bind and act on the channel, suggesting they also interact with the post-hydrolytic intermediate. Conversely, glibenclamide, a K_{ATP} antagonist, binds with high affinity to the inward facing pre-hydrolytic states (**A, B**), but with lowered affinity to (**C**). The binding affinity of glibenclamide to the pre-hydrolytic outward facing state (**D**) has not been established.

been shown to be important for cyanoguanidine and benzopyran openers (i.e., pinacidil and cromakalim), which bind to SUR1 and SUR2 with low and high affinity, respectively. Mutating these two residues in SUR1 to their SUR2 counterparts (M1290T and

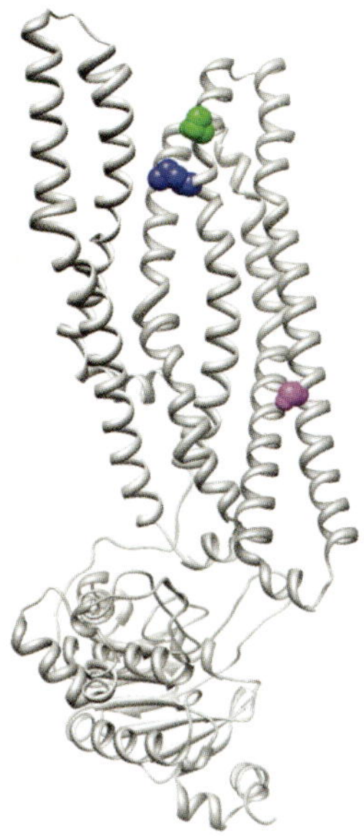

Fig. 4. TMD2 and NBD2 of SUR1 (homology model based on SAV1866.[47] Residue S1237 (magenta) on the cytoplasmic loop below TM helix 16 is important for glibenclamide binding and T1286L and M1290T (green and blue, respectively) at the top of helix 17 have been shown to be important for binding of pinacidil and cromakalim.

T1286L) confers SUR1 with higher sensitivity to these openers.[57,58] Figure 4 illustrates the location of these residues in relation to S1237, which has been shown to be important for glibenclamide binding.[59] Although its location is less clear, diazoxide appears to bind at a distinct site that includes TM8–11.[60,61] Since SUR1 is a member of the ABC superfamily, it is tempting to speculate that these drugs may mimic transport substrates and/or modulate ATP hydrolysis. The evidence that glibenclamide and diazoxide each preferentially bind distinct enzymatic intermediates of SUR1 suggests they may both inhibit ATPase activity, however their effects remain to be shown experimentally. The dissociation of the cyanoguanidine, P1075, from SUR2B is slow, estimated $t_{1/2} > 10$ minutes,[62,63] consistent with these potassium channel openers reducing the ATPase activity of K_{ATP} channels.

5. K_{ATP} Channels in the Pancreatic Islets of Langerhans

The islets of Langerhans are clusters of a few thousand cells, distributed all over the organ, that comprise the endocrine pancreas. The peptides glucagon, insulin, somatostatin and pancreatic polypeptide, secreted by the islet α-, β-, δ- and PP-cells, respectively, are the major hormones regulating glucose homeostasis. It is widely understood that the increase in the concentration of glucose in the blood stream during a meal prompts the secretion of insulin from islet β-cells. Insulin acts on multiple tissues, through binding to its cognate receptor, to increase the rates of glucose uptake and utilization and thus lower blood sugar levels following a meal. Similarly, glucagon is understood to have a reciprocal relation with insulin, counteracting its action and maintaining blood sugar levels between meals by increasing the production of glucose by the liver. Somatostatin is a potent inhibitor of the secretion of both insulin and glucagon,[64] but its role in glucose homeostasis is not yet as well defined. K_{ATP} channels have been described in α-, β- and δ-cells and implicated in the coordinated release of all three hormones in response to changes in blood glucose levels. The percentages of each islet cell type varies with species; human islets are ~70% β-cells, ~20% α-cells, <10% δ-cells and <5% PP-cells. Rat and mouse islets have a larger proportion of β-cells relative to α-cells and δ-cells. Small numbers of cells producing other glucostatic hormones, e.g., ε-cells capable of secreting ghrelin, a peptide hormone secreted by gastric cells during fasting to stimulate feeding, have been described,[65–67] but these findings are still controversial as GFP fluorescence was not detected in the pancreatic islets of ghrelin-GFP reporter mice.[68]

The regulation of islet hormone secretion is complex. Islets are highly innervated by both sympathetic and parasympathetic nerves; adrenaline acts as a potent stimulator of glucagon secretion and an inhibitor of insulin release, while acetylcholine stimulates insulin secretion.[69,70] For the purpose of this review on *ABCC8* we focused on local control within the islet, where K_{ATP} channels participate in regulating the secretion of insulin, glucagon and somatostatin.

The paracrine suppression of glucagon secretion by insulin during periods of hyperglycemia,[71] the "intra-islet insulin hypothesis", provides a rationale for the reciprocal relation between insulin and glucagon.[72–75] This idea was widely supported by early studies on the anatomy and flow of blood within rodent islets. The distribution of cells in a rodent islet is often quite striking with a core highly enriched with β-cells surrounded partially by a mantle of non-β-cells.[76,77] Elegant work on islet blood flow by Samols and colleagues[78,79] were consistent with a core-to-mantle flow ordered $\beta \rightarrow \alpha \rightarrow \delta$. Studies by Nyman *et al.*[80] using dynamic *in vivo* imaging techniques have confirmed the core-to-mantle flow and demonstrated an additional top-to-bottom flow. Recent studies from several laboratories[77,81,82] comparing rodent and human islets strongly suggest a more heterogeneous distribution of cells within human islets. These studies clearly show human islets don't have a distinct β-cell core, that islet cells tend to cluster about blood vessels, and that all three cell types are intermingled and in close contact. This organization suggests paracrine interactions with neighboring cells may play a greater role than the microcirculation in human islets. Additionally, studies by de Heer and Holst[83] and by de Heer *et al.*[84] using the isolated perifused rat pancreas imply that paracrine interactions in rodent islets may be more subtle than expected since stimulation of somatostatin release by the gut hormone, GLP-1, was able to suppress glucagon release, an effect blocked by somatostatin receptor antagonists.

6. K_{ATP} Channels in Insulin-Secreting β-Cells

K_{ATP} channels were identified in the early 1980s in cardiac myocytes by Noma[85] and by Trube and Hescheler[86] and in β-cells by Cook and Hales[87] as K^+ currents whose activity was reduced by application of ATP to the cytoplasmic side of a membrane patch. Subsequent studies showed that the β-cell channels were inhibited by sulfonylureas[88] and that under physiologic conditions, MgADP would antagonize the inhibitory action of ATP.[43,89] This suggested that K_{ATP} channels couple V_m with the cellular energy status via their response to the ATP/ADP ratio. A consensus view has developed to explain how a glucose stimulus is coupled with insulin secretion.[90–92] During euglycemia, when extracellular glucose is low (<5 mM) β-cells are "resting" and their cytosolic ATP, $[ATP]_c$, is estimated to be in the low millimolar range,[93,94] well above the low micromolar IC_{50} values for inhibition of K_{ATP} channels in isolated membrane patches.[19,95] The majority of K_{ATP} channels are silent, however openings of a small number of channels are considered sufficient, together with other currents, to maintain the membrane potential near –65 mV.[96] This is below the threshold, approximately –40 mV,[91] needed for the opening of L-type voltage-gated Ca^{2+} channels important for insulin secretion, thus cytoplasmic Ca^{2+} levels are low and insulin release is limited during hypo- or euglycemic periods. During a meal or snack blood glucose levels increase. Glucose is transported into β-cells via a glucose transporter, GLUT2 in rodents, and converted to glucose-6-phosphate by glucokinase.[97] Further metabolism increases the β-cell ATP/ADP ratio significantly.[93,98] This rise increases the inhibitory action of nucleotides on $K_{IR}6.2$ while decreasing stimulation via SUR1. The net effect is a reduction in K_{ATP} channel openings, which allows V_m to increase and voltage-gated L-type Ca^{2+} channels to open and stimulate insulin secretion. The consensus view is supported by both loss- and gain-of-function mutations in K_{ATP} subunits described below that result in HI and NDM, but presents an oversimplification that does not describe the complexity of β-cell responsiveness. The membrane depolarization produces Ca^{2+}-dependent action potentials and initiates slower oscillations in V_m, $[Ca^{2+}]_c$, and insulin release[99] that are coordinated among β-cells presumably by cell-to-cell interactions[100,101] and possibly paracrine factors.[102] K_{ATP} channels have been suggested to aid the repolarization of V_m and thus function during the slow oscillations,[103–107] however, β-cells from Sur1KO mice that lack SUR1/K_{IR}6.2-type K_{ATP} channels, exhibit membrane potential and $[Ca^{2+}]_c$ oscillations, even in low glucose when control β-cells are quiescent.[108] Further studies are necessary to determine whether there is a compensatory up-regulation of a repolarizing current in the SUR1KO islets. The consensus view implies a graded decrease in K_{ATP} channel activity over an extended range of blood sugar concentrations. In this regard, there are changes in the β-cell ATP/ADP ratio to ~10 mM, but little change under hyperglycemic conditions, 10-to-20 mM. Several studies suggest K_{ATP} channels should be largely closed above euglycemic conditions,[109] where the rate of insulin secretion is increased. The greater insulin secretion may reflect the augmentation of insulin secretion by glucose described by Henquin and others[90,92] but has also been suggested to reflect the activation of other metabolically sensitive currents, e.g., chloride currents.[110,111] Similarly, sulfonylureas stimulate insulin secretion during hyperglycemia when K_{ATP} channel activity is attenuated. The con-sensus view attributes this response to a further decrement in K_{ATP} channel activity, but activation of other depolarizing currents has been reported.[112]

7. K_{ATP} in Glucagon-Secreting α-Cells

SUR1/K_{IR}6.2-type K_{ATP} channels have been identified in the alpha TC-6 cell line[113,114] and α-cells in islets.[115,116] In contrast to the consensus view of the role K_{ATP} channels play in β-cells, their function in α-cells is controversial. Four hypotheses, three of which utilize K_{ATP} channels, have been put forward to account for the increase in glucagon secretion in response to hypoglycemia; (1) central modulation of glucagon secretion via parasympathetic and sympathetic signaling,[117] (2) the intra-islet insulin or paracrine hypothesis,[71,73] (3) the direct sensing of glucose by α-cells[115,118–120] and (4) autocrine stimulation via the co-release of glutamate with glucagon,[121] where glutamate activates AMPA/kainate-type ionotropic glutamate receptors (iGluRs), non-selective cation channels, resulting in membrane depolarization, and thus increased influx of Ca^{2+} to stimulate glucagon exocytosis.

7.1 *Central modulation of glucagon secretion*

A fall in peripheral blood glucose level below 3 mM is rapidly detected by glucose-depolarized neurons in the CNS. A reduction in firing rate produces a "counter-regulatory" response to mobilize the release of glucagon, catecholamines, and on a slower time scale, cortisol and growth hormone.[122–125] This counter-regulatory response enhances the production of glucose from the liver in an effort to restore euglycemia. Although our focus is on the islet, the same proteins, GLUT-2, glucokinase and SUR1/K_{IR}6.2-type channels present in β-cells are also found in glucose-excited (glucose-depolarized) neurons in the hypothalamus and are proposed to function in a manner that broadly parallels their function in islets.[126] During eu- and hyperglycemia, glucose uptake, metabolism and channel closure leads to hypothalamic neuronal activity, which must suppress glucagon release from pancreatic α-cells. Reducing glucose metabolism activates hyperpolarizing K_{ATP} channels, which reduces the activity of these glucose-excited neurons to trigger glucagon release. The observation that glucose-excited neurons in the ventromedial hypothalamus (VMH) of knockout mice lacking SUR1/K_{IR}6.2-type channels fail to hyperpolarize and can maintain their firing rate during periods of hypoglycemia supports this mechanism.[127]

7.2 *The intra-islet insulin or paracrine hypothesis*

Maruyama *et al.*[71] were one of the first to propose that the reciprocal relation between insulin and glucagon secretion might be a reflection of paracrine suppression of α-cells by β-cells. This idea was supported by studies on the distribution of islet cells and the islet microcirculation. Subsequent studies showed that several β-cell secretory products, i.e., insulin, Zn^{2+}, and gamma-amino butyric acid (GABA) could suppress glucagon secretion under different conditions. GABA, co-released with insulin, is reported to activate $GABA_A$ receptors, Cl^- channels, and hyperpolarize α-cells thus limiting glucagon secretion.[128] This effect is augmented by insulin, which up-regulates $GABA_A$ receptors to the α-cell surface.[129] Insulin itself is reported to suppress α-cell secretion via activation of K_{ATP} channels[130] and reported to activate K_{ATP} channels in selected neurons.[131,132] Zn^{2+}, co-released with insulin, is reported to suppress glucagon secretion under conditions of hypoglycemia. Wollheim and colleagues[133,134] studied the responsiveness of isolated rat α-cells and showed that when taken from the context of other islet cell types, increasing glucose metabolism stimulated the release of glucagon. Somewhat surprisingly, the glucose-stimulated glucagon secretion from isolated α-cells was analogous to glucose-stimulate insulin release from β-cells. Sulfonylureas stimulated and diazoxide, a K_{ATP} channel opener, inhibited glucagon secretion. Interestingly, the basal levels of glucagon secretion were significantly elevated, consistent with elimination of tonic suppression attributed to loss of junctional contacts.[135] The application of insulin had a small inhibitory effect on glucagon release, while the application of Zn^{2+} markedly inhibited release. The idea that Zn^{2+} was a potential regulator of glucagon release was supported by "switch-off" experiments using an *in vivo* perifused rat pancreas preparation where the simultaneous decrease in glucose and Zn^{2+}, but not glucose alone or Zn^{2+}-free insulin, prompted glucagon secretion.[136] In these experiments β-cells were destroyed by prior treatment with streptozotocin to eliminate endogenous insulin. It was suggested that the Zn^{2+} effect was specific to rats,[137] but subsequent work has confirmed the findings using mouse islets.[138,139] It is worth pointing out the requirement for a drop in either Zn^{2+} (or Zn^{2+}-insulin) *and* glucose to observe the increase in glucagon release suggests glucose exerts an additional suppressive action through an unspecified mechanism which is absent in the isolated α-cell experiments.[133,134] How Zn^{2+} acts to suppress glucagon release is unresolved. Zn^{2+} is reported to stimulate the activity of K_{ATP} channels partially inhibited by nucleotides.[140–142] Two histidine residues on the extracellular side of TM helix 6 of SUR1 are important for this Zn^{2+}-dependent activation of K_{ATP} (rodent residues, corresponding to His326 and Asp332 in human).[140] It is worth reiterating that, as with insulin exocytosis, glucagon release requires Ca^{2+} and the suppressive effects of GABA, insulin, and Zn^{2+} on α-cell function are all reported to involve membrane hyperpolarization that limits opening of Ca^{2+} channels, in short, it is this mechanism by which α-cells are *hyperpolarized* under hyperglycemic conditions.

7.3 *Direct sensing of glucose by α-cells*

In contrast to the glucose-mediated stimulation of glucagon release from isolated α-cells reported by Franklin *et al.*,[134] early reports demonstrated action potentials and oscillations in membrane potential and $[Ca^{2+}]_c$ in α-cells under hypoglycemic conditions that was suppressed by increasing the concentration of glucose[143,144] and multiple reports indicate glucose attenuates release from α-cells in isolated islets.[115,118–120,145] The mechanism of attenuation is unresolved. Gylfe and co-workers[118,119,146] have presented an extensive series of experiments supporting a role for a depolarizing, Ca^{2+}-store operated current. In this hypothesis, an increase in peripheral glucose and α-cell glucose metabolism results in sequestration of Ca^{2+} within the endoplasmic reticulum and inhibition of the depolarizing current as Ca^{2+} stores fill. This mechanism operates independently of paracrine effects from β- and δ-cells and K_{ATP} channels play a minor role. Lowering glucose presumably leads to depletion of Ca^{2+} stores with activation of the depolarizing current and subsequent activation of voltage-dependent Na^+ and Ca^{2+} channels, including L-type Ca^{2+} channels.[118]

Göpel *et al.*[115] and later MacDonald *et al.*[120] proposed an alternative model to explain glucose inhibition without paracrine action. In this model, action potentials and oscillations in V_m and $[Ca^{2+}]_c$ occur only within a limited "window" of membrane potentials. In contrast to β-cells, the ATP/ADP ratio of α-cells is approximately constant over a wide range of glucose concentrations[93] and is sufficient to reduce K_{ATP} channel activity. Hyperpolarizing α-cells by opening channels with diazoxide abolishes Ca^{2+} oscillations and reduces glucagon secretion.[115,120] Similarly, depolarizing α-cells by closing K_{ATP} channels with tolbutamide has a similar effect. This window behavior is suggested to result from the expression and usage of a different complement of ion channels in α- vs β-cells. For example, the N-type Ca^{2+} channels important for glucagon secretion are activated at lower potentials than L-type channels, but unlike L-type channels they inactivate at the depolarized membrane potentials of α-cells during hyperglycemia or in tolbutamide. In contrast to the paracrine model, in the window model α-cells are depolarized, but $[Ca^{2+}]_c$ is low because Ca^{2+} channels are inactivated. In support of this, opening channels, presumably in α-cells, with low concentrations of diazoxide stimulates glucagon secretion.[115,120] Given the apparent stability of the ATP/ADP ratio in α-cells[93,98] it is not understood in this model what intermediate of glucose metabolism acts to reduce K_{ATP} channel activity during hyperglycemia and thus depolarize α-cells sufficiently to reduce $[Ca^{2+}]_c$ and glucagon secretion. It is also worth noting that glucagon secretion, under hyperglycemic conditions, is strongly stimulated by amino acids. In a physiologic setting this is usually thought to counteract the hypoglycemic action of insulin during a meal and set a "floor" under blood sugar levels. The uptake of many amino acids is electrogenic via co-transport with Na^+ and thus would be expected to depolarize α-cells further. Arginine infusion, for example, is used clinically as a potent stimulator of glucagon secretion.

8. K_{ATP} in Somatostatin Secreting δ-Cells

Somatostatin is produced by endocrine cells in a variety of tissues, particularly the CNS and is central to the control of growth hormone regulation. Somatostatin is also produced and secreted locally by islet δ-cells in response to increased glucose metabolism. Somatostatin (SST) is a potent inhibitor of both insulin and glucagon release.[147–150] Both type 1 and type 2 diabetic patients have impaired glucose-stimulated SST secretion. $SST^{-/-}$ mice lacking SST show enhanced insulin and glucagon secretory responses upon stimulation with arginine. The lack of SST eliminates the inhibition of glucagon release by glucose suggesting somatostatin exerts a tonic inhibition on both insulin and glucagon secretion. K_{ATP} channels have been identified in δ-cells using electrophysiology[151,152] and by antibody staining.[153] Like insulin, SST secretion is stimulated by glucose and by sulfonylureas *in vivo*.[145,154,155] More recent studies indicate both K_{ATP} channel-dependent and independent pathways are involved.[156] The role of K_{ATP} channels in controlling SST secretion has been proposed to be similar to the consensus view for β-cells.[152] Secretion is increased by lowering channel activity with tolbutamide or rising glucose and is reduced by opening channels with diazoxide.[145,156] A recent study indicates that while K_{ATP} channels play a role in the control of SST secretion under hypoglycemic conditions, <3 mM glucose, and their closure is needed for exocytosis, at high glucose concentrations a Ca^{2+}-induced Ca^{2+} release mechanism predominates to augment secretion.[157]

9. Summary of Intra-Islet Regulation

Figure 5 summarizes the interactions between islet cells and where K_{ATP} channels are involved. Increases in glucose metabolism prompt closure of K_{ATP} channels in both β- and δ-cells stimulating secretion of both insulin and somatostatin. Both hormones have been proposed to act as paracrine suppressors of α-cells, reducing glucagon secretion when blood glucose levels rise. Insulin itself, and Zn^{2+}, co-secreted with insulin, have been proposed to hyperpolarize α-cells via activation of K_{ATP} channels. Somatostatin has been argued to activate K_{ATP} channels, but activation of G-protein coupled GIRK channels via binding to

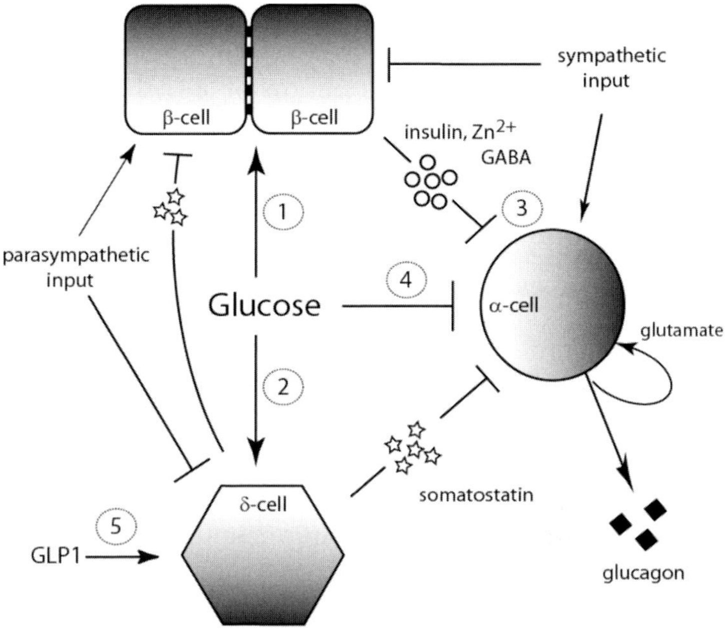

Fig. 5. Summary of intra-islet interactions. The autonomic system provides CNS input to maintain glucose homeostasis. (1, 2) Glucose metabolism stimulates closure of K_{ATP} channels in both β- and δ-cells to stimulate hormone release. (3) Zn^{2+} and insulin have both been suggested to *hyperpolarize* α-cells by *opening* channels and thus limiting glucagon release during hyperglycemia. (4) Alternatively, *closure* of K_{ATP} channels in high glucose, depolarization of α-cells and inactivation of Na^+ and Ca^{2+} channels are key components of the window hypothesis to explain inhibition of glucagon release during hyperglycemia. Somatostatin is a powerful inhibitor of both insulin and glucagon secretion. A recent report suggests GLP1 may indirectly reduce glucagon release by increasing somatostatin secretion. (5) GLP1 has been proposed to modulate K_{ATP} channels in other cell types including β-cells.[158]

somatostatin receptors is a more viable mechanism and long-lasting somatostatin analogs, e.g., octreotide, can suppress insulin release in cases of HI. GABA has been proposed to hyperpolarize α-cells via activation of $GABA_A$ chloride channels, a process amplified by the action of insulin to move $GABA_A$ receptors to the plasma membrane. K_{ATP} channels are also a key component of the "window" hypothesis, although in contrast to the paracrine suppression mechanisms, channels are closed in elevated glucose and α-cells are depolarized. A recent study by de Heer *et al.*[84] suggests that the action of GLP1 (glucagon-like peptide-1) to reduce glucagon secretion is mediated via an increase in somatostatin secretion and consequent reduction in glucagon release.

10. Disease States

Although monogenic forms of diabetes mellitus have genetically heterogeneous etiologies and are rare, accounting for only 1–2% of the total cases of diabetes, they have been extremely useful in adding to our understanding of the molecular mechanisms of abnormal pancreatic development, β-cell dysfunction and β-cell destruction or apoptosis. Knowing the genetic etiology in patients with Maturity Onset Diabetes of the Young (MODY), Mitochondrial Diabetes and Neonatal Diabetes (NDM) allows for the proper diagnosis and treatment, as well as screening of family members for genetic counseling.[159,160]

As shown in Fig. 6, loss of function mutations in SUR1/*ABCC8* and Kir6.2/*KCNJ11* reduce K_{ATP} channel activity. This results in closure of the channel, a persistent depolarization of the β-cell membrane and persistent openings of L-type voltage-operated Ca^{2+} channels that causes unregulated insulin secretion. Conversely, gain-of-function mutations that result in overactive K_{ATP} channels are a common cause of NDM. These mutations exert the opposite effect, K_{ATP} channels are more difficult to close in response to metabolically generated changes in adenine nucleotides. Thus β-cells remain hyperpolarized when blood glucose levels rise, L-type voltage-operated Ca^{2+} channels remain closed and Ca^{2+} influx and insulin secretion are limited.

11. Hyperinsulinemic Hypoglycemia of Infancy

Although congenital hyperinsulinism (CHI) is the most frequent cause of hypoglycemia presenting in the first few hours, days or

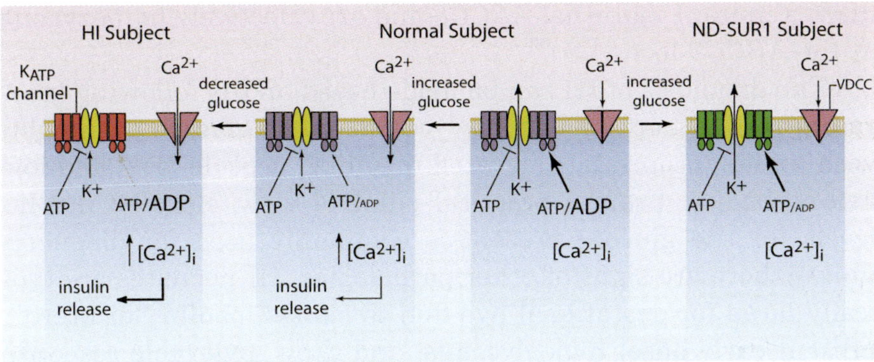

Fig. 6. Loss-of-function mutations (HI subject) cause more K_{ATP} channels to remain closed leading to dysregulation of insulin secretion. Gain-of-function mutations (ND subject) cause more K_{ATP} channels to remain open resulting in decreased insulin secretion.

weeks after birth, it is a rare autosomal recessive or dominant disease with a world prevalence of 1:30,000 to 1:50,000, that can be as frequent as 1:2,500 in some consanguineous populations, and as low as 1 in 70,000 as reported by Sandal,[161] based on mutational screening in Norway over the last decade. HI is characterized by excess insulin secretion relative to the degree of hypoglycemia. Even though other genes have been identified as causes of HI, e.g., glucokinase, glutamate dehydrogenase and the mitochondrial enzyme short-chain 3-hydroxyacyl-CoA dehydrogenase, mutations in *ABCC8* and *KCNJ11* account for ~50% of HI cases. To prevent brain damage due to recurrent hypoglycemia prompt diagnosis and treatment is important and requires a team of physicians, surgeons and other health care specialists. More than 150 mutations in *ABCC8* have been reported.[162] These mutations cause loss of channel function by reducing the number of K_{ATP} channels in the plasma membrane secondary to abnormal assembly and/or trafficking[16,17] by reducing channel activation by MgADP and/or MgATP[163,164] or by loss of heterogeneity at the paired *ABCC8*/*KCNJ11* locus. In the latter cases the paternal *ABCC8* allele is lost through a separate mutation or chromosomal rearrangement. Affected individuals carry a mutant maternal *ABCC8* and are effectively homozygous at the *ABCC8* locus.[165,166]

The diagnosis of HI can be made based on the following laboratory results: recurrent hypoglycemia (blood glucose <3 mmol/l) with an inappropriately elevated insulin/C-peptide level and low ketone bodies. From a practical point of view, elevated insulin levels for the glucose present is commonly used for diagnosis unless there are signs of other pathologies. HI neonates are typically large for gestational age due to excess insulin "in utero", present early onset hypoglycemia, and show a variable response to the clinical treatment that is available using K_{ATP} channel openers that result in inhibition of insulin release.

The high degree of hypoglycemia requires high rates of intravenous glucose infusion (>15 mg/kg/min) to maintain normoglycemia. The first choice of treatment is diazoxide in an attempt to stimulate K_{ATP} channel openings. Non-responsive patients are candidates for a second treatment using the long-lasting somatostatin analog, octreotide in an effort to suppress insulin release via activation of G-protein coupled K^+ channels. Infants who fail to respond to either diazoxide or octreotide become candidates for surgical intervention, a focal resection or a sub-total pancreatectomy, or are fed a high carbohydrate diet to maintain euglycemia.[167–171]

HI cases resulting from loss-of-heterozygosity (LOH) manifest distinctive pancreas histology versus those due to diffuse loss of K_{ATP} channel function. In the LOH or "focal" form, patients present a paternal mutation and loss of heterozygosity of the 11p15 region that includes the maternal alleles of ABCC8, IGF2, and CDKN1C. This focal β-cell expansion occurs in 40–60% of HI cases and produces foci of adenomatous hyperplasia within the normal pancreas. Islet cells within the focal lesion have abnormal or enlarged islet cell nuclei missing in adjacent tissue. In comparison, in the diffuse case, all of the islets throughout the pancreas are affected and are characterized by hypertrophied β-cells with large nuclei.[172,173] The histological distinction between focal and diffuse cases is important because the surgical management is completely different.[170] Successful excision of the focal lesion in cases with localized hyperplasia cures the patient and restores normal function, whereas patients with the diffuse form who do not respond to medical therapy may require a total or sub-total pancreatectomy that results in a form of insulin dependent diabetes mellitus. Although genetic analysis of the *ABCC8/KCNJ11* locus can be used to distinguish diffuse, i.e., homozygous/compound heterozygous mutations in *ABCC8* and *KCNJ11*, from focal lesions,[165,174] it is important to be aware that dominant forms can also result in more severe phenotypes.

Ribeiro *et al.*[175] used ^{18}F-fluoro-L-DOPA-PET as a non-invasive technique to distinguish between focal and diffuse forms of HI. Patients with a focal lesion can benefit from this technique,[176–179] which allows a limited resection of the lesion without the long-term risk of diabetes. Additionally, surgical resection by laparoscopy has been successful with excellent outcome and faster recovery.[168,171,180]

12. Neonatal Diabetes Mellitus

Neonatal diabetes mellitus (NDM) is defined as hyperglycemia occurring within the first six to nine months of life that lasts for at least two weeks and requires insulin therapy.[181] NDM is rare, an incidence of approximately 1/300,000–1/500,000 live births.[182–184] Recently, referral rates for diagnostic testing show an incidence of permanent neonatal diabetes in three European countries of ~1 in 260,000 live births.[185] Clinically these cases can be transient (TNDM) or permanent (PNDM). Approximately 50% of the transient cases enter remission and maintain euglycemia, while the other half relapse after being in remission for several months and become permanent. The permanent cases do not go into remission[181,184,186] and require treatment with insulin for life. Mutations in both *KCNJ11*[162,187] and *ABCC8*[188,189] that cause hyperactive channels and thus lead to reduced insulin secretion and neonatal diabetes can be classified in broad categories, those that: desensitize $K_{IR}6.2$ to inhibitory ATP, cause a SUR1 MgATP-dependent hyperstimulation of the channel or increase the channel open probability (P_0). This is an active area of investigation, which we[186] and others[190] have reviewed recently.

13. Diagnosis and Implications for Therapy

Type 1 diabetes mellitus can present with the same symptoms as NDM at early ages and thus there is the possibility of misdiagnosis. Because it had been initially reported that early-onset diabetes reported before 180 days of age showed a greater tendency to have protective HLA antigens and less frequent autoimmune markers associated with type 1 diabetes,[191] six months of age was used as a provisional boundary until recently where older patients have been diagnosed. Making the correct diagnosis has immediate therapeutic benefits. Until recently patients with neonatal diabetes were treated with insulin to establish good metabolic control. This changed when Gloyn *et al.*[187] and Sagen *et al.*[192] identified the initial mutations and showed that NDM patients responded to oral sulfonylureas. This observation has now been confirmed by many researchers[188,193–196] and treatment with sulfonylureas has been extended to transient cases during the initial period with good results.[197] Although more than 95% of NDM cases respond to variable doses of the oral agents, there are some syndromic cases that do not respond.[186,196] However, large-scale studies with mutations in *KCNJ11*[196] and *ABCC8*[198] have shown improvement in glycemic control and it has also been reported that higher dose sulfonylurea treatment improves CNS symptoms.[199–201]

In conclusion, the knowledge we have recently acquired in the understanding of how basic cellular processes can be disrupted and result in diseases like neonatal diabetes and hyperinsulinemic hypoglycemia is one of the best examples of translational research. The identification of the specific β-cell defects which define the molecular genetic etiology in these patients, have not only enabled a new therapeutic approach that improves their quality of life in the case of neonatal diabetes but has given new information on the function that the high affinity sulfonylurea receptor plays as the regulatory subunit of the K_{ATP} channel.

References

1. Aguilar-Bryan, L. and Bryan, J. (1999) *Endocr. Rev.* **20**, 101.
2. Bryan, J. and Aguilar-Bryan, L. (1999) *Biochim. Biophys. Acta* **1461**, 285.
3. Bryan, J., Crane, A., Vila-Carriles, W.H., Babenko, A.P., and Aguilar-Bryan, L. (2005) *Curr. Pharm. Des.* **11**, 2699.
4. Bryan, J., Munoz, A., Zhang, X., Dufer, M., Drews, G., Krippeit-Drews, P., and Aguilar-Bryan, L. (2007) *Pflugers Arch.* **453**, 703.
5. Aittoniemi, J., Fotinou, C., Craig, T.J., de Wet, H., Proks, P., and Ashcroft, F.M. (2009) *Philos. Trans. R. Soc. Lond., B, Biol. Sci.* **364**, 257.
6. Seino, S. (2003) *J. Diabetes Complicat.* **17**, 2.
7. Seino, S. and Miki, T. (2003) *Prog. Biophys. Mol. Biol.* **81**, 133.
8. Yokoshiki, H., Sunagawa, M., Seki, T., and Sperelakis, N. (1998) *Am. J. Physiol.* **274**, C25.
9. Ashcroft, F.M. and Gribble, F.M. (1998) *Trends Neurosci.* **21**, 288.
10. Tusnády, G.E., Bakos, E., Varadi, A., and Sarkadi, B. (1997) *FEBS Lett.* **402**, 1.
11. Raab-Graham, K.F., Cirilo, L.J., Boettcher, A.A., Radeke, C.M., and Vandenberg, C.A. (1999) *J. Biol. Chem.* **274**, 29122.
12. Babenko, A.P. and Bryan, J. (2003) *J. Biol. Chem.* **278**, 41577.
13. Chan, K.W., Zhang, H., and Logothetis, D.E. (2003) *EMBO J.* **22**, 3833.
14. Doyle, D.A., Morais Cabral, J., Pfuetzner, R.A., Kuo, A., Gulbis, J.M., Cohen, S.L., Chait, B.T., and MacKinnon, R. (1998) *Science* **280**, 69.
15. Zerangue, N., Schwappach, B., Jan, Y.N., and Jan, L.Y. (1999) *Neuron* **22**, 537.
16. Crane, A. and Aguilar-Bryan, L. (2004) *J. Biol. Chem.* **279**, 9080.
17. Sharma, N., Crane, A., Clement, J.Pt., Gonzalez, G., Babenko, A.P., Bryan, J., and Aguilar-Bryan, L. (1999) *J. Biol. Chem.* **274**, 20628.
18. Tucker, S.J., Gribble, F.M., Zhao, C., Trapp, S., and Ashcroft, F.M. (1997) *Nature* **387**, 179.

19. Babenko, A.P., Gonzalez, G., Aguilar-Bryan, L., and Bryan, J. (1999) *FEBS Lett.* **445**, 131.
20. Tanabe, K., Tucker, S.J., Matsuo, M., Proks, P., Ashcroft, F.M., Seino, S., Amachi, T., and Ueda, K. (1999) *J. Biol. Chem.* **274**, 3931.
21. Antcliff, J.F., Haider, S., Proks, P., Sansom, M.S., and Ashcroft, F.M. (2005) *EMBO J.* **24**, 229.
22. Babenko, A.P. (2005) *J. Mol. Cell. Cardiol.* **39**, 79.
23. Hilgemann, D.W. (1997) *Annu. Rev. Physiol.* **59**, 193.
24. Fan, Z. and Makielski, J.C. (1999) *J. Gen. Physiol.* **114**, 251.
25. Shyng, S.L. and Nichols, C.G. (1998) *Science* **282**, 1138.
26. Baukrowitz, T. and Fakler, B. (2000) *Biochem. Pharmacol.* **60**, 735.
27. Baukrowitz, T., Schulte, U., Oliver, D., Herlitze, S., Krauter, T., Tucker, S.J., Ruppersberg, J.P., and Fakler, B. (1998) *Science* **282**, 1141.
28. Branstrom, R., Corkey, B.E., Berggren, P.O., and Larsson, O. (1997) *J. Biol. Chem.* **272**, 17390.
29. Haider, S., Tarasov, A.I., Craig, T.J., Sansom, M.S., and Ashcroft, F.M. (2007) *EMBO J.* **26**, 3749.
30. Schulze, D., Krauter, T., Fritzenschaft, H., Soom, M., and Baukrowitz, T. (2003) *J. Biol. Chem.* **278**, 10500.
31. Schulze, D., Rapedius, M., Krauter, T., and Baukrowitz, T. (2003) *J. Physiol. (Lond.)* **552**, 357.
32. Logothetis, D.E., Jin, T., Lupyan, D., and Rosenhouse-Dantsker, A. (2007) *Pflugers Arch.* **455**, 83.
33. Linton, K.J. and Higgins, C.F. (2007) *Pflugers Arch.* **453**, 555.
34. Ueda, K., Inagaki, N., and Seino, S. (1997) *J. Biol. Chem.* **272**, 22983.
35. Ueda, K., Komine, J., Matsuo, M., Seino, S., and Amachi, T. (1999) *Proc. Natl. Acad. Sci. U.S.A.* **96**, 1268.
36. Matsuo, M., Kimura, Y., and Ueda, K. (2005) *J. Mol. Cell. Cardiol.* **38**, 907.
37. de Wet, H., Mikhailov, M.V., Fotinou, C., Dreger, M., Craig, T.J., Venien-Bryan, C., and Ashcroft, F.M. (2007) *FEBS J.* **274**, 3532.
38. Mikhailov, M.V., Campbell, J.D., de Wet, H., Shimomura, K., Zadek, B., Collins, R.F., Sansom, M.S., Ford, R.C., and Ashcroft, F.M. (2005) *EMBO. J.* **24**, 4166.
39. Babenko, A.P. (2008) *J. Biol. Chem.* **283**, 8778.
40. Bienengraeber, M., Olson, T.M., Selivanov, V.A., Kathmann, E.C., O'Cochlain, F., Gao, F., Karger, A.B., Ballew, J.D., Hodgson, D.M.,

Zingman, L.V., Pang, Y.P., Alekseev, A.E., and Terzic, A. (2004) *Nat. Genet.* **36**, 382.
41. Zingman, L.V., Alekseev, A.E., Bienengraeber, M., Hodgson, D., Karger, A.B., Dzeja, P.P., and Terzic, A. (2001) *Neuron* **31**, 233.
42. Zingman, L.V., Hodgson, D.M., Bienengraeber, M., Karger, A.B., Kathmann, E.C., Alekseev, A.E., and Terzic, A. (2002) *J. Biol. Chem.* **277**, 14206.
43. Dunne, M.J. and Petersen, O.H. (1986) *FEBS Lett.* **208**, 59.
44. Hopkins, W.F., Fatherazi, S., Peter-Riesch, B., Corkey, B.E., and Cook, D.L. (1992) *J. Membr. Biol.* **129**, 287.
45. de Wet, H., Proks, P., Lafond, M., Aittoniemi, J., Sansom, M.S., Flanagan, S.E., Pearson, E.R., Hattersley, A.T., and Ashcroft, F.M. (2008) *EMBO Rep.* **9**, 648.
46. de Wet, H., Rees, M.G., Shimomura, K., Aittoniemi, J., Patch, A.M., Flanagan, S.E., Ellard, S., Hattersley, A.T., Sansom, M.S., and Ashcroft, F.M. (2007) *Proc. Natl. Acad. Sci. U.S.A.* **104**, 18988.
47. Dawson, R.J. and Locher, K.P. (2007) *FEBS Lett.* **581**, 935.
48. Ward, A., Reyes, C.L., Yu, J., Roth, C.B., and Chang, G. (2007) *Proc. Natl. Acad. Sci. U.S.A.* **104**, 19005.
49. Aller, S.G., Yu, J., Ward, A., Weng, Y., Chittaboina, S., Zhuo, R., Harrell, P.M., Trinh, Y.T., Zhang, Q., Urbatsch, I.L., and Chang, G. (2009) *Science* **323**, 1718.
50. Schwanstecher, M., Loser, S., Brandt, C., Scheffer, K., Rosenberger, F., and Panten, U. (1992) *Br. J. Pharmacol.* **105**, 531.
51. Schwanstecher, M., Loser, S., Rietze, I., and Panten, U. (1991) *Naunyn Schmiedebergs Arch. Pharmacol.* **343**, 83.
52. Hambrock, A., Loffler-Walz, C., and Quast, U. (2002) *Br. J. Pharmacol.* **136**, 995.
53. Matsuo, M., Kioka, N., Amachi, T., and Ueda, K. (1999) *J. Biol. Chem.* **274**, 37479.
54. Mikhailov, M.V., Mikhailova, E.A., and Ashcroft, S.J. (2001) *FEBS Lett.* **499**, 154.
55. Vila-Carriles, W.H., Zhao, G., and Bryan, J. (2007) *FASEB J.* **21**, 18.
56. Schwanstecher, M., Sieverding, C., Dorschner, H., Gross, I., Aguilar-Bryan, L., Schwanstecher, C., and Bryan, J. (1998) *EMBO J.* **17**, 5529.
57. Moreau, C., Jacquet, H., Prost, A.L., D'Hahan, N., and Vivaudou, M. (2000) *EMBO J.* **19**, 6644.

58. Hambrock, A., Kayar, T., Stumpp, D., and Osswald, H. (2004) *Diabetes* **53**(3), S128.
59. Ashfield, R., Gribble, F.M., Ashcroft, S.J., and Ashcroft, F.M. (1999) *Diabetes* **48**, 1341.
60. Babenko, A.P., Gonzalez, G., and Bryan, J. (2000) *J. Biol. Chem.* **275**, 717.
61. Dabrowski, M., Ashcroft, F.M., Ashfield, R., Lebrun, P., Pirotte, B., Egebjerg, J., Bondo Hansen, J., and Wahl, P. (2002) *Diabetes* **51**, 1896.
62. Manley, P.W., Loffler-Walz, C., Russ, U., Hambrock, A., Moenius, T., and Quast, U. (2001) *Br. J. Pharmacol.* **133**, 275.
63. Hambrock, A., Loffler-Walz, C., Kurachi, Y., and Quast, U. (1998) *Br. J. Pharmacol.* **125**, 577.
64. Schuit, F.C., Derde, M.P., and Pipeleers, D.G. (1989) *Diabetologia* **32**, 207.
65. Stefan, Y., Orci, L., Malaisse-Lagae, F., Perrelet, A., Patel, Y., and Unger, R.H. (1982) *Diabetes* **31**, 694.
66. Rahier, J., Goebbels, R.M., and Henquin, J.C. (1983) *Diabetologia* **24**, 366.
67. Wierup, N., Svensson, H., Mulder, H., and Sundler, F. (2002) *Regul. Pept.* **107**, 63.
68. Sakata, I., Nakano, Y., Osborne-Lawrence, S., Rovinsky, S.A., Lee, C.E., Perello, M., Anderson, J.G., Coppari, R., Xiao, G., Lowell, B.B., Elmquist, J.K., and Zigman, J.M. (2009) *Regul. Pept.* **155**, 91.
69. Dunning, B.E. and Taborsky, G.J., Jr. (1991) *Adv. Exp. Med. Biol.* **291**, 107.
70. Ahren, B. (2000) *Diabetologia* **43**, 393.
71. Maruyama, H., Hisatomi, A., Orci, L., Grodsky, G.M., and Unger, R.H. (1984) *J. Clin. Invest.* **74**, 2296.
72. Samols, E. and Stagner, J.I. (1988) *Am. J. Med.* **85**, 31.
73. Banarer, S., McGregor, V.P., and Cryer, P.E. (2002) *Diabetes* **51**, 958.
74. Gosmanov, N.R., Szoke, E., Israelian, Z., Smith, T., Cryer, P.E., Gerich, J.E., and Meyer, C. (2005) *Diabetes Care* **28**, 1124.
75. Raju, B. and Cryer, P.E. (2005) *Diabetes* **54**, 757.
76. Bonner-Weir, S. and Orci, L. (1982) *Diabetes* **31**, 883.
77. Kharouta, M., Miller, K., Kim, A., Wojcik, P., Kilimnik, G., Dey, A., Steiner, D.F., and Hara, M. (2009) *Diabetes Res. Clin. Pract.* **85**, 252.
78. Samols, E., Stagner, J.I., Ewart, R.B., and Marks, V. (1988) *J. Clin. Invest.* **82**, 350.

79. Stagner, J.I., Samols, E., and Bonner-Weir, S. (1988) *Diabetes* **37**, 1715.
80. Nyman, L.R., Wells, K.S., Head, W.S., McCaughey, M., Ford, E., Brissova, M., Piston, D.W., and Powers, A.C. (2008) *J. Clin. Invest.* **118**, 3790.
81. Brissova, M., Shostak, A., Shiota, M., Wiebe, P.O., Poffenberger, G., Kantz, J., Chen, Z., Carr, C., Jerome, W.G., Chen, J., Baldwin, H.S., Nicholson, W., Bader, D.M., Jetton, T., Gannon, M., and Powers, A.C. (2006) *Diabetes* **55**, 2974.
82. Cabrera, O., Berman, D.M., Kenyon, N.S., Ricordi, C., Berggren, P.O., and Caicedo, A. (2006) *Proc. Natl. Acad. Sci. U.S.A.* **103**, 2334.
83. de Heer, J. and Holst, J.J. (2007) *Diabetes* **56**, 438.
84. de Heer, J., Rasmussen, C., Coy, D.H., and Holst, J.J. (2008) *Diabetologia* **51**, 2263.
85. Noma, A. (1983) *Nature* **305**, 147.
86. Trube, G. and Hescheler, J. (1984) *Pflugers Arch.* **401**, 178.
87. Cook, D.L. and Hales, C.N. (1984) *Nature* **311**, 271.
88. Sturgess, N.C., Ashford, M.L., Cook, D.L., and Hales, C.N. (1985) *Lancet* **2**, 474.
89. Misler, S., Falke, L.C., Gillis, K., and McDaniel, M.L. (1986) *Proc. Natl. Acad. Sci. U.S.A.* **83**, 7119.
90. Henquin, J.C. (2000) *Diabetes* **49**, 1751.
91. Satin, L.S. (2000) *Endocrine* **13**, 251.
92. Henquin, J.C. (2009) *Diabetologia* **52**, 739.
93. Detimary, P., Dejonghe, S., Ling, Z., Pipeleers, D., Schuit, F., and Henquin, J.C. (1998) *J. Biol. Chem.* **273**, 33905.
94. Kennedy, H.J., Pouli, A.E., Ainscow, E.K., Jouaville, L.S., Rizzuto, R., and Rutter, G.A. (1999) *J. Biol. Chem.* **274**, 13281.
95. Babenko, A.P., Aguilar-Bryan, L., and Bryan, J. (1998) *Annu. Rev. Physiol.* **60**, 667.
96. Cook, D.L., Satin, L.S., Ashford, M.L., and Hales, C.N. (1988) *Diabetes* **37**, 495.
97. Matschinsky, F.M. (2005) *Curr. Diab. Rep.* **5**, 171.
98. Quoix, N., Cheng-Xue, R., Mattart, L., Zeinoun, Z., Guiot, Y., Beauvois, M.C., Henquin, J.C., and Gilon, P. (2009) *Diabetes* **58**, 412.
99. Gilon, P., Shepherd, R.M., and Henquin, J.C. (1993) *J. Biol. Chem.* **268**, 22265.
100. Gopel, S., Kanno, T., Barg, S., Galvanovskis, J., and Rorsman, P. (1999) *J. Physiol. (Lond.)* **521(3)**, 717.

101. Charollais, A., Gjinovci, A., Huarte, J., Bauquis, J., Nadal, A., Martin, F., Andreu, E., Sanchez-Andres, J.V., Calabrese, A., Bosco, D., Soria, B., Wollheim, C.B., Herrera, P.L., and Meda, P. (2000) *J. Clin. Invest.* **106**, 235.
102. Hellman, B., Dansk, H., and Grapengiesser, E. (2004) *Am. J. Physiol. Endocrinol. Metab.* **286**, E759.
103. Krippeit-Drews, P., Dufer, M., and Drews, G. (2000) *Biochem. Biophys. Res. Commun.* **267**, 179.
104. Tornheim, K. (1997) *Diabetes* **46**, 1375.
105. Detimary, P., Gilon, P., and Henquin, J.C. (1998) *Biochem. J.* **333**(2), 269.
106. Kindmark, H., Kohler, M., Brown, G., Branstrom, R., Larsson, O., and Berggren, P.O. (2001) *J. Biol. Chem.* **276**, 34530.
107. Rolland, J.F., Henquin, J.C., and Gilon, P. (2002) *Diabetes* **51**, 376.
108. Dufer, M., Haspel, D., Krippeit-Drews, P., Aguilar-Bryan, L., Bryan, J., and Drews, G. (2004) *Diabetologia* **47**, 488.
109. Best, L. (2002) *J. Membr. Biol.* **185**, 193.
110. Best, L. (2002) *Pflugers Arch.* **445**, 97.
111. Best, L. (2005) *J. Physiol. (Lond.)* **568**, 137.
112. Kinard, T.A. and Satin, L.S. (1995) *Diabetes* **44**, 1461.
113. Rajan, A.S., Aguilar-Bryan, L., Nelson, D.A., Nichols, C.G., Wechsler, S.W., Lechago, J., and Bryan, J. (1993) *J. Biol. Chem.* **268**, 15221.
114. Ronner, P., Matschinsky, F.M., Hang, T.L., Epstein, A.J., and Buettger, C. (1993) *Diabetes* **42**, 1760.
115. Gopel, S.O., Kanno, T., Barg, S., Weng, X.G., Gromada, J., and Rorsman, P. (2000) *J. Physiol. (Lond.)* **528**, 509.
116. Gromada, J., Ma, X., Hoy, M., Bokvist, K., Salehi, A., Berggren, P.O., and Rorsman, P. (2004) *Diabetes* **53**(3), S181.
117. Ahren, B. and Taborsky, G.J., Jr. (1986) *Endocrinology* **118**, 1551.
118. Vieira, E., Salehi, A., and Gylfe, E. (2007) *Diabetologia* **50**, 370.
119. Salehi, A., Vieira, E., and Gylfe, E. (2006) *Diabetes* **55**, 2318.
120. MacDonald, P.E., De Marinis, Y.Z., Ramracheya, R., Salehi, A., Ma, X., Johnson, P.R., Cox, R., Eliasson, L., and Rorsman, P. (2007) *PLoS Biol.* **5**, e143.
121. Cabrera, O., Jacques-Silva, M.C., Speier, S., Yang, S.N., Kohler, M., Fachado, A., Vieira, E., Zierath, J.R., Kibbey, R., Berman, D.M., Kenyon, N.S., Ricordi, C., Caicedo, A., and Berggren, P.O. (2008) *Cell Metab.* **7**, 545.

122. Amiel, S.A., Simonson, D.C., Tamborlane, W.V., DeFronzo, R.A., and Sherwin, R.S. (1987) *Diabetes* **36**, 518.
123. Cryer, P.E. (1981) *Diabetes* **30**, 261.
124. Taborsky, G.J., Jr., Ahren, B., and Havel, P.J. (1998) *Diabetes* **47**, 995.
125. Steffens, A.B., Strubbe, J.H., Balkan, B., and Scheurink, J.W. (1990) *Neurosci. Biobehav. Rev.* **14**, 305.
126. Levin, B.E. (2001) *Int. J. Obes. Relat. Metab. Disord.* **25**(5), S68.
127. Miki, T., Liss, B., Minami, K., Shiuchi, T., Saraya, A., Kashima, Y., Horiuchi, M., Ashcroft, F., Minokoshi, Y., Roeper, J., and Seino, S. (2001) *Nat. Neurosci.* **4**, 507.
128. Rorsman, P., Berggren, P.O., Bokvist, K., Ericson, H., Mohler, H., Ostenson, C.G., and Smith, P.A. (1989) *Nature* **341**, 233.
129. Xu, E., Kumar, M., Zhang, Y., Ju, W., Obata, T., Zhang, N., Liu, S., Wendt, A., Deng, S., Ebina, Y., Wheeler, M.B., Braun, M., and Wang, Q. (2006) *Cell Metab.* **3**, 47.
130. Nunemaker, C.S., Zhang, M., and Satin, L.S. (2004) *Diabetes* **53**, 1765.
131. Mirshamsi, S., Laidlaw, H.A., Ning, K., Anderson, E., Burgess, L.A., Gray, A., Sutherland, C., and Ashford, M.L. (2004) *BMC Neurosci.* **5**, 54.
132. Konner, A.C., Janoschek, R., Plum, L., Jordan, S.D., Rother, E., Ma, X., Xu, C., Enriori, P., Hampel, B., Barsh, G.S., Kahn, C.R., Cowley, M.A., Ashcroft, F.M., and Bruning, J.C. (2007) *Cell Metab.* **5**, 438.
133. Ishihara, H., Maechler, P., Gjinovci, A., Herrera, P.L., and Wollheim, C.B. (2003) *Nat. Cell Biol.* **5**, 330.
134. Franklin, I., Gromada, J., Gjinovci, A., Theander, S., and Wollheim, C.B. (2005) *Diabetes* **54**, 1808.
135. Gromada, J., Franklin, I., and Wollheim, C.B. (2007) *Endocr. Rev.* **28**, 84.
136. Zhou, H., Zhang, T., Harmon, J.S., Bryan, J., and Robertson, R.P. (2007) *Diabetes* **56**, 1107.
137. Ravier, M.A. and Rutter, G.A. (2005) *Diabetes* **54**, 1789.
138. Gyulkhandanyan, A.V., Lu, H., Lee, S.C., Bhattacharjee, A., Wijesekara, N., Fox, J.E., MacDonald, P.E., Chimienti, F., Dai, F.F., and Wheeler, M.B. (2008) *J. Biol. Chem.* **283**, 10184.
139. Slucca, M., Harmon, J.S., Oseid, E.A., Bryan, J., and Robertson, R.P. (2009) *Diabetes*.

140. Bancila, V., Cens, T., Monnier, D., Chanson, F., Faure, C., Dunant, Y., and Bloc, A. (2005) *J. Biol. Chem.* **280**, 8793.
141. Bloc, A., Cens, T., Cruz, H., and Dunant, Y. (2000) *J. Physiol. (Lond.)* **529**(3), 723.
142. Prost, A.L., Bloc, A., Hussy, N., Derand, R., and Vivaudou, M. (2004) *J. Physiol. (Lond.)* **559**, 157.
143. Berts, A., Ball, A., Gylfe, E., and Hellman, B. (1996) *Biochim. Biophys. Acta* **1310**, 212.
144. Nadal, A., Quesada, I., and Soria, B. (1999) *J. Physiol. (Lond.)* **517**(1), 85.
145. Quesada, I., Nadal, A., and Soria, B. (1999) *Diabetes* **48**, 2390.
146. Liu, Y.J., Hellman, B., and Gylfe, E. (1999) *Endocrinology* **140**, 5524.
147. Sakurai, H., Dobbs, R., and Unger, R.H. (1974) *J. Clin. Invest.* **54**, 1395.
148. Taborsky, G.J., Jr. (1983) *Am. J. Physiol.* **245**, E598.
149. Brunicardi, F.C., Wen, D., Bradley, J.C., Elahi, D., Miller, C., and Hanks, J. (2000) *Int. J. Surg. Investig.* **1**, 381.
150. Hauge-Evans, A.C., King, A.J., Carmignac, D., Richardson, C.C., Robinson, I.C., Low, M.J., Christie, M.R., Persaud, S.J., and Jones, P.M. (2009) *Diabetes* **58**, 403.
151. Berts, A., Ball, A., Dryselius, G., Gylfe, E., and Hellman, B. (1996) *Endocrinology* **137**, 693.
152. Gopel, S.O., Kanno, T., Barg, S., and Rorsman, P. (2000) *J. Physiol. (Lond.)* **528**, 497.
153. Guiot, Y., Stevens, M., Marhfour, I., Stiernet, P., Mikhailov, M., Ashcroft, S.J., Rahier, J., Henquin, J.C., and Sempoux, C. (2007) *Diabetologia* **50**, 1889.
154. Ipp, E., Dobbs, R.E., Arimura, A., Vale, W., Harris, V., and Unger, R.H. (1977) *J. Clin. Invest.* **60**, 760.
155. Efendic, S., Enzmann, F., Nylen, A., Uvnas-Wallensten, K., and Luft, R. (1979) *Proc. Natl. Acad. Sci. U.S.A.* **76**, 5901.
156. Braun, M., Ramracheya, R., Amisten, S., Bengtsson, M., Moritoh, Y., Zhang, Q., Johnson, P.R., and Rorsman, P. (2009) *Diabetologia* **52**, 1566.
157. Zhang, Q., Bengtsson, M., Partridge, C., Salehi, A., Braun, M., Cox, R., Eliasson, L., Johnson, P.R., Renstrom, E., Schneider, T., Berggren, P.O., Gopel, S., Ashcroft, F.M., and Rorsman, P. (2007) *Nat. Cell Biol.* **9**, 453.

158. Leech, C.A., Holz, G.G., and Habener, J.F. (1996) *Ann. N. Y. Acad. Sci.* **805**, 81.
159. Murphy, R., Ellard, S., and Hattersley, A.T. (2008) *Nat. Clin. Pract. Endocrinol. Metab.* **4**, 200.
160. Vaxillaire, M., Bonnefond, D.A., and Froguel, P. (2009) *Pediatr. Endocrinol. Rev.* **6**, 405.
161. Sandal, T., Laborie, L.B., Brusgaard, K., Eide, S.A., Christesen, H.B., Sovik, O., Njolstad, P.R., and Molven, A. (2009) *Clin. Genet.* **75**, 440.
162. Flanagan, S.E., Clauin, S., Bellanne-Chantelot, C., de Lonlay, P., Harries, L.W., Gloyn, A.L., and Ellard, S. (2009) *Hum. Mutat.* **30**, 170.
163. Nichols, C.G., Shyng, S.L., Nestorowicz, A., Glaser, B., Clement, J.P.t., Gonzalez, G., Aguilar-Bryan, L., Permutt, M.A., and Bryan, J. (1996) *Science* **272**, 1785.
164. Shyng, S.L., Ferrigni, T., Shepard, J.B., Nestorowicz, A., Glaser, B., Permutt, M.A., and Nichols, C.G. (1998) *Diabetes* **47**, 1145.
165. Verkarre, V., Fournet, J.C., de Lonlay, P., Gross-Morand, M.S., Devillers, M., Rahier, J., Brunelle, F., Robert, J.J., Nihoul-Fekete, C., Saudubray, J.M., and Junien, C. (1998) *J. Clin. Invest.* **102**, 1286.
166. Fournet, J.C. and Junien, C. (2004) *Endocr. Pathol.* **15**, 233.
167. Mazor-Aronovitch, K., Landau, H., and Gillis, D. (2009) *Pediatr. Endocrinol. Rev.* **6**, 424.
168. Al-Shanafey, S. (2009) *J. Pediatr. Surg.* **44**, 957.
169. Adzick, N.S., Thornton, P.S., Stanley, C.A., Kaye, R.D., and Ruchelli, E. (2004) *J. Pediatr. Surg.* **39**, 270.
170. Kapoor, R.R. James, C., and Hussain, K. (2009) *Endocr. Dev.* **14**, 95.
171. Bax, K.N. and van der Zee, D.C. (2007) *Semin. Pediatr. Surg.* **16**, 245.
172. Lyonnet, S., Bonnefont, J.P., Saudubray, J.M., Nihoul-Fekete, C., and Brunelle, F. (1989) *Lancet* **2**, 671.
173. Goossens, A., Gepts, W., Saudubray, J.M., Bonnefont, J.P., Nihoul-Fekete, C., Heitz, P.U., and Kloppel, G. (1989) *Am. J. Surg. Pathol.* **13**, 766.
174. Christesen, H.B., Brusgaard, K., Alm, J., Sjoblad, S., Hussain, K., Fenger, C., Rasmussen, L., Hovendal, C., Otonkoski, T., and Jacobsen, B.B. (2007) *Horm. Res.* **67**, 184.
175. Ribeiro, M.J., De Lonlay, P., Delzescaux, T., Boddaert, N., Jaubert, F., Bourgeois, S., Dolle, F., Nihoul-Fekete, C., Syrota, A., and Brunelle, F. (2005) *J. Nucl. Med.* **46**, 560.

176. Otonkoski, T., Nanto-Salonen, K., Seppanen, M., Veijola, R., Huopio, H., Hussain, K., Tapanainen, P., Eskola, O., Parkkola, R., Ekstrom, K., Guiot, Y., Rahier, J., Laakso, M., Rintala, R., Nuutila, P., and Minn, H. (2006) *Diabetes* **55**, 13.
177. Mohnike, K., Blankenstein, O., Christesen, H.T., De Lonlay, J., Hussain, K., Koopmans, K.P., Minn, H., Mohnike, W., Mutair, A., Otonkoski, T., Rahier, J., Ribeiro, M., Schoenle, E., and Fekete, C.N. (2006) *Horm. Res.* **66**, 40.
178. Mohnike, K., Blankenstein, O., Minn, H., Mohnike, W., Fuchtner, F., and Otonkoski, T. (2008) *Horm. Res.* **70**, 65.
179. Mohnike, K., Blankenstein, O., Pfuetzner, A., Potzsch, S., Schober, E., Steiner, S., Hardy, O.T., Grimberg, A., and van Waarde, W.M. (2008) *Horm. Res.* **70**, 59.
180. Blakely, M.L., Lobe, T.E., Cohen, J., and Burghen, G.A. (2001) *Surg. Endosc.* **15**, 897.
181. von Muhlendahl, K.E. and Herkenhoff, H. (1995) *N. Engl. J. Med.* **333**, 704.
182. Shield, J.P., Gardner, R.J., Wadsworth, E.J., Whiteford, M.L., James, R.S., Robinson, D.O., Baum, J.D., and Temple, I.K. (1997) *Arch. Dis. Child. Fetal Neonatal Ed.* **76**, F39.
183. Polak, M. and Cave, H. (2007) *Orphanet J. Rare Dis.* **2**, 12.
184. Polak, M. and Shield, J. (2004) *Pediatr. Endocrinol. Rev.* **2**, 193.
185. Slingerland, A.S., Shields, B.M., Flanagan, S.E., Bruining, G.J., Noordam, K., Gach, A., Mlynarski, W., Malecki, M.T., Hattersley, A.T., and Ellard, S. (2009) *Diabetologia* **52**, 1683.
186. Aguilar-Bryan, L. and Bryan, J. (2008) *Endocr. Rev.* **29**, 265.
187. Gloyn, A.L., Pearson, E.R., Antcliff, J.F., Proks, P., Bruining, G.J., Slingerland, A.S., Howard, N., Srinivasan, S., Silva, J.M., Molnes, J., Edghill, E.L., Frayling, T.M., Temple, I.K., Mackay, D., Shield, J.P., Sumnik, Z., van Rhijn, A., Wales, J.K., Clark, P., Gorman, S., Aisenberg, J., Ellard, S., Njolstad, P.R., Ashcroft, F.M., and Hattersley, A.T. (2004) *N. Engl. J. Med.* **350**, 1838.
188. Babenko, A.P., Polak, M., Cave, H., Busiah, K., Czernichow, P., Scharfmann, R., Bryan, J., Aguilar-Bryan, L., Vaxillaire, M., and Froguel, P. (2006) *N. Engl. J. Med.* **355**, 456.
189. Proks, P., Arnold, A.L., Bruining, J., Girard, C., Flanagan, S.E., Larkin, B., Colclough, K., Hattersley, A.T., Ashcroft, F.M., and Ellard, S. (2006) *Hum. Mol. Genet.* **15**, 1793.

190. Hattersley, A.T. and Ashcroft, F.M. (2005) *Diabetes* **54**, 2503.
191. Iafusco, D., Stazi, M.A., Cotichini, R., Cotellessa, M., Martinucci, M.E., Mazzella, M., Cherubini, V., Barbetti, F., Martinetti, M., Cerutti, F., and Prisco, F. (2002) *Diabetologia* **45**, 798.
192. Sagen, J.V., Raeder, H., Hathout, E., Shehadeh, N., Gudmundsson, K., Baevre, H., Abuelo, D., Phornphutkul, C., Molnes, J., Bell, G.I., Gloyn, A.L., Hattersley, A.T., Molven, A., Sovik, O., and Njolstad, P.R. (2004) *Diabetes* **53**, 2713.
193. Zung, A., Glaser, B., Nimri, R., and Zadik, Z. (2004) *J. Clin. Endocrinol. Metab.* **89**, 5504.
194. Proks, P., Girard, C., and Ashcroft, F.M. (2005) *Hum. Mol. Genet.* **14**, 2717.
195. Tonini, G., Bizzarri, C., Bonfanti, R., Vanelli, M., Cerutti, F., Faleschini, E., Meschi, F., Prisco, F., Ciacco, E., Cappa, M., Torelli, C., Cauvin, V., Tumini, S., Iafusco, D., and Barbetti, F. (2006) *Diabetologia* **49**, 2210.
196. Pearson, E.R., Flechtner, I., Njolstad, P.R., Malecki, M.T., Flanagan, S.E., Larkin, B., Ashcroft, F.M., Klimes, I., Codner, E., Iotova, V., Slingerland, A.S., Shield, J., Robert, J.J., Holst, J.J., Clark, P.M., Ellard, S., Sovik, O., Polak, M., and Hattersley, A.T. (2006) *N. Engl. J. Med.* **355**, 467.
197. Loomba-Albrecht, L.A., Glaser, N.S., Styne, D.M., and Bremer, A.A. (2009) *Clin. Ther.* **31**, 816.
198. Rafiq, M., Flanagan, S.E., Patch, A.M., Shields, B.M., Ellard, S., and Hattersley, A.T. (2008) *Diabetes Care* **31**, 204.
199. Shimomura, K., Horster, F., de Wet, H., Flanagan, S.E., Ellard, S., Hattersley, A.T., Wolf, N.I., Ashcroft, F., and Ebinger, F. (2007) *Neurology* **69**, 1342.
200. Slingerland, A.S., Nuboer, R., Hadders-Algra, M., Hattersley, A.T., and Bruining, G.J. (2006) *Diabetologia* **49**, 2559.
201. Mlynarski, W., Tarasov, A.I., Gach, A., Girard, C.A., Pietrzak, I., Zubcevic, L., Kusmierek, J., Klupa, T., Malecki, M.T., and Ashcroft, F.M. (2007) *Nat. Clin. Pract. Neurol.* **3**, 640.

Chapter 12

The ABCD Subfamily: Peroxisomal Transporters in Health and Disease

Aurora Pujol and Patrick Aubourg

1. Introduction

First described by Baudhuin and DeDuve in the mid-1950s, peroxisomes are single membrane-bound, typically spherical, organelles ranging in size from 0.1–1 microns in diameter and numbering from a few hundred to a few thousand in mammalian cells.[1] In the late 1970, it has been shown that deficiencies of peroxisome structure and/or in peroxisomal enzyme activities are associated with inherited metabolic diseases in humans that involve mostly the central nervous system (CNS). During the last two decades, important new insights have been gained on the mechanisms by which proteins are targeted to peroxisomes, the peroxisome structures in disease states and, more globally, the functions of peroxisomes that are indispensable in particular for the development and maintenance of the CNS.

Peroxisomes play a key role in lipid homeostasis and the first peroxisomal "ATP-binding cassette" (ABC) transporter that was identified turned out to be a membrane protein of 70 kDA (originally called PMP70, but now renamed ABCD3 the product of the *ABCD3* gene) owing to its large abundance in liver peroxisomes.[2] The typical nucleotide binding fold (NMD) motif present in PMP70 was however overlooked until the gene responsible for X-adrenoleukodystrophy (X-ALD), the most frequent (1:17,000 males) inherited peroxisomal disorder, was identified by positional cloning.[3] The ALD protein (encoded by the *ABCD1* gene and now called ABCD1) also shows a typical ABC transporter architecture. The cloning of the *ABCD1* gene led to the rapid identification of two other peroxisomal ABC transporters in mammals, the ALD related protein (formerly ALDRP and now known as ABCD2) and the PMP70 related (formerly PMP70R or PMP69, and now known as ABCD4) proteins encoded respectively by the *ABCD2* and *ABCD4* genes.[4–6]

Most eukaryotic ABC transporters are full transporters composed of two homologous halves, which each half made up of a hydrophobic transmembrane domain containing multiple alpha helices, and a hydrophilic nucleotide binding domain (NBD) with Walker A and B consensus motifs. In contrast, the peroxisomal ABCD transporters are half-transporters, with only one single hydrophobic and hydrophilic domain and consequently need to homo or heterodimerize to become functional.

The biochemical hallmark of X-ALD is the accumulation of very-long-chain fatty acids (VLCFAs) in blood and tissues, so when the X-ALD gene (*ABCD1*) was cloned it was immediately suspected that the ABCD1 protein as well as the ABCD4 could be involved in the transport of fatty acids across the peroxisomal membrane. The localization of the ABCD transporters within the peroxisomal membrane and the technical difficulties encountered in purifying peroxisomes has hampered for many years our ability to make significant progresses in the physiological functions of these ABC transporters. However, the situation has changed lately owing to the development of mouse models with deficiencies in ABCD transporters and the expression of mammalian ABCD transporters in yeast.

2. X-Linked Adrenoleukodystrophy (X-ALD): A Complex Disease Due to Loss of Function of ABCD1 Gene

X-ALD was first described by Siemerling and Creutzfeld in 1923 in a 6-year-old boy who rapidly developed cerebral demyelination with neuro-inflammation. In 1976, Budka and Griffin reported that X-ALD may manifest in young adults as a progressive spastic paraplegia without cerebral demyelination (the "adrenomyeloneuropathy" phenotype, AMN). A key observation was the demonstration of the characteristic accumulation of VLCFAs in the brain of X-ALD patients.

X-ALD is characterized by two distinct phenotypes: adrenomyeloneuropathy (AMN) and cerebral demyelinating form of X-ALD (CDA). AMN and CDA occur frequently within the same family but there is no correlation between X-ALD phenotypes and mutations of the *ABCD1* gene.

AMN is the most frequent (65%) phenotype of X-ALD with onset of neurologic symptoms between the second and fourth decade of life. The disease involves the spinal cord and presents with slowly progressive stiffness and weakness of legs, impaired vibration sense, sphincter disturbances and impotence. Clinical signs of peripheral nerve involvement are sometimes present. More than 60% of heterozygous women for X-ALD develop also AMN in the fourth to fifth decade of life. Epidemiologically, motor disability and progression is less severe than in males. Adrenal insufficiency (Adisson's disease) is present in two-thirds of males and less than 1% of heterozygous women. Brain magnetic resonance imaging (MRI) is roughly normal in males and heterozygous women with AMN, but 35% of adult males with AMN also develop cerebral demyelination between the ages of 25 and 40 years. This risk is nearly zero after the age of 45 years. The outcome of cerebral ALD in adult males is the same as in boys: it results in a vegetative stage or death a few years after the onset of cerebral demyelination. Heterozygous women with AMN never develop the cerebral demyelinating form of ALD.

A male newborn with a mutation in the *ABCD1* gene has a 35% risk to develop CDA between five to 12 years of age, before developing AMN. Boys are completely unaffected until they develop cerebral demyelination visible by brain MRI. The lesions involve either the white matter of the parieto-occipital or frontal lobes. At this stage, the demyelinating lesions are not inflammatory (with no evidence of blood-brain barrier disruption demonstrated by an enhancement of demyelinating lesions after peripheral injection of gadolinium) and progress slowly. The patients have virtually no neurologic symptoms, excepting mild cognitive dysfunctions. The disease then becomes inflammatory (with evidence of enhancement after peripheral injection of gadolinium) and the patients enter into active phase of neurologic and cognitive degradation that include emotional lability, hyperactive behavior, school failure in children, visuospatial impairment or frontal syndrome according to the sites of demyelinating lesions. This evolves to the rapid development of motor symptoms (hemiplegia or quadriparesis), impaired central auditory discrimination, quadranopsia and then loss of vision, and often seizures. The disease progresses to a vegetative state within two to four years and to death at varying intervals thereafter.

Adrenal insufficiency may be the first manifestation of X-ALD and 10% of X-ALD males have only primary adrenal insufficiency without evidence of nervous system involvement at some time of their life (either in childhood or adulthood). These patients are, however, at risk of developing either AMN or CDA. The AMN phenotype is fully penetrant in males before the age of 55 years. Adult X-ALD males may develop testicular insufficiency due to the involvement of Sertoli and Leydig cells. Heterozygous women usually present with no endocrine problems, although <1% show evidence of adrenal insufficiency.

The diagnosis of X-ALD relies upon the demonstration of increased levels of plasma VLCFAs; this is 100% reliable for detection of affected males, but detects only 90–95% of X-ALD heterozygous women.

More than 450 different mutant *ABCD1* alleles have been reported and catalogued on the regularly-updated X-ALD mutation database (www.x-ald.nl).[7] Of these, 50% are missense mutations and the remainder are nonsense mutations, amino acid insertions and deletions, frame shift mutations and deletions of one or more exons. 40% of missense mutations and 100% of the other mutations result in the absence of detectable ABCD1 in fibroblasts or white blood cells. There are no real hotspot mutations and many *ABCD1* gene mutations are private. The percentage of *de novo* mutation is less than 8%. As mentioned above, there is no correlation between the clinical phenotype and the *ABCD1* gene mutations, which suggests that environmental factors, "modifier" genes, epigenetic and even stochastic factors might play a crucial role in the onset of cerebral demyelination.

3. The ABCs of Peroxisomes

ABCDs are thought to import fatty acids for α- and β-oxidation in peroxisomes. Defects in either pathway are causative for a range of fatal inherited human disorders, such as adult Refsum disease (ARD) or X-linked adrenoleukodystrophy (X-ALD) to cite but two.[8]

Peroxisomes contain a fatty acid β-oxidation system which allows the chain-shortening of a large variety of different fatty acids including saturated and monosaturated very-long-chain fatty acid (VLCFAs) and branched fatty acids such as pristanic, di- and trihydroxycholestanoic acids. The peroxisomal β-oxidation system has the same basic structure as that in mitochondria comprising four linked steps; dehydrogenation, hydration, a second dehydrogenation and thiolytic cleavage (see Fig. 1). The ultimate result is that the first two carbon atoms of the fatty acyl-CoA ester are released as acetyl-CoA leaving a shortened acyl-CoA ester which can undergo subsequent rounds of β-oxidation. Since peroxisomes are not able to degrade fatty acyl-CoAs to completion, the chain-shortened acyl-CoAs are shuttled to mitochondria as a carnitine ester for full β-oxidation to acetyl-CoA units.

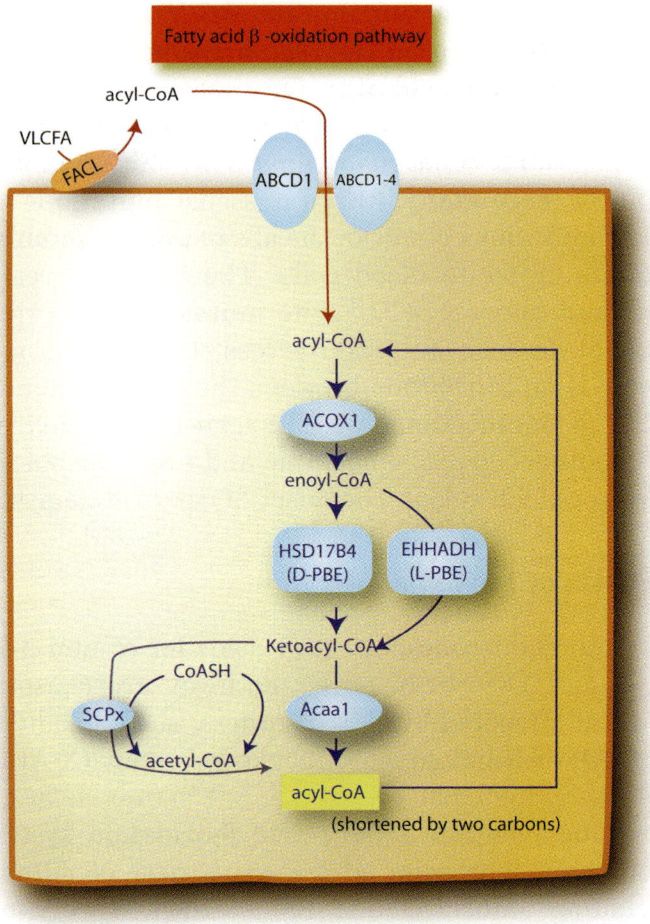

Fig. 1. Peroxisomal β-oxidation pathway, adapted from www.peroxisomedb.org. Dehydrogenation is carried out by ACOX1; bifunctional enzymes D-PBE and L-PBE catalyze the hydratation and a second dehydrogenation and the thiolases ACCA1 and SCPx catalyze the thiolytic cleavage and chain shortening.

In order to catalyze the β-oxidation of these different fatty acids, peroxisomes are equipped with the full enzymatic machinery required to oxidize a large range of different fatty acids such as straight-chain fatty acids, mono- and poly-unsaturated fatty acids, 2-methyl branched-chain fatty acids, 2-hydroxy fatty acids, etc.

With respect to the peroxisomal β-oxidation *per se* it has been clearly established that peroxisomes contain two acyl-CoA oxidases (ACOX1 and ACOX2) to catalyze the first step, two bifunctional enzymes, L- and D-bifunctional enzymes (or L-PBE and D-PBE, also called multifunctional protein 1 and 2, MFP2 and MFP2), catalyzing the second and third steps, and two peroxisomal thiolases (Acaa1 or PTH1 and PTH2/Sterol carrier protein X [SCPx]), catalyzing the final step.[9] Studies in recent years have resolved the question which of these enzymes is required for the oxidation of each particular fatty acid. For instance, C26:0 β-oxidation can only take place via the concerted action of ACOX1, D-PBE, and either PTH1 (Acaa1) or PTH2 (SCPx), whereas oxidation of pristanic acid and di- and trihydroxycholestanoic acid requires the active participation of ACOX2, D-PBE, and PTH2 (SCPx).[9]

Phytanic acid (3,7,11,15-tetramethylhexadecanoic acid) is a branched chain fatty acid, present in the human diet. The 3-methyl group of phytanic acid prevents degradation by β-oxidation, and instead, the terminal carboxyl group is first removed by alpha-oxidation. The first step of the alpha-oxidation pathway involves the 2-hydroxylation of phytanoyl-CoA to 2-hydroxyphytanoyl-CoA by phytanoyl-CoA hydroxylase (PhyH/Pahx) which is localized in peroxisomes and deficient in patients with adult Refsum disease. The second step involves the decarboxylation of 2-hydroxyphytanoyl-CoA by 2-hydroxyphytanoyl-CoA lyase enzyme. The last step involves the conversion of pristanal to pristanic acid. This reaction is carried out by an NAD(P) dependent fatty aldehyde dehydrogenase (ALDH3A2, encoded by the ALDH3A2 gene) which is defective in patients suffering from Sjögren Larsson syndrome, but not in patients with ARD. ALDH3A2, which is not localized in peroxisomes, also oxidizes fatty aldehyde substrates arising from the metabolism of fatty alcohols, leukotriene B4, ether glycerolipids and other potential sources such as sphingolipids.[10]

Recognition of inherited disorders of peroxisome assembly (referred to as "peroxisome biogenesis disorders", PBDs) such as the Zellweger spectrum (ZS) and Rhizomelic chondrodysplasia punctata type 1 (RCDP1) had led to major progresses in the field of peroxisome biogenesis. Synthesized on free cytosolic ribosomes, peroxisomal matrix proteins or enzymes are targeted posttranslationally to peroxisomes by specific cytosolic receptors that recognize *cis*-acting sequences (peroxisomal targeting signals [PTSs]) in the primary peptide sequence. Most matrix proteins are targeted by peroxisomal targeting sequence 1 (PTS1), a C-terminal Serine-Lysine-Leucine-COOH (SKL) or conservative variant thereof. A few matrix proteins are targeted by PTS2, a degenerate sequence (-R/KX5Q/HL-) located near the N terminus. The PTS1 and 2 receptors have been cloned and characterized. The former, peroxin 5 which is encoded by the PEX5 gene, is a tetratricopeptide repeat protein; the latter, peroxin 7 which is encoded by the PEX7 gene, is a WD40 repeat protein. In mammalian cells, one isoform of peroxin 5 interacts with and is necessary for peroxin 7 function.[9]

Like matrix proteins, peroxisomal membrane proteins (PMPs) are synthesized on free cytosolic ribosomes and targeted to the organelle by *cis*-acting targeting sequences [membrane PTS (mPTS)]. In contrast to the discrete, well-defined, single targeting sequences found in matrix proteins, the model emerging for several PMPs includes at least two nonoverlapping targeting segments for each polypeptide, either of which is sufficient for localization and insertion into the peroxisome membrane. In most cases these targeting segments are relatively long and include one or two transmembrane segments. The largely cytosolic peroxin 19, encoded by the PEX19 gene, is required for the topogenesis of PMPs and is supposed to escort PMPs from the cytosol to the peroxisomal membrane. The peroxin 19 interacts with all PMPs tested so far.[11,12]

Seventeen different PEX genes encoding peroxins associated or embedded within the peroxisomal membrane have been identified in humans. These include: PEX1, PEX2, PEX3, PEX4, PEX5, PEX6, PEX7, PEX10, PEX11, PEX12, PEX13, PEX14, PEX15, PEX16, PEX19, and PEX26. Autosomal recessive mutations of twelve out those seventeen PEX genes have been shown to be responsible for PBDs.

Peroxins 3 and 16 are involved in the initial biogenesis of peroxisomes from pre-existing pre-peroxisomes. Peroxin 19 is involved with peroxin 3 in the import of PMPs into the peroxisomal membrane. Mutations in the Pex3, 16 and 19 genes result in the virtual absence of any peroxisomes in cells. Peroxins 11 (there are three peroxin 11 genes in human coding for different isoforms) are involved in the division and proliferation of peroxisomes that have already imported peroxisomal membrane proteins or enzymes.

The peroxin 5 binds its cargo proteins in the cytosol and targets them to docking machinery at the peroxisomal membrane that involves the peroxins 14 and 13. Then, the peroxin 5 forms with the peroxins 2, 10 and 12 a pore within the peroxisomal membrane, releases the cargo proteins into the peroxisomal lumen and shuttles back to the cytosol. Dislocation of the peroxin 5 from the peroxisomal membrane to the cytosol after cargo release is performed by the peroxins 1 and 6 that have an AAA motive. Recycling of the peroxin 5 in the cytosol involves its mono-ubiquitination by the peroxins 4 and 22 that are localized in the peroxisomal membrane. Destruction of the peroxin 5 by the 26S proteasome in the cytosol involves the poly-ubiquitination of peroxin 5 by the peroxin 15.[13,14] This system shows striking mechanistic similarities to the ERAD (endoplasmic reticulum associated degradation) machinery engines. The use of molecular phylogenetic tools has identified sequence homologies between both systems and has thus allowed tracing the evolutionary origin of peroxisomes to the ER membranes.[15] Comprehensive integrative peroxisomal pathways can be found on www.peroxisomedb.org.[16]

The number and matrix enzyme content of peroxisomes is dynamic and varies with the metabolic state of the resident cells and organism. Certain naturally occurring lipids and xenobiotics increase peroxisome number and induce expression of genes encoding peroxisome matrix and membrane proteins. This coordinated regulation is mediated primarily by activation of peroxisome proliferator-activated receptor α (PPAR α), a member of the nuclear receptor superfamily. Activated PPAR α heterodimerizes with a second member of the nuclear receptor superfamily, retinoid X receptor (RXR), to form an active transcription factor that binds *cis*-acting elements, called peroxisome proliferator responsive elements (PPRE), with a consensus sequence of two direct repeats of -AGGA/TCA- separated by a single base pair in the promoters of its target genes. An increase in peroxisome number is achieved usually by maturation, enlargement, and division of preexisting peroxisomes but *de novo* synthesis is also possible.[17,18]

4. Fatty Acid Transport Across the Peroxisomal Membrane: The Role of ABCD-Family Transporters

The peroxisomal membrane, as other cellular membranes, is selectively permeable to solute and macromolecules owing to the presence of specific transporters. Whilst hydrophobic and apolar molecules can cross the lipid bilayer by diffusion, the transport of polar, hydrophilic and large molecules require specific transport systems. There is a pH gradient between the cytosol and peroxisomal matrix, suggesting that transport of protons is regulated. Available evidence holds that fatty acids taken up by peroxisomes are not in the form of carnitine esters, the form utilized by mitochondria. Indeed, $C_{26:0}$-CoA and the CoA ester of the bile acid precursor trihydroxycholestanoic acid, which are exclusively β-oxidized in peroxisomes, are not converted into the respective carnitine esters by mitochondrial CPT1 (carnitine palmitoyltransferase I),[8] or by any other acyltransferase. It is currently believed that fatty acids can enter the peroxisome either as non-esterified fatty acid or directly as a CoA ester through ABCD transporters, depending on chain length. ABCD transporters are thought to be involved in the transport of fatty acids across the peroxisomal membrane, based on three principal observations: (i) peroxisomal membranes are not freely permeable, suggesting that the entrance and exit of metabolites requires the presence of transporter proteins; (ii) the two peroxisomal ABC transporters of *S. cerevisiae*, Pxa1p and Pxa2p, form functional heterodimers that transports long-chain fatty acyl-CoAs into the peroxisome;[19–21] (iii) loss of ABCD1 gene function causes X-ALD, which is biochemically characterized by the pathognomonic accumulation of saturated VLCFAs, mainly hexacosanoic acid (C26:0), in plasma and tissues.

5. The ABCDs of Baker's Yeast

Most of our current understanding of the import of fatty acids into peroxisomes comes from studies of the yeast *S. cerevisiae*. The peroxisomal membrane in *S. cerevisiae* contains only two members of the superfamily of ABC transporters, Pxa1p and Pxa2p.[19,22] The β-oxidation of oleic acid ($C_{18:1}$) is impaired in *pxa1* and *pxa2* gene-disruption mutants, but normal in cell lysates.[23] This strongly suggests a function of Pxa1p and Pxa2p in a transport process required for peroxisomal β-oxidation. Moreover, since the β-oxidation of lauric acid ($C_{12:0}$) was found to be completely normal in these knockouts, it was concluded that the defect must occur upstream of the β-oxidation process, presumably at the level of fatty acid import. These results suggest the existence of at least two distinct pathways for the peroxisomal import of fatty acids: (i) a Pxa1p/Pxa2p-independent pathway, which is used by lauric acid, and (ii) a Pxa1p/Pxa2p-dependent pathway, by which oleic acid is actively imported.[21,23] This led to a model which proposes that lauric acid is imported as a non-esterified fatty acid and then activated to its CoA ester in the peroxisomes, whereas oleic acid is activated in the cytosol and imported as a CoA ester into peroxisomes. Most fatty acids appear to be imported through both pathways to various extents. Two factors are probably decisive for which pathway is utilized by a specific fatty acid. Firstly, the physiochemical properties of the fatty acid, in particular its solubility in a non-polar solvent. Secondly, the substrate specificity and subcellular localization of the acyl-CoA synthetases present.

6. The ABCDs of Mammals

Four ABC transporters have been identified at the peroxisomal membrane in mammals (*ABCD1* to *ABCD4* genes, encoding homologous proteins). The 70 kD ABCD3 (PMP70),[2] comprises 659 amino acids with an ATP binding domain on the cytoplasmic side of the peroxisomal membrane. The second mammalian ABC transporter identified, by Mosser *et al*. in 1993[3] using a positional cloning strategy in the search for the gene responsible for X-ALD was named ALDP (ALD protein) and is encoded by the *ABCD1* gene. Identification of the X-ALD causing gene as a peroxisomal transporter came as a surprise, as the disease is associated with defective β-oxidation of VLCFAs. It was therefore thought that the ALD gene should encode a β-oxidation enzyme or a synthetase responsible for activating VLCFAs into CoA derivatives. Topology studies have shown that the hydrophilic carboxy-terminal half containing the ATP-binding domain of ABCD1 also faces the cytosol. Shortly thereafter, the third peroxisomal ABC transporter was identified[5] and called the ALD-related protein (ALDR) as the similarity with ABCD1 was marked (66% amino acid identity and 88% homology). ALDRP is encoded by the *ABCD2* gene. Finally, a fourth mammalian ABCD transporter was identified by two groups simultaneously (34,79), and named PMP69 or PMP70R as its closest homolog was the PMP70. The PMP70R protein is encoded by the *ABCD4* gene and its identity to ABCD1 is much lower (25% amino acid identity). A phylogenetic tree describing the relationships of the family is depicted in Fig. 2.

Localization of the ABC transporters to the peroxisomal membrane is dependent on an amino-terminal targeting motif and is mediated by the peroxin 19.[24] For ABCD1, amino acids 68–82 function as the binding site for the peroxin 19 and two membrane α-helices are necessary for the correct intracellular localization.[25] Amino acids 1–124 have been shown to be necessary for the peroxisomal localization of ABCD4, and the binding site for the peroxin 19 is included in the first 60 amino acids.[26]

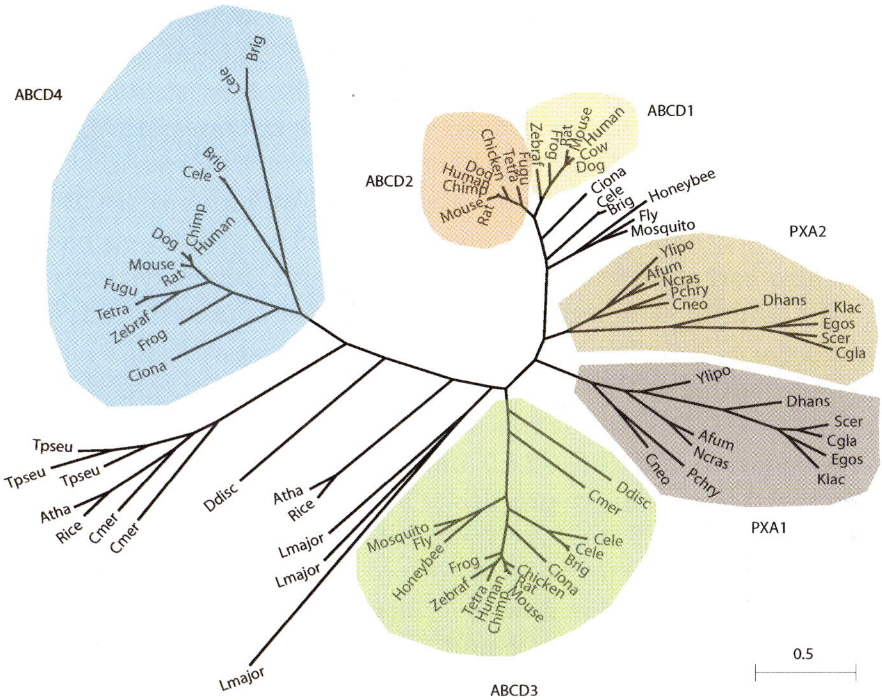

Fig. 2. Phylogenetic tree of ABCDs, adapted from Fourcade *et al.*[28]

Photoaffinity labeling studies, and a coupled assay for nucleotide hydrolysis, have shown that ABCD1 and ABCD3 specifically hydrolyze ATP.[27] Mutations of conserved residues of the NBD (G478R or S572I for ABCD3, respectively; and G512S or S606L for ABCD1, respectively) decrease or abolish the ATPase activity of ABCD1 and ABCD3 without influencing their homodimerization or heterodimerization.

7. The Function of ABCD1 (ALDP) Transporter

The β-oxidation of VLCFAs, including C26:0 and C24:0, is impaired in X-ALD fibroblasts, suggesting that the observed β-oxidation defect in X-ALD should be secondary to the impaired transport of VLCFAs across the peroxisomal membrane.[3,29] By analogy with Pxalp and Pxa2p in yeast, it has been postulated that ABCD1 transports the CoA esters of VLCFAs across the peroxisomal membrane in mammals. Recent studies have demonstrated the ability of human ABCD1 to transport VLCFA-CoA and other fatty-CoA esters across the peroxisomal membrane in *S. cerevisiae*.[20] Expression of human ABCD1 in *pxa1/pxa2Δ* (deficient) cells not only restored growth on oleate, but also corrected the acyl-CoA ester profile, albeit partially on both accounts. The oxidation rate of C26:0 is extremely low in wild-type *S. cerevisiae*, therefore it was not possible to measure the β-oxidation of this fatty acid in *pxa1/pxa2Δ* yeast cells rescued with ABCD1, however it was shown that these cells were capable of β-oxidation of C18:1, C22:0 and C24:6 fatty acids and that they generated the requisite profiles of acyl-CoA esters. Nevertheless, definitive proof of the functional role of ABCD1 is still missing and awaits successful reconstitution of ABCD1 into liposomes, for the direct demonstration that fatty acid CoA esters, including VLCFA-CoA esters, bind to, and are transported by, the protein.

The exact function of ABCD1 has evident implications in the physiopathogenesis of X-ALD. Cerebral demyelination with neuroinflammation is only one aspect of X-ALD. It is, in fact, the end-stage of X-ALD disease. Although cerebral demyelination with neuroinflammation may appear very early in life (between 5–12 years of age), adrenomyeloneuropathy (AMN) requires at least two decades to manifest. AMN is characterized by an axonopathy with microgliosis but without significant myelin changes and neuro-inflammation in the spinal cord.[30] The latter are also the only abnormalities apparent in the ABCD1 knockout mice that develop, at 16 months, a mild phenotype that resembles AMN.[31] ABCD1 is markedly expressed in oligodendrocytes and studies in human patients and in mice have shown that oligodendrocyte impairment may result in demyelination within the CNS, or axonal damage without demyelination.[32,33] X-ALD illustrates the fact that a loss of the ABCD1 transporter function in oligodendrocytes results primarly in axonal damage in spinal cord without demyelination. Importantly, lipoxidative damage can be seen as early as three months of age in the spinal cord of ALD mice, i.e. twelve months before the onset of any neuropathological or abnormalities of motor function.[34] This highlights that disturbance of fatty-acyl CoA metabolism in peroxisomes, including the accumulation of VLCFA-CoA esters, might trigger production of reactive oxygen species and hamper natural defences against oxidative stress in oligodendrocytes.[34] Compared to other brain cell types, oligodendrocytes are more susceptible to oxidative stress possibly because of their high content of iron and reduced levels of glutathione. A discussion of the additional factors that trigger cerebral demyelination and then neuroinflammation in X-ALD is out the scope of this chapter, but recent studies indicate that secondary peroxisomal dysfunctions in oligodendrocytes and specific dysfunction of microglial cells that also express ABCD1 might play a role in the cerebral demyelinating process that occurs in X-ALD.[35–37]

The observation that, at least in yeast, ABCD1 plays a role in the import of a wide range of fatty-CoA esters, and not only of VLCFA-CoA esters, raises important and unresolved issues: what is the functional role of peroxisomes in fatty acid metabolism in mammals? It remains unexplained why the degradation of fatty acids in mammalian cells is partitioned in two different organelles, the mitochondria and the peroxisomes, whereas in yeast, β-oxidation takes place mainly in peroxisomes? The answer cannot be merely that mitochondria and peroxisomes handle different classes of fatty acids as there is significant overlap in substrate specificity. This issue awaits additional and crucial *in vivo* and *in vitro* experiments to assess the physiological role of peroxisomes in fatty acid metabolism in mammals.

8. Do Mammalian Peroxisomal ABCD Half-Transporter Act as Homo- or Heterodimers?

ABCD proteins are half-size ABC transporters which require dimerization as a prerequisite for activity, but the question whether the peroxisomal ABC half-transporters function as a homodimer and/or a heterodimer is not completely resolved. The majority of eukaryotic ABC half-transporters identified, like TAP1 (*ABCB2*) and TAP2 (*ABCB3*), and *ABCG5* and *ABCG8,* form a functional full-transporter by heterodimerization. The only mammalian example of a full-transporter that is formed through homodimerization of two half transporters is the breast cancer resistance protein (ABCG2, see Chapter 4).[38]

In *S. cerevisiae*, the two peroxisomal half transporters Pxa1p and Pxa2p form a functional heterodimer that transports long-chain fatty acyl-CoAs into the peroxisome,[20,39] but the situation can be different in other species. Using co-immunoprecipitation assays and the yeast two-hybrid system, evidence for homodimerization and heterodimerization of the carboxyl-terminal halves of ABCD1, ABCD2 and ABCD3 has been demonstrated.[40] However, formation of heterodimers *in vivo*, i.e. in tissues, has never been shown. In contrast, isolation of protein complexes from mouse liver indicate that neither PMP70 nor ABCD2 co-precipitates with ABCD1.[41] FRET microscopy in intact living cells demonstrated that ABCD1 and ABCD3 form homodimers and might form ABCD1/ABCD3 heterodimers, although ABCD1 homodimers prevail.[42] Supporting the hypothesis that ABCD1 forms mostly homodimers is the observation that a knock-in ALD mouse (i.e. a mouse expressing a stable but deficient ABCD1) shows the same phenotype as the knock-out ALD mouse that does not express any ABCD1 (K. Nave, unpublished). The subunit complexity of ABCD4 transporters has not been investigated, mainly due to the unavailability of suitable antibodies.

Finally, indirect evidence supporting the hypothesis that ABCD1 and other peroxisomal ABCD transporters form homodimers comes from analysis of the ABCD1 gene mutations and expression of these two ABCD transporters in tissue. After careful analysis of the *ABCD1* gene, all X-ALD patients have so far been found to carry a mutation of the ABCD1 gene (http://www.x-ald.nl).[7] If ABCD1 would only function as a heterodimer, X-ALD patients without a mutation in the *ABCD1* gene should exist. These X-ALD patients would carry a mutation in the *ABCD2*,

ABCD3 or *ABCD4* gene instead. This notion is supported by mutation analysis in patients with sitosterolemia which is caused by a mutation in either the *ABCG5* or the *ABCG8* gene. Moreover, ABCD1 and ABCD2 have strikingly different patterns of expression in mouse tissues.[43,44] In mouse brain, ABCD2 is expressed at high levels in the neurons from the cerebral cortex, in the pyramidal and granular cell layers of the dentate gyrus of the hypoccampus, and in Purkinje cells and granular cells in the cerebellum. In the brain and spinal cord, ABCD1 is not expressed in neurons but in glial cells alongside ABCD2, and also in endothelial cells which do not express ABCD2. In the adrenal gland, ABCD2 is expressed in the medulla and ABCD1 in the adrenal cortex. Langmann and coworkers[45] generated expression profiles of all known human ABC transporters in 20 different human tissues using real-time RT–PCR analysis and showed that the four peroxisomal half-transporters exhibit distinct expression patterns. ABCD1 expression is highest in lung, testis, placenta and uterus. ABCD2 is most abundant in brain, trachea and uterus, ABCD3 in trachea, testis and prostate and ABCD4 is expressed at high levels in many tissues, including trachea, lung, testis, uterus, prostate, placenta and spleen. During mouse brain development, the four ABCD genes are also regulated differently: *ABCD1* mRNA is most abundant in embryonic brain and gradually decreases during maturation; *ABCD2* and *ABCD4* mRNA increases in the early postnatal period; and the amount of *ABCD3* transcript increases slightly during the second and third postnatal week.[46]

Taken together, these data make it less likely that ABCD1, ABCD2 and ABCD3 act as obligate partners and imply that the proteins are likely to function as homodimers rather than heterodimers.

9. ABCD2, ABCD3 and ABCD4 Transporters and Their Overlapping Roles in Fatty Acid Metabolism

The function of the other three mammalian peroxisomal ABC transporters has remained largely unresolved (Fig. 3). Up to now,

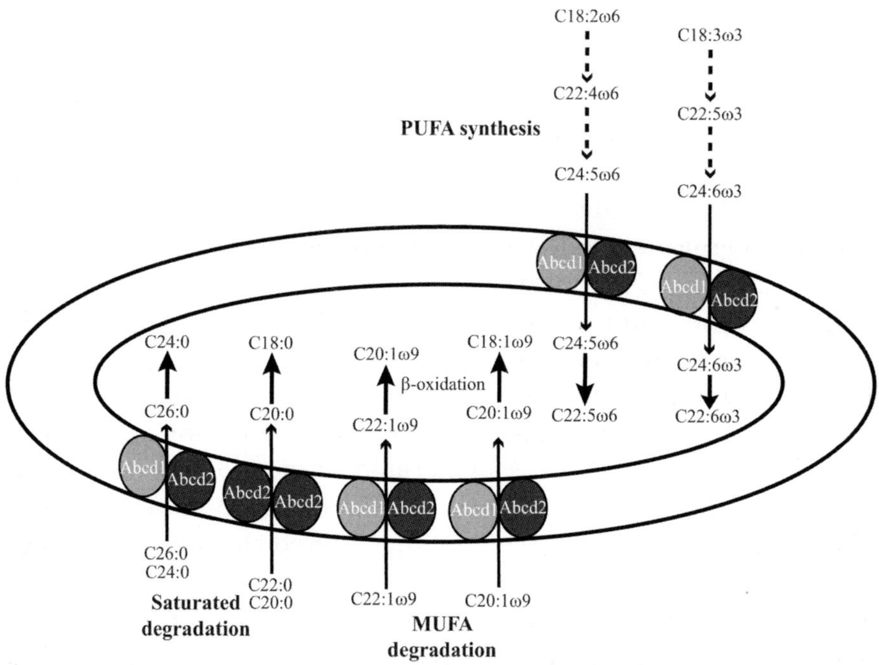

Fig. 3. Schematic depiction of the putative substrates of Abcd1 and Abcd2 transporters.

no mutations of ABCD2, ABCD3 and ABCD4 genes have been found to be associated with a human disease. The four human ABCD transporters share a high sequence homology, suggesting similarity of function to a certain degree. ABCD2 shares 66%, ABCD3 35%, and ABCD4 25% amino acid identity with ABCD1,[4,28] which suggest potential functional redundancy or at least overlapping roles. Even though the exact functions of ABCD2, ABCD3 and ABCD4 remain to be precisely identified, it is established that ABCD2 and ABCD3 are at least partially functionally redundant and have overlapping substrate specificities with ABCD1.[47–49] Over-expression of ABCD3 in X-ALD fibroblasts partially corrects the deficiency in VLCFA β-oxidation, and over-expression of ABCD2 results in complete correction of VLCFA degradation. Most importantly, we have reported a functional overlap between ABCD1 and ABCD2 transporters on the metabolism of VLCFAs *in vivo*. Indeed, ubiquitous overexpression of *Abcd2* gene fully normalizes not only VLCFA profiles in the nervous system and adrenal glands from *Abcd1* knock-out mice, but also results in a striking prevention of axonal degeneration in the spinal cord.[50] Additional evidence of redundant function between ABCD1 and ABCD2 can be found in the fact that *Abcd2* knockout mice accumulate hexacosanoic acid (C26:0), the main fatty acid accumulated in Abcd1 knockouts, in a specific cell population that does not express Abcd1 to detectable levels: the sensitive neurons of dorsal root ganglia.[43]

However, the differences in sequence, expression patterns,[44] and phenotypes of *Abcd1* and *Abcd2* knock-out mice suggest specific and distinct roles for ABCD2 in fatty acid homeostasis. Indeed, *Abcd2–/–* mice develop a late-onset cerebellar and sensory ataxia, areflexia, loss of cerebellar Purkinje cells and dorsal root ganglia cell degeneration that are absent in *Abcd1–/–* mice.[43] This indicates that ABCD2 has a substrate specificity which is likely to be different from that of ABCD1 despite a certain degree of overlap to metabolize C26:0, and/or that some cell types are most sensitive to demise upon loss of function given the difference in the expression patterns of *Abcd1* and *Abcd2* genes.[44]

Very recent findings indicate that ABCD2 plays a role in the degradation of long-chain saturated and ω9-monounsaturated fatty acids since these fatty acids accumulate in the CNS, adrenal glands and adipose tissue of *Abcd2–/–* mice[28,51] Moreover, β-oxidation assays performed *ex-vivo* on cerebral cortex slices confirmed the role of ABCD2 in degradation of C26:0, and interestingly of C24:6 ω3, the precursor of docosahexacosanoic acid (DHA).[28] DHA is the predominant omega-3 polyunsaturated fatty acid (PUFA) in the mammalian nervous system, reaching up to 15% concentration of the total fatty acids in brain.[52]

The functions of the other two mammalian peroxisomal ABC transporters, ABCD3 and ABCD4 have remained largely unresolved. *Abcd3*-knockout mice were found to accumulate pristanic acid and the bile acid precursors THCA (trihydroxycholestanoic acid) and DHCA (dihydroxycholestanoic acid)[53] and Wanders personal communication), strongly suggesting that ABCD3 could be involved in the import of these branched fatty acids into the peroxisomes, presumably in the form of their corresponding CoA esters.

The function of the fourth and most enigmatic member of the ABCD family, ABCD4, remains unknown at this time. Even the intracellular localization of ABCD4 remains controversial as conflicting results localize it to peroxisomes[6] or the endoplasmic reticulum[54] in overexpression experiments. Molecular phylogenetic analyses indicate that ABCD4 is the ancestor of the ABCD subfamily.[55] In this scenario, it is tantalizing to speculate that ABCD4 could not have moved from the original membrane of the endoplasmic reticulum, the source that gave birth to the first peroxisomes in early eukaryotes.[15] Of note, the NBD of ABCD4 fails to heterodimerize with the NBD of any other ABCD transporters using the yeast two-hybrid system (Aubourg *et al.*, unpublished). The function of ABCD4 could therefore be quite different from that of other peroxisomal ABC transporters. The generation of an *abcd4* knockout mouse should help to resolve this issue.

10. Inducible Expression of the Mammalian ABCD Proteins and Relevance for X-ALD

The various degrees of functional redundancy among the ABCD transporters hold therapeutic promise for X-ALD, favoring intense investigation on the matter. The expression of two ABCD half-transporters, ABCD2 and ABCD3 can be increased by the exposure to peroxisome proliferators which are synthetic ligands of peroxisome proliferator-activated receptor (PPAR) α. Many genes involved in lipid metabolism, most notably those involved in peroxisomal β-oxidation, contain a peroxisome proliferator response element (PPRE) in their promoter. Peroxisome proliferation is a process mediated in rodents by PPARα and one isoform of the peroxin 11 encoded by the

PEX11α gene. PPARα is predominantly expressed in tissues that metabolize high amounts of fatty acids, such as liver, kidney, heart, and muscle, and to a lesser extent in the brain.[56] Peroxisome proliferators include plasticizers, like DEHP,[57] and fibrates, like ciprofibrate and fenofibrate.[58,59] While treatment of mice with fenofibrate results in a 10-fold increase in the levels of ABCD2 in liver, and ABCD3 in liver and intestine, the expression of *ABCD1* and *ABCD4* genes remain unchanged. The induction of *ABCD2* and *ABCD3* gene expression is PPARα dependent, as no induction of these genes can be demonstrated in fenofibrate treated *PPARα* knockout mice. The *ABCD2* promoter does contain putative PPRE elements, which turned out to be non functional.[60,61] Instead, it appears that the induction of ABCD2 gene expression by fenofibrate is mediated via PPARα but through an indirect mechanism, which might involve the sterol regulatory element binding protein (SREBP) pathway.[62] Unfortunately, fenofibrate treatment does not increase the expression of the *ABCD2* in the brain,[46,49] likely owing to the action of the multidrug-transporter P-glycoprotein (MDR1, *ABCB1*; see Chapter 4) that actively pumps out fenofibrate from the brain.[46] The findings that (i) expression of *Aldr* gene is inducible by pharmacological agents like fibrates, (ii) ABCD2 can functionally compensate for the absence of ABCD1 and (iii) ABCD2 over-expression can prevent the onset of clinical symptoms in *Abcd1* mice, have made *ABCD2* an attractive candidate gene for a pharmacological gene therapy approach in X-ALD. Several pharmacological compounds have been identified that can up-regulate ABCD2 expression both in human and mouse fibroblasts, including thyroid hormone (T_3), fenofibrate and 4-phenylbutyrate (4-PBA). Treatment of rat oligodendroglial CG4 cells and human and mouse ABCD1-deficient fibroblasts with T_3 results in an increased expression of ABCD2, and in correction of the VLCFA levels in fibroblasts.[63] However, treatment of rats with T_3 increases the expression of *Aldr* gene in liver but failed to induce *Aldr* in brain.[63]

Initial experiments with 4-PBA were more promising. Treatment of fibroblasts from X-ALD patients and *Abcd1* knockout mice with 4-PBA resulted in the correction of the peroxisomal β-oxidation of VLCFA, a reduction of the VLCFA levels in peripheral tissues, increased expression of the *ABCD2* gene and peroxisome proliferation.[48] In contrast to that observed with T_3 and fenofibrate, dietary treatment of *Abcd1* knockout mice with 4-PBA resulted in correction of VLCFA levels in the adrenals, but also in brain.[48] These results suggested that the effect of 4-PBA might be mediated through *ABCD2* overexpression or peroxisome proliferation. A more recent publication by Gondcaille *et al.*[64] suggests that 4-PBA may exert its action in brain via a different, yet unclear, mechanism. Treatment of rats with 4-PBA for six weeks with an equivalent dose to the one used for mice,[48] resulted in a two-fold increase in peroxisome abundance and a 1.5–1.9-fold increase in the expression of *Abcd2* gene in the liver. No gene induction or peroxisome proliferation, however, could be observed in kidney, testis, spleen or brain. Interestingly, these effects in liver, including the proliferation of peroxisomes, does not involve the PPARα pathway but the PEX11 genes, which plays a role in peroxisome biogenesis and division.[64,65] To test whether this could be due to the rapid metabolism of 4-PBA by the liver, 4-PBA was perfused directly into the cerebral fourth ventricle, but no increase of Abcd2 expression or peroxisome proliferation was found. 4-PBA is also a weak inhibitor of histone deacetylase 1 (HDAC1) with an IC50 in the millimolar range.[66] McGuinness *et al.*[67] compared 4-PBA with one of the most potent HDAC inhibitors known, trichostatin A (TSA), and demonstrated that treatment of mouse X-ALD fibroblasts with either 5 mM 4-PBA or 50 nM TSA induces both mitochondrial and peroxisomal VLCFA β-oxidation. Interestingly, while 4-PBA treatment results in increased ABCD2 expression, TSA did not affect the expression of *ABCD2* gene, and its effects on brain and adrenal VLCFA levels were not reported.

Because of the promising results obtained with 4-PBA in the cell cultures and the *Abcd1*-deficient mice, a small clinical trial was conducted to investigate the effect of 4-PBA in seven patients with AMN.[68] No side effects were reported other than the inconvenience of taking 40 pills per day. Brain magnetic resonance spectroscopy studies demonstrated that 4-PBA could pass the blood–brain barrier as it was detectable in cerebrospinal fluid and the brain white matter. However, VLCFA measurements in plasma, red blood cells, white blood cells and platelets showed that C22:0 and C24:0 levels were only reduced in platelets, with no changes C26:0 levels. In addition, treatment did not result in increased levels of ABCD2 in white blood cells.[68]

A more potent HDAC inhibitor than 4-PBA, the well known antiepileptic drug valproic acid, has been shown able to upregulate *ALDR* gene levels *ex-vivo*, in rodent and human brain slices, through an HDAC inhibition mechanism.[69] A small pilot clinical trial undertaken has demonstrated that after six months of treatment, VPA is able to increase the protein levels of ABCD2 in peripheral mononuclear cells.[70] Most strikingly, VPA normalizes oxidative lesions to proteins in this specific cell population, thus providing therapeutic hope and warranting the engagement of larger numbers of patients over longer periods of administration in order to evaluate potential therapeutic benefit.

11. Conclusion

Peroxisomes perform a broad range of different functions, dependent upon organism, tissue type or developmental stage, many of which are connected with lipid metabolism and more specifically to β-oxidation of fatty acids. Peroxisomes are selectively permeable to a limited number of solutes and macromolecules and ABCD transporters play a major role in the transport of fatty acids across the peroxisomal membrane. In mammalian cells, the differential expansion of the ABCD family reflects not only functional diversification but also both tissue and cell specificity of peroxisomes. This is illustrated in the central nervous system where peroxisomes exhibit different functions in neurons, oligodendrocytes, astrocytes and microglia with different expression of ABCDs in those cells. This raises a first question of how several apparently different roles for ABCDs are controlled and co-ordinated.

Many biochemical details for peroxisomal import of fatty acids by ABCDs are yet to be determined. Unequivocal demonstration of fatty acid or fatty CoA ester transport will require the reconstitution of mammalian peroxisomal ABC proteins into lipid vesicles, a challenging task given the difficulties of purifying membrane proteins, the limited water solubility of putative substrates and the heterodimeric nature of ABCDs. An additional layer of complexity could be added to the picture by the potential requirement for accessory proteins which may deliver lipophilic substrates to the peroxisome or interact with the transporter on the lumenal side of the membrane.

Then, questions will remain about the mechanisms and the physiological purposes for which peroxisomal ABC transporters mediate and regulate the transport of fatty acids through the peroxisomal membrane. The evidence for the yeast Pxa1p/Pxa2p is direct since β-oxidation is confined to peroxisomes, as in plants. In mammalian cells, the answer is less clear since β-oxidation of fatty acids is shared with the mitochondria. Such studies are key to our understanding of the ABCD subfamily. X-linked adrenoleukodystrophy illustrates best the pivotal role of one ABCD member in the maintenance of axonal and myelin integrity, thus justifying the fascination exerted by this group of ABC transporters.

Acknowledgments

The work was supported by funds from INSERM, the European Leukodystrophy Association (ELA 2008–040C4), the Association Française contre les Myopathies (AFM, Project n° 9315), the Fondo de Investigación Sanitaria (FIS, Fondo de Investigación Sanitaria del Instituto de Salud Carlos III, PI051118 and PI080991 grants). The CIBER de Enfermedades Raras is an initiative of the ISCIII. The work was developed under the COST action BM0604 to which both authors belong.

References

1. De Duve, C. and Baudhuin, P. (1996) *Physiol. Rev.* **46**, 323.
2. Kamijo, K., Taketani, S., Yokota, S., Osumi, T., and Hashimoto, T. (1990) *J. Biol. Chem.* **265**, 4534.
3. Mosser, J., *et al.* (1993) *Nature* **361**, 726.
4. Holzinger, A., Kammerer, S., and Roscher, A.A. (1997) *Biochem. Biophys. Res. Commun.* **237**, 152.
5. Lombard-Platet, G., Savary, S., Sarde, C.O., Mandel, J.L., and Chimini, G. (1996) *Proc. Natl. Acad. Sci. U.S.A.* **93**, 1265.
6. Shani, N., Jimenez-Sanchez, G., Steel, G., Dean, M., and Valle, D. (1997) *Hum. Mol. Genet.* **6**, 1925.
7. Kemp, S., *et al.* (2001) *Hum. Mutat.* **18**, 499.
8. Wanders, R.J., *et al.* (2001) *Biochem. Soc. Trans.* **29**, 250.
9. Wanders, R.J. and Waterham, H.R. (2006) *Annu. Rev. Biochem.* **75**, 295.
10. Rizzo, W.B. (2007) *Mol. Genet. Metab.* **90**, 1.
11. Fujiki, Y., Matsuzono, Y., Matsuzaki, T., and Fransen, M. (2006) *Biochim. Biophys. Acta* **1763**, 1639.
12. Pinto, M.P., *et al.* (2006) *J. Biol. Chem.* **281**, 34492.
13. Eckert, J.H. and Erdmann, R. (2003) *Rev. Physiol. Biochem. Pharmacol.* **147**, 75.
14. Steinberg, S.J., *et al.* (2006) *Biochim. Biophys. Acta.* **1763**, 1733.
15. Schluter, A., *et al.* (2006) *Mol. Biol. Evol.* **23**, 838.
16. Schluter, A., *et al.* (2007) *Nucleic Acids Res.* **35**, D815.
17. Platta, H.W. and Erdmann, R. (2007) *Trends Cell Biol.* **17**, 474.
18. Schrader, M. and Fahimi, H.D. (2006) *Biochim. Biophys. Acta.* **1763**, 1755.
19. Shani, N., Watkins, P.A., and Valle, D. (1995) *Proc. Natl. Acad. Sci. U.S.A.* **92**, 6012.
20. van Roermund, C.W., *et al.* (2008) *FASEB J.* **22**, 4201.
21. Verleur, N., Hettema, E.H., van Roermund, C.W., Tabak, H.F., and Wanders, R.J. (1997) *Eur. J. Biochem.* **249**, 657.
22. Shani, N. and Valle, D. (1996) *Proc. Natl. Acad. Sci. U.S.A.* **93**, 11901.
23. Hettema, E.H., *et al.* (1996) *EMBO J.* **15**, 3813.
24. Gloeckner, C.J., *et al.* (2000) *Biochem. Biophys. Res. Commun.* **271**, 144.
25. Mayerhofer, P.U., Kattenfeld, T., Roscher, A.A., and Muntau, A.C. (2002) *Biochem. Biophys. Res. Commun.* **291**, 1180.

26. Biermanns, M. and Gartner, J. (2001) *Biochem. Biophys. Res. Commun.* **285**, 649.
27. Roerig, P., Mayerhofer, P., Holzinger, A., and Gartner, J. (2001) *FEBS Lett.* **492**, 66.
28. Fourcade, S., *et al.* (2009) *Am. J. Physiol. Endocrinol. Metab.* **296**, E211.
29. Lazo, O., *et al.* (1988) *Proc. Natl. Acad. Sci. U.S.A.* **85**, 7647.
30. Powers, J.M., DeCiero, D.P., Ito, M., Moser, A.B., and Moser, H.W. (2000) *J. Neuropathol. Exp. Neurol.* **59**, 89.
31. Pujol, A., *et al.* (2002) *Hum. Mol. Genet.* **11**, 499.
32. Griffiths, I., *et al.* (1998) *Science* **280**, 1610.
33. Lappe-Siefke, C., *et al.* (2003) *Nat. Genet.* **33**, 366.
34. Fourcade, S., *et al.* (2008) *Hum. Mol. Genet.* **17**, 1762.
35. Brites, P., Mooyer, P.A., El Mrabet, L., Waterham, H.R., and Wanders, R.J. (2009) *Brain* **132**, 482.
36. Kassmann, C.M., *et al.* (2007) *Nat. Genet.* **39**, 969.
37. Paintlia, M.K., Paintlia, A.S., Contreras, M.A., Singh, I., and Singh, A.K. (2008) *Exp. Neurol.* **210**, 560.
38. Kage, K., *et al.* (2002) *Int. J. Cancer* **97**, 626.
39. Hettema, E.H., *et al.* (1998) *J. Cell Biol.* **142**, 421.
40. Liu, L.X., *et al.* (1999) *J. Biol. Chem.* **274**, 32738.
41. Guimaraes, C.P., *et al.* (2004) *Biochim. Biophys. Acta* **1689**, 235.
42. Hillebrand, M., *et al.* (2007) *J. Biol. Chem.* **282**, 26997.
43. Ferrer, I., *et al.* (2005) *Hum. Mol. Genet.* **14**, 3565.
44. Troffer-Charlier, N., *et al.* (1998) *Eur. J. Cell Biol.* **75**, 254.
45. Langmann, T., *et al.* (2003) *Clin. Chem.* **49**, 230.
46. Berger, J., *et al.* (1999) *Eur. J. Biochem.* **265**, 719.
47. Imanaka, T., *et al.* (1999) *J. Biol. Chem.* **274**, 11968.
48. Kemp, S., *et al.* (1998) *Nat. Med.* **4**, 1261.
49. Netik, A., *et al.* (1999) *Hum. Mol. Genet.* **8**, 907.
50. Pujol, A., *et al.* (2004) *Hum. Mol.Genet.* **13**, 2997.
51. Liu, J., *et al.* (2010) *J. Lipid Res.* **51**, 162.
52. Salem, N.J., Kim, H.Y., and Yergey, J.A. (1986) Docosahexaenoic acid: membrane function and metabolism, in: Simopoulos, A., Kifer, R., Martin, R. (eds.) *Health Effects of Polyunsaturated Fatty Acids in Seafoods,* 2nd edition. Academic Press, Orlando, FL, USA, p. 263.
53. Silva-Zolezzi, I., *et al.* (2003) Paper presented at *The 53rd Annual Meeting of the American Society of Human Genetics*, Los Angeles, California, November 4–8, 2003.

54. Kashiwayama, Y., *et al.* (2009) *Exp. Cell Res.* **315**, 190.
55. Schlüter, A., *et al.* (2010) *Nucleic Acids Res.* **38**, D800.
56. Auboeuf, D., *et al.* (1997) *Diabetes* **46**, 1319.
57. Wilcke, M. and Alexson, S.E. (2001) *Biochim. Biophys. Acta* **1544**, 358.
58. Albet, S., *et al.* (1997) *FEBS Lett.* **405**, 394.
59. Fourcade, S., *et al.* (2001) *Eur. J. Biochem.* **268**, 3490.
60. Pujol, A., Troffer-Charlier, N., Metzger, E., Chimini, G., and Mandel, J.L. (2000) *Genomics* **70**, 131.
61. Rampler, H., *et al.* (2003) *Mol. Genet. Metab.* **80**, 398.
62. Weinhofer, I., Forss-Petter, S., Zigman, M., and Berger, J. (2002) *Hum. Mol. Genet.* **11**, 2701.
63. Fourcade, S., *et al.* (2003) *Mol. Pharmacol.* **63**, 1296.
64. Gondcaille, C., *et al.* (2005) *J. Cell Biol.* **169**, 93.
65. Li, X., *et al.* (2002) *Mol. Cell Biol.* **22**, 8226.
66. de Ruijter, A.J., van Gennip, A.H., Caron, H.N., Kemp, S., and van Kuilenburg, A.B. (2003) *Biochem. J.* **370**, 737.
67. McGuinness, M.C., Zhang, H.P., and Smith, K.D. (2001) *Mol. Genet. Metab.* **74**, 256.
68. Moser, H.W., Smith, K.D., Watkins, P.A., Powers, J., and Moser, A.B. (2001) X-linked adrenoleukodystrophy, in: Scriver, C.R., Beaudet, A.L., Sly, W.S., Valle, D. (eds.) *The Metabolic and Molecular Bases of Inherited Disease*, Vol. II. McGraw-Hill, New York, 2001, p. 3303.
69. Pujol, A., *et al.* (2005) *ESHG meeting, Amsterdam.*
70. Fourcade, S., *et al.* (submitted).

Chapter 13

ABCG5/8 and Its Role in Sterol Homeostasis

Ronald P. J. Oude Elferink and Coen C. Paulusma

1. Introduction

Cholesterol homeostasis is a crucial aspect of human pathophysiology. Cholesterol is an essential building block of membranes and the precursor to all steroid hormones. Not only is plasma cholesterol one of the most important risk factors for the development of cardiovascular disease, the formation of cholesterol gallstones is also a highly frequent health problem in humans. The function of biliary cholesterol secretion is actually not quite clear but it certainly adds another level of regulation to body cholesterol homeostasis. The maintenance of body cholesterol homeostasis comprises a highly complex regulatory network that acts at transcriptional, translational and posttranslational levels. In this review we will limit ourselves to the role of the ABC transporter couple ABCG5/ABCG8 in cholesterol homeostasis. Other ABC transporters involved in cholesterol homeostasis will be discussed elsewhere (see Chapter 2). The ABC heterodimer ABCG5/8 is the key transporter for sterol excretion from the liver into bile. Expression of the same transporter in the small intestine prevents the uptake of plant sterols in the circulation and reduces intestinal absorption of cholesterol.

2. The ABC Transporter Couple ABCG5/ABCG8

Like all other members of the ABC G-subfamily, ABCG5 and ABCG8 consist of a single ATP binding cassette in the amino terminus followed by six putative transmembrane helices; to become functionally active they have to dimerize, and hence they are referred to as half-transporters. In contrast to the other G-family members, these two half-transporters represent a heterodimer and no evidence exists that homodimerization of either of the two gives rise to active transport of a substrate across the plasma membrane of cells. The strongest evidence that ABCG5 and ABCG8 act as obligate heterodimers comes from genetic data: sitosterolemia, the disease caused by deficiency of functional sterol transport is caused by mutations in both alleles of either *ABCG5* or *ABCG8*, and nonetheless lead to indistinguishable phenotypes.[1,2] Both half-transporters appear to function as a chaperone for their partner; if only ABCG5 or only ABCG8 was expressed in cultured cells, neither could exit from the endoplasmic reticulum (ER) of cells as witnessed by the absence of terminal glycosylation.[3] In this condition small amounts (in the case of ABCG5) or significant amounts (in the case of ABCG8) of homodimers were formed which, however, cannot be functional since they did not reach the plasma membrane.[3] Similarly, dimerization of ABCG5 or ABCG8 with other members of the G-family did not give rise to functional transporters. When co-expressed, ABCG1, ABCG2, and ABCG4 can be co-immunoprecipitated with ABCG5, and ABCG4 co-immunoprecipitated with G8, but in these cases too, the putative dimers were retained in the ER.[4] Hence, obligate heterodimerization of ABCG5 and ABCG8 occurs in the ER and is necessary for proper trafficking of the functional transporter to the plasma membrane. In polarized (WIF-B9) cells the heterodimer trafficked to the apical (canalicular) membrane.[3]

Wang *et al.* have purified and reconstituted the human heterodimer from tranduced insect cells,[5] as well as from mouse liver.[6] They developed an *in vitro* assay for cholesterol translocation; this is not a simple and straightforward experiment because cholesterol is insoluble in water. To solve this problem, the authors made use of a two vesicle system, preparing proteoliposomes into which the heterodimer (isolated from insect cells overexpressing both half-transporters) was reconstituted, and also unilamelar liposomes containing a mixture of phospholipids and radioactive cholesterol.[5] These preparations were mixed and after incubation the two types of vesicles were separated by density ultracentrifugation and the amount of radioactivity in the acceptor membrane vesicles was measured (because the donor liposomes contain no protein, they have a lower buoyant density). Apparently, there is sufficient transfer of cholesterol from the donor liposomes to the acceptor membrane vesicles to supply enough cholesterol for translocation towards the intravesicular leaflet of these proteoliposomes. Clearly, it was impossible to perform kinetic studies with this assay because the entire process entails several additional transfer steps that may be more rate-limiting than the translocation via ABCG5/8. Nevertheless, the results showed convincingly that the overall transfer of cholesterol was enhanced by co-expression of ABCG5 and ABCG8. In the presence of both proteins as well as ATP and Mg^{2+} transfer of cholesterol was about four-fold higher than in the absence of any of these components. Membranes from cells overexpressing only a single half-transporter were inactive. Hence, both ABCG5 and ABCG8 had to be present, indicating that homodimers are not capable of cholesterol transport. In addition, the process depended entirely on the presence of Mg^{2+} and ATP (GTP, UTP and dCTP produced much lower rates) and the process could effectively be inhibited by vanadate or beryllium fluoride, indicating the requirement for continued ATP hydrolysis. ABCG5/8-mediated cholesterol transfer was not accompanied by transfer of phosphatidylcholine. Interestingly, the rate of transfer with sitosterol was as high as with cholesterol. Since plant sterols are pumped out of the enterocytes much more efficiently than cholesterol it might have been expected that the process is more efficient with plant sterols than with cholesterol. This was not the case, which suggests that the capture of cholesterol by ACAT-mediated esterification within the enterocytes determines the different behavior of cholesterol vs. plant sterols. The group of Urbatsch[7] also purified and reconstituted both half-transporters upon expression in the yeast *Pichia pastoris*. They also showed that the two proteins stabilize each other. In addition, they found that heterodimerization increased the ATPase activity of the transporter couple albeit with a high K_m for Mg^{2+}/ATP (about 4 mM). Strikingly, the presence of cholesterol in the reconstituted lipid mixture had no influence on the ATPase activity of the complex.[7]

The half-transporters ABCG5 and ABCG8 are encoded by two genes that are arranged in a head-to-head configuration on chromosome 2. The genes are separated by a short 374 basepair intergenic sequence, which acts as a bidirectional promoter. The genes are predominantly expressed in liver, intestine and gallbladder.[1,8] Expression of ABCG5 and (to a lesser extent) ABCG8 is regulated by the oxysterol-sensing nuclear receptor LXRα although the responsive element has not been identified.[9] In mice a high cholesterol diet as well as administration of the artificial LXR agonist T0901317 increases *Abcg5/8* expression in the liver and increases biliary cholesterol secretion.[10] The intergenic promoter region also contains elements for LRH-1,[11] as well as GATA4 and HNF4α, which seem to synergistically activate transcription of both genes.[12]

3. Sitosterolemia

The role of the transporter couple ABCG5/8 was discovered by the fact that in patients with sitosterolemia, one of either gene is mutated. Sitosterolemia is a very rare inherited disorder, characterized by up to 100-fold increased serum levels of plant sterols (phytosterols). There are more than 20 different plant sterols which differ from cholesterol (which is not present in plants) by their side-chain configuration. The most common plant sterol is sitosterol which bears an additional ethyl group on the side chain.

The presence of plant sterols in the plasma of sitosterolemia patients is mainly caused by defective extrusion of these sterols in the gut and in the liver. The accumulation of plant sterols has several pathological effects, the main being hemolytic anemia. Other symptoms are thrombocytopenia, hypersplenism, formation of xanthomas, premature cardiovascular disease and atherosclerosis. These symptoms can be largely explained by the accumulation of dietary sterols. For example, in one of the mouse models for sitosterolemia, the *Abcg5*$^{-/-}$ mouse, it could be shown that the thrombocytopenia is caused by the circulating plant sterols: when *Abcg5*$^{-/-}$ mice were lethally irradiated and transplanted with bone marrow from wild type mice, they still developed thrombocytopenia and this could be reversed by administration of ezetimibe, which inhibits the intestinal uptake of plant sterols.[13]

The group of Hobbs[14] showed in *Abcg5/8*$^{-/-}$ mice that selected dietary plant sterols disrupt cholesterol homeostasis in the adrenal gland. The adrenal glands of the *Abcg5/8*$^{-/-}$ mice were grossly abnormal in appearance (brown, not white) due to a 91% reduction in cholesterol content. Adrenal cholesterol levels returned to near-normal levels in mice treated with ezetimibe, which blocks phytosterol absorption. More recently, Mushtaq *et al.*[15] reported on adrenal insufficiency in a family with sitosterolemia, suggesting that accumulation of plant sterols also may affect adrenal function in humans.

Presently, administration of ezetimibe is the treatment of choice in patients with sitosterolemia. Ezetimibe effectively inhibits NPC1L1 (Niemann Pick C 1-like 1 protein), the intestinal transporter that represents the main entry mechanism for sterols in the gut. Lütjohann *et al.*[16] performed a placebo-controlled study in which patients with sitosterolemia (who have a mutation in either *ABCG5* or *ABCG8*) were treated with ezetimibe for two years. Their conclusion was that this regimen was effective in reducing plasma plant sterol concentrations with an overall favorable safety and tolerability profile.

Interestingly, liver transplantation in a patient with sitosterolemia strongly reduced serum plant sterol levels to values only slightly higher than normal,[17] indicating that in the absence of intestinal ABCG5/8 function the liver can ensure sufficient biliary plant sterol secretion to maintain low systemic plant sterol concentrations.

4. Intestinal Sterol Absorption and Excretion

Cholesterol absorption is an important determinant of plasma LDL cholesterol levels. In the past this has led to the conclusion that dietary intake of cholesterol may strongly influence plasma cholesterol levels. However, strong restriction of dietary cholesterol intake has not resulted in significant decreases in LDL cholesterol. The most likely explanation for this lack of effect is the fact that the majority of cholesterol in the gut is derived from biliary cholesterol and not from dietary cholesterol. Only about 25% of the cholesterol that enters the gut is derived from the diet (about 400 mg/day) whereas 75% comes from the liver via bile (about 1.2 gr/day). This means that cholesterol synthesis by the liver to a large extent determines the levels of cholesterol in the intestinal lumen.

Multiple processes determine the net cholesterol absorption in the intestine. Quantitatively, the most important factor is probably uptake in the small intestine via NPC1L1, which mediates facilitated diffusion across the membrane. The exact mechanism by which NPC1L1 mediates this uptake is still unclear. The importance of NPC1L1 is based on two crucial observations. Firstly, mice with a gene disruption for *Npc1l1* have strongly reduced cholesterol absorption.[18,19] Secondly, the development of ezitimibe, a potent inhibitor of intestinal cholesterol uptake and its binding to NPC1L1 showed that this protein plays a crucial role in absorption.[20–22] SR-B1 has also been postulated to play a role in intestinal cholesterol absorption; however, mice with a disrupted Sr-b1 gene do not display reduced intestinal cholesterol absorption.[23,24]

The absorption of cholesterol (and other lipids) depends on solubilization of the lipid by bile salts that are secreted by the liver into bile (see Chapter 3). Mixed micelles of these lipids with bile salts are the vehicle that carries lipid through the unstirred mucin layer to the apical membrane of enterocytes, where the lipid can be presented to uptake transporters. The mechanism by which NPC1L1 transports cholesterol across the membrane is still poorly understood and may involve internalization of cholesterol with NPC1L1 into cells through clathrin/AP2-mediated endocytosis.[25]

NPC1L1 does not only mediate cholesterol absorption but also facilitates plant sterol absorption. This is clear from the fact that hyperabsorption of plant sterols in patients with sitosterolemia can be inhibited by ezetimibe.[26] Our diet thus contains considerable amounts of plant sterols; yet, only very small amounts of these sterols enter the blood circulation. More recently, it has been shown that in mice lacking Abcg5/8 the systemic accumulation of plant sterols could be largely prevented by also disrupting the Npc1l1 gene.[27] These triple knockout mice have almost normal (low) plant sterol levels in their serum and demonstrate that selective sterol absorption is governed by the combined action of uptake via Npc1l1, and efflux via Abcg5/8 (and intestinal esterification).

In contrast to other inherited forms of hypercholesterolemia, sitosterolemia is characterized by increased sensitivity of serum cholesterol to dietary cholesterol.[28] In normal individuals large changes in dietary cholesterol are associated with modest changes in serum cholesterol. In contrast, in sitosterolemic patients diets containing high cholesterol levels elicit severe hypercholesterolemia. Conversely, restriction of dietary cholesterol is able to normalize the serum cholesterol level in these patients. These observations indicate that intestinal ABCG5/8 functions to limit the amount of absorbed cholesterol as it does for plant sterols. The transporter appears to have equal affinity for cholesterol and plant sterols. Yet, the effect of ABCG5/8 function on net absorption of plant sterols is much stronger than on cholesterol; although a normal western diet contains only twice the amount of cholesterol compared to plant sterols, more than

50% of the ingested cholesterol is absorbed whereas <5% of the plant sterols are absorbed. This is mainly caused by the fact that cholesterol is avidly esterified upon entry into enterocytes by the action of acyl-CoA:cholesterol acyltransferase 2 (ACAT2). In contrast, plant sterols are poorly recognized by this enzyme,[29] remain unesterified, and can therefore be pumped back into the intestinal lumen by ABCG5/8. Hence, by its esterification, cholesterol escapes secretion by ABCG5/8 back into the intestinal lumen, whereas this is not the case for the unesterified plant sterols. The esterified cholesterol is efficiently packed into chylomicrons which transport it to liver and peripheral tissues.

It is well established that sloughing of intestinal cells contributes substantially to net cholesterol excretion by the intestine. Since this process occurs along the entire intestinal tract and sterol absorption is most intense in the proximal small intestine, a large fraction of cholesterol from sloughed cells cannot be reabsorbed and will be excreted via the feces. It was recently demonstrated, however, that another, as yet unidentified, mechanism of active cholesterol excretion may take place in the small intestine; van de Velde *et al.*[30,31] unmasked an active mechanism of cholesterol excretion in the proximal small intestine that depends on the presence of phospholipid vesicles as acceptors.

5. Biliary Cholesterol Secretion

Although it has now been firmly established that a large part of biliary cholesterol secretion is driven by the action of the transporter ABCG5/8, biliary cholesterol secretion is a complex process that is still only partly understood. However, several aspects are clear.

Firstly, it is established that biliary lipid secretion is driven by bile salts. In the complete absence of bile salts there is no biliary phospholipid or cholesterol output and there is a curvilinear relationship between biliary bile salt and lipid secretion.[32] In line with this, Vrins *et al.* showed that cholesterol efflux from polarized epithelial cells transduced with ABCG5/8 depends on the presence of bile salts in the medium.[33]

Secondly, the rate of cholesterol secretion into bile strongly depends on the expression level of *Abcg5/8*. In *Abcg5/8*$^{-/-}$ mice cholesterol secretion is reduced by about 80%, in heterozygotes it is reduced by about 40%.[34,35] The Abcg5/8-independent cholesterol secretion may at least in part be mediated by SR-B1.[36] In mice overexpressing ABCG5/8 biliary cholesterol secretion is strongly induced; biliary cholesterol levels varied directly with ABCG5/8 copy number and hepatic mRNA levels over a >16-fold range.[37]

Thirdly, the rate of cholesterol secretion into bile depends on the nature of acceptor micelles or vesicles in bile. Pure bile salt micelles are poor cholesterol solubilizers and solubilization is greatly enhanced by the presence of phospholipids. In *Abcb4*$^{-/-}$ mice cholesterol secretion is very low because of the complete absence of biliary phospholipids.[32] A minimal amount of phospholipids is required to allow full cholesterol secretion.[38] Also in cultured epithelial cells this aspect of ABCG5/8 mediated cholesterol efflux could be replicated.[33]

For a long time it was assumed that bile salts extract phospholipids and cholesterol from the membrane in the form of vesicles. This model became untenable, however, when it became clear that secretion is completely (in the case of phospholipids) or largely (in the case of cholesterol) dependent on protein-mediated translocation of these lipid molecules within the membrane and that deficiency of either of these transport proteins leads to selective absence of the corresponding lipid in bile. Thus, by knocking out the phospholipid translocator Abcb4 in mice, phospholipid secretion was completely abolished, whereas (under the appropriate conditions) cholesterol secretion can be normal.[32] Conversely, the absence of Abcg5/8 reduces cholesterol secretion by about 80% but leaves phospholipid secretion unchanged.[34,35] If biliary lipid secretion would entail the pinching off of pieces of canalicular membrane, the secreted lipid would mirror the composition of the membrane domain from which it originated and phospholipid and cholesterol secretion would always be coupled at least to some extent. It is therefore much more likely that lipid secretion from the outer leaflet of the canalicular membrane occurs in a molecular fashion rather than by extraction of larger membrane domains or even intact vesicles. The exact mechanism by which cholesterol is extracted from the membrane remains elusive. The fact that an ABC transporter (ABCG5/8) is necessary for the large majority of cholesterol secretion strongly suggests that cholesterol cannot be extracted from the membrane nonspecifically. It may be assumed that the canalicular membrane does have substantial amounts cholesterol in its outer leaflet even in the absence of the translocator ABCG5/8. Cholesterol is essential to make the membrane sufficiently resistant towards bile salts which are strong detergents. Membranes that lack cholesterol are highly sensitive to detergents[39] and the fact that patients with sitosterolemia as well as mouse models lacking Abcg5/8 do not have severe liver damage strongly indicates that the outer leaflet of the canalicular membrane does contain cholesterol even in the absence of Abcg5/8. The amount of cholesterol in the outer leaflet may, however, to some extent depend on Abcg5/8 activity, particularly under conditions of high cholesterol secretion rates. The question remains why a translocator such as Abcg5/8 is necessary for the efflux of cholesterol if sufficient cholesterol is already present in the outer leaflet and bile salts are present in the canalicular lumen to extract the cholesterol. In a seminal paper, Small[40] hypothesized that ABCG5/8 is not so much necessary for the actual translocation step (although it most likely functions as a translocator), but rather to subsequently partially extrude the molecule from the lipid bilayer so that it can be accepted by bile salt micelles. In a membrane bilayer mostly composed of sphingomyelin, cholesterol is deeply buried in the outer leaflet and it requires substantial activation energy to lift it out of the leaflet. Hence, after translocation ABCG5/8 may lift cholesterol from the bilayer and the latter step may actually be the crucial rate-controlling step in secretion. These considerations certainly do not exclude the possibility that at higher cholesterol secretion rates Abcg5/8 is necessary to maintain sufficient cholesterol in the outer membrane leaflet. Indeed, Kosters *et al.*[41] observed that when

Abcg8$^{-/-}$ mice were subjected to high bile salt infusion rates (leading to high biliary bile salt secretion rates) these mice were more sensitive to the development of bile salt-induced cholestasis than wild type animals. Under these conditions it could be shown that the cholesterol content of the *Abcg8*$^{-/-}$ canalicular membrane was significantly reduced as compared to wild type.[41] We hypothesize that this represents non-specifically extracted cholesterol. It was recently shown that reduction of the cholesterol content of this membrane impairs the function of the ABC transporters Abcb11 (Bsep) and Abcc2 (Mrp2) and probably other ABC transporters as well.[42] This impaired transporter function by reduced cholesterol is probably the causative factor of bile salt-induced cholestasis.

Notwithstanding the important role of ABCG5/8 in biliary cholesterol secretion it is clear from various reports that a substantial fraction of biliary cholesterol secretion occurs independent of ABCG5/8.[10,34,35] In mice lacking Abcg5 or Abcg8 or both, cholesterol secretion is decreased by 75–80% indicating that 20–25% of the secretion occurs independent of Abcg5/8. The physicochemical state of the canalicular membrane appears to be an important parameter for this independent cholesterol extraction. The canalicular membrane harbors a phospholipid flippase, ATP8B1, that increases detergent resistance of the membrane.[43] ATP8B1 is a P-type ATPase that flips phosphatidylserine from the outer to the inner leaflet of the canalicular membrane. Absence of this flippase leads to severe liver damage and mutations in the *ATP8B1* gene are the cause of a severe inherited liver disease called Progressive Familial Intrahepatic Cholestasis type 1.[44] In mice lacking Atp8b1 there is significantly increased biliary secretion of phosphatidylserine, cholesterol and ectoenzymes from the canalicular membrane. Hence, in the absence of inward PS flipping the canalicular membrane is more sensitive to bile salt-induced membrane damage. We hypothesize that this is caused by extraction from the membrane due to a reduced relative sphingomyelin content of the outer leaflet of the canalicular membrane, indirectly caused by the absence of inward PS flipping. Groen *et al.*[45] recently showed that the observed increase in biliary cholesterol output in Atp8b1-deficient mice was completely independent of Abcg5/8 function. This is in line with the hypothesis that in these mice the canalicular membrane is less resistant towards bile salt-mediated extraction. These studies demonstrate that with reduced lipid asymmetry substantial amounts of cholesterol can be extracted from the canalicular membrane without translocation via Abcg5/8. Our hypothesis on a reduced membrane resistance against detergents in the absence of PS flipping is based on the physicochemical properties of membranes with different phospholipid composition (Fig. 1). At body temperature,

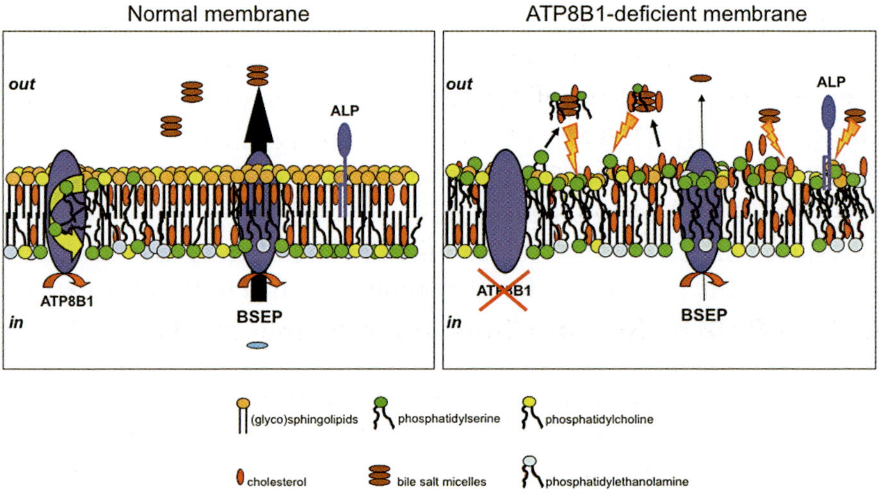

Fig. 1. Hypothetical model that explains the cholestasis in ATP8B1 deficiency. The figure displays a highly schematical representation of bile formation at the canalicular membrane of a healthy subject (left) and of a patient with ATP8B1 deficiency (right). In the ATP8B1-deficient membrane there is no inward flipping of phophatidylserine, which as a result of this is exposed in the exoplasmic leaflet of the canalicular membrane. This increases the amount of phospholipids with unsaturated fatty acylchains and induces a more liquid disordered state which makes the membrane more vulnerable to extraction of cholesterol and ectoenzymes by bile salt micelles. Cholesterol extraction from the membrane leads to impairment of BSEP activity, which causes the cholestasis. ATP8B1, phosphatidylserine flippase; BSEP, the major bile salt transporter; ALP, alkaline phosphatase.

membranes composed of glycerophospholipids (phosphatidylcholine, phosphatidylserine and phosphatidylethanolamine) are in the "liquid-disordered" phase (also referred to as liquid crystalline), in which the lipids are rather loosely packed. The loose packing is mainly caused by the high content of unsaturated fatty acyl chains in glycerophospholipids.[46] Such lipid bilayers are highly prone to intercalation of detergents and subsequent solubilization. However, addition of sphingomyelin and cholesterol to membranes induces a much more rigid membrane structure that is in the "liquid-ordered" phase. This phase represents tighter packing of lipids, mainly owing to the long, saturated fatty acyl chains of sphingomyelins, which combine quite well with the flat structure of the cholesterol molecule. These phases coexist in the membrane and are most likely in a dynamic equilibrium. Membranes consisting of lipids forming a liquid-ordered phase were found to be resistant to the detergent Triton X-100.[47] Unfortunately, elaborate studies of this kind have not been performed with bile salts. Some studies do, however, suggest that a similar behavior does occur with bile salts.[39]

6. The Role of Bile in Cholesterol Homeostasis

Recent reports on mice overexpressing human ABCG5/8 in the liver[48,49] have underscored that cholesterol homeostasis critically depends on the balance between biliary cholesterol secretion and intestinal cholesterol absorption. Specific overexpression of ABCG5/8 in the liver (2- to 3-fold) increased biliary cholesterol secretion about 1.7-fold when mice were fed a cholesterol-rich diet.[49] This had, however, no effect on plasma cholesterol or triglyceride levels, suggesting that the excess cholesterol in the gut was efficiently reabsorbed. When these experiments were performed in mice that also lacked the LDL-receptor or apoE (which induces symptoms of atherosclerosis) there was no difference in the number of aortic lesions. However, when the same experiments were done with simultaneous feeding of ezetimide to inhibit intestinal sterol absorption, there was a significant decrease in plasma cholesterol as well as in the number of aortic lesions.[48] Wilund *et al.*[50] found very different results when the ABCG5/8 transporter was overexpressed in both liver and intestine (and other tissues): in these mice overexpression against a background of LDL receptor deficiency did protect against aortic lesions (even without administration of ezetimibe). This result is not contradictory to that described above[49] as simultaneous expression of ABCG5/8 in liver and intestine increases biliary cholesterol secretion but also prevents reabsorption in the gut. Indeed, in these mice fractional absorption of cholesterol was about 60% of that in non-overexpressing animals, whereas in mice with hepatic overexpression fractional cholesterol absorption was equal to that of the wild type. Hence, biliary cholesterol secretion on its own does not regulate cholesterol homeostasis and an increase in biliary cholesterol secretion should be accompanied by a decreased intestinal absorption to achieve a reduction of the cholesterol pool.

7. ABCG5/8 and Gallstones

Cholesterol gallstones are a major clinical problem considering the fact that about 25% of the western population develops gallstones at some point in life. Although in many cases this does not lead to clinical complications, cholecystectomy needs to be performed frequently. The underlying problem of gallstone formation is the virtual insolubility of cholesterol in water. The solubility of cholesterol is critically dependent on the amount of bile salts and phospholipids in bile. The average bile composition of western individuals is such that cholesterol concentration is around its point of saturation and this is expressed by the cholesterol saturation index. This index is the result of a polynomial equation that depends on the relative bile salt, phospholipid and cholesterol concentration in a given bile sample. Basically, increased cholesterol concentrations and decreased phospholipid concentrations increase the risk of precipitation of cholesterol from the solution. This leads to the formation of cholesterol crystals in bile which are the basis of stone formation. Cholesterol gallstones are composed of cholesterol crystals glued together by proteins present in bile. Not only the biliary lipid composition but also gallbladder motility and biliary protein content are factors that contribute to gallstone formation.[51] As mentioned above, low phospholipid and high cholesterol concentrations lead to cholesterol crystallization; this means that increased cholesterol secretion or decreased phospholipid secretion into bile predisposes to gallstone formation. Therefore, the activity of the corresponding canalicular transporters, ABCG5/8 and ABCB4 respectively, need to be tightly regulated in order to prevent this pathological process. In recent years it has become clear that mutations or variants of these genes are, indeed, associated with gallstone formation. Indeed, it could be shown that impaired biliary phospholipid secretion in *Abcb4*$^{-/-}$ mice gives rise to gallstone formation because insufficient phospholipid is present to properly solubilized cholesterol.[52] Meanwhile, a large set of mutations in the human *ABCB4* gene has been recognized that leads to what now is called "Low phospholipid associated cholelithiasis".[53] These mutations most likely impair expression or activity of ABCB4 resulting in reduced amounts of phospholipid in bile. This leads to a reduced solubilizing capacity of bile for cholesterol and crystal formation. These mutations are not strong enough to cause complete absence of phospholipids in bile because in that case a much more severe phenotype is found (Progressive Familial Intrahepatic Cholestasis type 3).[53]

Conversely, overexpression of the cholesterol efflux system, i.e. ABCG5/8, may be expected to cause cholesterol crystallization as well. In mice, cholesterol secretion into bile is relatively low, so crystals are not easily formed even if mice are fed a "lithogenic diet" (0.5% cholate, 1% cholesterol and 15% fat). Indeed, in *Abcg5/8* overexpressing mice, cholesterol secretion into bile was five-fold increased compared to wild type animals without detection of cholesterol crystals.[54] However, in liver-specific insulin receptor knockout (LIRKO) mice that were fed a lithogenic diet, *Abcg5/8* expression was only 3-fold increased and yet gallstone formation was observed.[55] This was attributed to the fact that besides increased cholesterol secretion these mice also produced a more "lithogenic" bile salt composition containing more hydrophobic bile salt.[55] Humans secrete much higher amounts of cholesterol (near the point of saturation) and any increase in *ABCG5/8* expression or activity may predispose to cholesterol crystal formation. Interestingly, several studies have screened for association between polymorphisms in ABCG5 and ABCG8 on the one hand and gallstone disease on the other. One polymorphism in ABCG8 (D19H) was consistently found to occur more frequently in gallstone patient populations as compared with controls.[56–58] On the basis of this association it may be postulated that the D19H polymorphism leads to increased cholesterol secretion into bile. In line with this contention lower serum plant sterol levels were observed in carriers of the D19H polymorphism compared to others.[59] In addition, reduced cholesterol absorption was reported to be associated with the D19H polymorphism.[60] Thus far, it has been impossible to perform quantitative analysis of ABCG5/8 activity, so it cannot be decided whether the D19H polymorphism lead to increased activity, stability or expression of ABCG8. Other polymorphisms, like E604Q in ABCG5 and T400K and C54Y in ABCG8 were also found to correlate with gallstone disease albeit less significantly and/or consistently. Interestingly, a recent large genome-wide association study reported that a polymorphism in ABCG5 (R50C, rs6756629) associates with total and LDL cholesterol as well as with the risk for coronary heart disease.[61] Finally, other genetic factors that cause increased *ABCG5/8* expression will contribute to gallstone formation. In this context it should be noted that in a small Chinese cohort of gallstone patients a higher expression level of LXRα was observed as compared with gallstone-free subjects. A significant correlation between LXRα and ABCG5 expression was found which is in line with the fact that LXR regulates expression of *ABCG5/8*.[62] Further discussion of the role of nuclear hormone receptors on ABC gene expression can be found in Chapter 14.

The aforementioned observations underscore the importance for maintaining a tight regulation of cholesterol homeostasis. Although the role of ABCG5/8 in this homeostasis is indisputable, it is still poorly understood how ABCG5/8 functions at the molecular level. Is it required for translocation of cholesterol within the bilayer or does it primarily function as a "liftase"? Are these two independent activities and how is ABCG5/8 molecular activity regulated? Are there co-factors involved? All questions that need to be addressed in future research.

References

1. Berge, K.E., Tian, H., Graf, G.A., Yu, L., Grishin, N.V., Schultz, J., Kwiterovich, P., Shan, B., Barnes, R., and Hobbs, H.H. (2000) *Science* **290**, 1771.
2. Lu, K., Lee, M.H., Hazard, S., Brooks-Wilson, A., Hidaka, H., Kojima, H., Ose, L., Stalenhoef, A.F., Mietinnen, T., Bjorkhem, I., *et al.* (2001) *Am. J. Hum. Genet.* **69**, 78.
3. Graf, G.A., Li, W.P., Gerard, R.D., Gelissen, I., White, A., Cohen, J.C., and Hobbs, H.H., (2002) *J. Clin. Invest.* **110**, 659.
4. Graf, G.A., Yu, L., Li, W.P., Gerard, R., Tuma, P.L., Cohen, J.C., and Hobbs, H.H. (2003) *J. Biol. Chem.* **278**, 48275.
5. Wang, J., Sun, F., Zhang, D.W., Ma, Y., Xu, F., Belani, J.D., Cohen, J.C., Hobbs, H.H., and Xie, X.S. (2006). *J. Biol. Chem.* **281**, 27894.
6. Wang, J., Zhang, D.W., Lei, Y., Xu, F., Cohen, J.C., Hobbs, H.H., and Xie, X.S. (2008). *Biochemistry* **47**, 5194.
7. Wang, Z., Stalcup, L.D., Harvey, B.J., Weber, J., Chloupkova, M., Dumont, M.E., Dean, M., and Urbatsch, I.L. (2006). *Biochemistry* **45**, 9929.
8. Tauscher, A. and Kuver, R. (2003) *Biochem. Biophys. Res. Commun.* **307**, 1021.
9. Repa, J.J., Berge, K.E., Pomajzl, C., Richardson, J.A., Hobbs, H., and Mangelsdorf, D.J. (2002) *J. Biol. Chem.* **277**, 18793.
10. Plösch,T., van der Veen, J.N., Havinga, R., Huijkman, N.C., Bloks, V.W., and Kuipers, F. (2006) *Am. J. Physiol. Gastrointest. Liver Physiol.* **291**, G414.
11. Freeman, L.A., Kennedy, A., Wu, J., and Bark, S. (2004) *J. Lipid Res.* **45**, 1197.

12. Sumi, K., Tanaka, T., Uchida, A., Magoori, K., Urashima, Y., Ohashi, R., Ohguchi, H., Okamura, M., Kudo, H., Daigo, K., *et al.* (2007) *Mol. Cell Biol.* **27**, 4248.
13. Kruit, J.K., Drayer, A.L., Bloks, V.W., Blom, N., Olthof, S.G., Sauer, P.J., de, H.G., Kema, I.P., Vellenga, E., and Kuipers, F. (2008) *J. Biol. Chem.* **283**, 6281.
14. Yang, C., Yu, L., Li, W., Xu, F., Cohen, J.C., and Hobbs, H.H. (2004) *J. Clin. Invest.* **114**, 813.
15. Mushtaq, T., Wales, J.K., and Wright, N.P. (2007) *Eur. J. Endocrinol.* **157**(Suppl 1), S61.
16. Lutjohann, D., von, B.K., Sirah, W., Macdonell, G., Johnson-Levonas, A.O., Shah, A., Lin, J., Sapre, A., and Musliner, T. (2008) *Int. J. Clin. Pract.* **62**, 1499.
17. Miettinen, T.A., Klett, E.L., Gylling, H., Isoniemi, H., and Patel, S.B. (2006) *Gastroenterology* **130**, 542.
18. Altmann, S.W., Davis, H.R. Jr., Zhu, L.J., Yao, X., Hoos, L.M., Tetzloff, G., Iyer, S.P., Maguire, M., Golovko, A., Zeng, M., *et al.* (2004) *Science* **303**, 1201.
19. Davis, H.R. Jr., Zhu, L.J., Hoos, L.M., Tetzloff, G., Maguire, M., Liu, J., Yao, X., Iyer, S.P., Lam, M.H., Lund, E.G., *et al.* (2004) *J. Biol. Chem.* **279**, 33586.
20. Rosenblum, S.B., Huynh, T., Afonso, A., Davis, H.R. Jr., Yumibe, N., Clader, J.W., and Burnett, D.A. (1998) *J. Med. Chem.* **41**, 973.
21. Garcia-Calvo, M., Lisnock, J., Bull, H.G., Hawes, B.E., Burnett, D.A., Braun, M.P., Crona, J.H., Davis, H.R. Jr., Dean, D.C., Detmers, P.A., *et al.* (2005) *Proc. Natl. Acad. Sci. U.S.A.* **102**, 8132.
22. Dujovne, C.A., Ettinger, M.P., McNeer, J.F., Lipka, L.J., LeBeaut, A.P., Suresh, R., Yang, B., and Veltri, E.P. (2002) *Am. J. Cardiol.* **90**, 1092.
23. Altmann, S.W., Davis, H.R. Jr., Yao, X., Laverty, M., Compton, D.S., Zhu, L.J., Crona, J.H., Caplen, M.A., Hoos, L.M., Tetzloff, G., *et al.* (2002) *Biochim. Biophys. Acta.* **1580**, 77.
24. Mardones, P., Quinones, V., Amigo, L., Moreno, M., Miquel, J.F., Schwarz, M., Miettinen, H.E., Trigatti, B., Krieger, M., VanPatten, S., *et al.* (2001) *J. Lipid Res.* **42**, 170.
25. Ge, L., Wang, J., Qi, W., Miao, H.H., Cao, J., Qu, Y.X., Li, B.L., and Song, B.L. (2008) *Cell Metab.* **7**, 508.
26. Salen, G., von, B.K., Lutjohann, D., Kwiterovich, P., Kane, J., Patel, S.B., Musliner, T., Stein, P., and Musser, B. (2004) *Circulation* **109**, 966.

27. Tang, W., Ma, Y., Jia, L., Ioannou, Y.A., Davies, J.P., and Yu, L. (2009) *J. Lipid Res.* **50**, 293.
28. Belamarich, P.F., Deckelbaum, R.J., Starc, T.J., Dobrin, B.E., Tint, G.S., and Salen, G. (1990) *Pediatrics* **86**, 977.
29. Temel, R.E., Gebre, A.K., Parks, J.S., and Rudel, L.L. (2003) *J. Biol. Chem.* **278**, 47594.
30. van der Velde, A.E., Vrins, C.L., van den, O.K., Seemann, I., Oude Elferink, R.P., van, E.M., Kuipers, F., and Groen, A.K. (2008) *Am. J. Physiol. Gastrointest. Liver Physiol.* **295**, G203.
31. van der Velde, A.E., Vrins, C.L., van den, O.K., Kunne, C., Oude Elferink, R.P., Kuipers, F., and Groen, A.K. (2007) *Gastroenterology* **133**, 967.
32. Oude Elferink, R.P., Ottenhoff, R., van Wijland, M., Frijters, C.M., van Nieuwkerk, C., and Groen, A.K. (1996) *J. Lipid Res.* **37**, 1065.
33. Vrins, C., Vink, E., Vandenberghe, K.E., Frijters, R., Seppen, J., and Groen, A.K. (2007) *FEBS Lett.* **581**, 4616.
34. Klett, E.L., Lu, K., Kosters, A., Vink, E., Lee, M.H., Altenburg, M., Shefer, S., Batta, A.K., Yu, H., Chen, J., *et al.* (2004) *BMC. Med.* **2**, 5.
35. Yu, L., Hammer, R.E., Li-Hawkins, J., Von Bergmann, K., Lutjohann, D., Cohen, J.C., and Hobbs, H.H. (2002) *Proc. Natl. Acad. Sci. U.S.A.* **99**, 16237.
36. Wiersma, H., Gatti, A., Nijstad, N., Oude Elferink, R.P., Kuipers, F., and Tietge, U.J. (2009) *Hepatology* **50**, 1263.
37. Yu, L., Gupta, S., Xu, F., Liverman, A.D., Moschetta, A., Mangelsdorf, D.J., Repa, J.J., Hobbs, H.H., and Cohen, J.C. (2004) *J. Biol. Chem.* **280**, 8742.
38. Smith, A.J., de Vree, J.M., Ottenhoff, R., Oude Elferink, R.P., Schinkel, A.H., and Borst, P. (1998) *Hepatology* **28**, 530.
39. Oude Elferink, R.P., Paulusma, C.C., and Groen, A.K. (2006) *Gastroenterology* **130**, 908.
40. Small, D.M. (2003) *Proc. Natl. Acad. Sci. U.S.A.* **100**, 4.
41. Kosters, A., Kunne, C., Looije, N., Patel, S.B., Oude Elferink, R.P., and Groen, A.K. (2006) *J. Lipid Res.* **47**, 1959.
42. Paulusma, C.C., de Waart, D.R., Kunne, C., Mok, K.S., and Elferink, R.P. (2009) *J. Biol. Chem.* **284**, 9947.
43. Paulusma, C.C., Groen, A., Kunne, C., Ho-Mok, K.S., Spijkerboer, A.L., Rudi de, W.D., Hoek, F.J., Vreeling, H., Hoeben, K.A., van, M.J., *et al.* (2006) *Hepatology* **44**, 195.

44. Paulusma, C.C. and Oude Elferink, R.P. (2005) *Biochim. Biophys. Acta.* **1741**, 11.
45. Groen, A., Kunne, C., Jongsma, G., van den, O.K., Mok, K.S., Petruzzelli, M., Vrins, C.L., Bull, L., Paulusma, C.C., and Oude Elferink, R.P. (2008) *Gastroenterology* **134**, 2091.
46. Simons, K. and Vaz, W.L. (2004) *Annu. Rev. Biophys. Biomol. Struct.* **33**, 269.
47. Schroeder, R.J., Ahmed, S.N., Zhu, Y., London, E., and Brown, D.A. (1998) *J. Biol. Chem.* **273**, 1150.
48. Basso, F., Freeman, L.A., Ko, C., Joyce, C., Amar, M.J., Shamburek, R.D., Tansey, T., Thomas, F., Wu, J., Paigen, B., *et al.* (2007) *J. Lipid Res.* **48**, 114.
49. Wu, J.E., Basso, F., Shamburek, R.D., Amar, M.J., Vaisman, B., Szakacs, G., Joyce, C., Tansey, T., Freeman, L., Paigen, B.J., *et al.* (2004) *J. Biol. Chem.* **279**, 22913.
50. Wilund, K.R., Yu, L., Xu, F., Hobbs, H.H., and Cohen, J.C. (2004) *J. Lipid Res.* **45**, 1429.
51. Van Erpecum, K.J. (2005) *Biol. Cell.* **97**, 815.
52. Lammert, F., Wang, D.Q., Hillebrandt, S., Geier, A., Fickert, P., Trauner, M., Matern, S., Paigen, B., and Carey, M.C. (2004) *Hepatology* **39**, 117.
53. Oude Elferink, R.P. and Paulusma, C.C. (2007) *Pflugers Arch.* **453**, 601.
54. Yu, L., Li-Hawkins, J., Hammer, R.E., Berge, K.E., Horton, J.D., Cohen, J.C., and Hobbs, H.H. (2002) *J. Clin. Invest.* **110**, 671.
55. Biddinger, S.B., Haas, J.T., Yu, B.B., Bezy, O., Jing, E., Zhang, W., Unterman, T.G., Carey, M.C., and Kahn, C.R. (2008) *Nat. Med.* **14**, 778.
56. Buch, S., Schafmayer, C., Volzke, H., Becker, C., Franke, A., von Eller-Eberstein, H., Kluck, C., Bassmann, I., Brosch, M., Lammert, F., *et al.* (2007) *Nat. Genet.* **39**, 995.
57. Kuo, K.K., Shin, S.J., Chen, Z.C., Yang, Y.H., Yang, J.F., and Hsiao, P.J. (2008) *Br. J. Surg.* **95**, 1005.
58. Grunhage, F., Acalovschi, M., Tirziu, S., Walier, M., Wienker, T.F., Ciocan, A., Mosteanu, O., Sauerbruch, T., and Lammert, F. (2007) *Hepatology* **46**, 793.
59. Berge, K.E., von, B.K., Lutjohann, D., Guerra, R., Grundy, S.M., Hobbs, H.H., and Cohen, J.C. (2002) *J. Lipid Res.* **43**, 486.

60. Gylling, H., Hallikainen, M., Pihlajamaki, J., Agren, J., Laakso, M., Rajaratnam, R.A., Rauramaa, R., and Miettinen, T.A. (2004) *J. Lipid Res.* **45**, 1660.
61. Aulchenko, Y.S., Ripatti, S., Lindqvist, I., Boomsma, D., Heid, I.M., Pramstaller, P.P., Penninx, B.W., Janssens, A.C., Wilson, J.F., Spector, T., *et al.* (2009) *Nat. Genet.* **41**, 47.
62. Jiang, Z.Y., Parini, P., Eggertsen, G., Davis, M.A., Hu, H., Suo, G.J., Zhang, S.D., Rudel, L.L., Han, T.Q., and Einarsson, C. (2008) *J. Lipid Res.* **49**, 464.

Chapter 14

Hot Topics, Recent Advances and Novel Approaches for Clinical Intervention

Irwin M. Arias, Saskia W. C. van Mil, Heidrun Potschka, Alan S. Verkman, Uta Griesenbach, Eric W. F. W. Alton, Balazs Sarkadi and Gergely Szakacs

Introduction

The following five sections illustrate two important emerging themes in translational research concerning ABC transporters: treatment by gene replacement or small molecules, and basic research to elucidate new therapeutic targets. Although each article focuses on a specific disease or group, it is likely that the results have broad application to other inherited and acquired ABC transporter diseases. Research in inherited ABC transporter diseases initially focused on identification of specific genetic mutations which result in the observed phenotype. Subsequent inquiry has expanded to consider cell biologic aspects, including regulation of hepatobiliary ABC transporter synthesis, processing, trafficking, recycling, membrane insertion, degradation and turnover. The individual components involved in these processes are usually not distinct for ABC transporters; however, acquired or inheritable defects in these components can produce the same phenotype as observed with transcriptional ABC transporter mutations. The concept that acquired diseases, including infectious, immune- or drug-mediated, or neoplastic, can manifest ABC transporter phenotypes by virtue of altered expression and regulation is relatively new and worthy of more extensive study. An example is described by Potschka in Sec. 2 in which multidrug-resistance ABCB1 abnormalities in epilepsy are treated with Cox-2 inhibitors.

Misfolding of polytopic ABC transporters, particularly CFTR (ABCC7), focuses attention on cellular processes for proper folding and quality control of membrane proteins in the ER-Golgi system. Advances in understanding the complex process of proper protein folding have led to new therapeutic efforts directed to the basic cell biologic problem (i.e. protein processing) rather than the primary genetic defect. The article by Verkman (Sec. 3) elaborates this theme and reviews the current status of efforts to modify CFTR by mitigating the misfolding of delta F508 mutated CFTR. Similar studies have been described involving alpha trypsin inhibitor deficiency and other non-ABC transporter diseases.

Discovery of inheritable ABC genetic defects prompted interest in gene and cell therapy to correct the dysfunction. Initial enthusiasm for gene therapy, particularly in CF, has waned due to unanticipated problems in gene delivery, targeting, persistence and regulation. The article by Griesenbach and Alton reviews progress in these efforts (Sec. 4). The recent discovery that virtually any nucleated mammalian cell can be reprogrammed into a developmental pathway by expression of only four specific genes followed by culture in specific growth medium and creation of iPS cells dramatically increases opportunities for possible cell-based therapy in selected inheritable diseases. Phenotypically normal hepatocytes have been shown to correct albeit transiently, Crigler-Najjar syndrome resulting from mutations in UDP glucuronyl transferase; however, similar attempts have not been reported in patients with other liver-specific ABC transporter mutations, such as PFIC2, PFIC3 or the Dubin-Johnson syndrome.

Over 15 years ago, Igor Roninson described that progenitor hematopoietic cells express ABCB1. This was the first observation indicating that progenitor cells, including stem cells, express one or more ABC transporters. Subsequent studies revealed that virtually every normal organ has "side population" cells, which are defined as manifesting exclusion of a Hoechst dye which is eliminated when ABC transporter inhibitors are added. Whether all of these cells express ABC multidrug transporters is not established. The recent focus on cancer stem cells has extended the original observation and resulted in efforts to target these cells with therapeutic agents to ameliorate the cancer. The concept has emerged that failure to eradicate cancer stem cells only palliates but cannot cure. The current hypothesis is that certain ABC transporters, particularly ABCG2 and ABCB1, protect vulnerable stem (and other progenerative) cells from death due to multiple metabolic or environmental factors. Alternative proposals include the possibility that ABC transporters may secrete factors essential for cell and organ development. To date, the only established example in biology for this alternative possibility is secretion of A factor, which is required for mating, by an ABC transporter in budding yeast. The article by Sarkadi and Szakacs (Sec. 5) reviews the status of cancer stem cells from the perspective of ABC transporters.

Each article reviews the current status and proposes new directions for research. The bottom line is that considerably more study is required in all of these areas, each of which holds considerable promise for translational research and new therapies.

1. The ABC of Hepatobiliary ABC-Transporter Regulation by Nuclear Receptors

The expression of several hepatobiliary ABC transporters is under tight transcriptional regulation. In the past decade, the superfamily of Nuclear Receptors have been found to play an important role in this regulation.

Nuclear receptors (NRs) are ligand-activated transcription factors that regulate transcription of many target genes in a coordinated fashion. NRs function by transactivation of target genes via binding to consensus NR-response elements (NR-RE). Upon ligand binding, the nuclear receptor releases corepressor (CoR) complexes and interacts with multiprotein coactivator complexes, which modify chromatin structure, yielding access to the promoter DNA for general transcription factors and RNA polymerase II. Transcription is subsequently initiated. As such, NRs mediate a variety of physiological processes, including reproduction, metabolism, development and differentiation. NRs are unique transcription factors because their activity is regulated by specific ligands that easily pass biological membranes, making them ideal drug targets. The most important and extensively studied nuclear receptors relevant in transcriptional regulation of hepatobiliary ABC transporters are FXR (NR1H4), the farnesoid X receptor, LXR (NR1H2/3), the liver X receptor, and PXR/SXR (NR1I2), the pregnane/steroid X receptor (expressed in rodents and humans, respectively) and CAR (NR1I3), the Constitutive Androgen Receptor. All function as obligate heterodimers with the retinoid X receptor (RXR) as a common partner. In this chapter, we will highlight the most important NR regulation mechanisms of hepatobiliary ABC transporters and evaluate the therapeutic potential of those NRs.

1.1 *The Bile Salt Sensor Farnesoid X Receptor*

Farnesoid X receptor (FXR, NR1H4) is a nuclear receptor activated by endogenous bile salts, which are hormone-like molecules that circulate between the liver and intestine and are crucial for the uptake of dietary fats and fat-soluble vitamins from the intestinal lumen. However, bile salts are highly toxic. FXR is the master regulator of bile salt concentrations in cells by regulating the inward and outward bile salt transport systems, bile salt metabolism and the *de novo* synthesis in the liver.[1] FXR is expressed most prominently in liver and small intestine, but also in stomach, kidney, adrenal glands, adipose tissue and immune cells. In the majority of cases, FXR modulates the expression of its target genes by binding as a heterodimer with

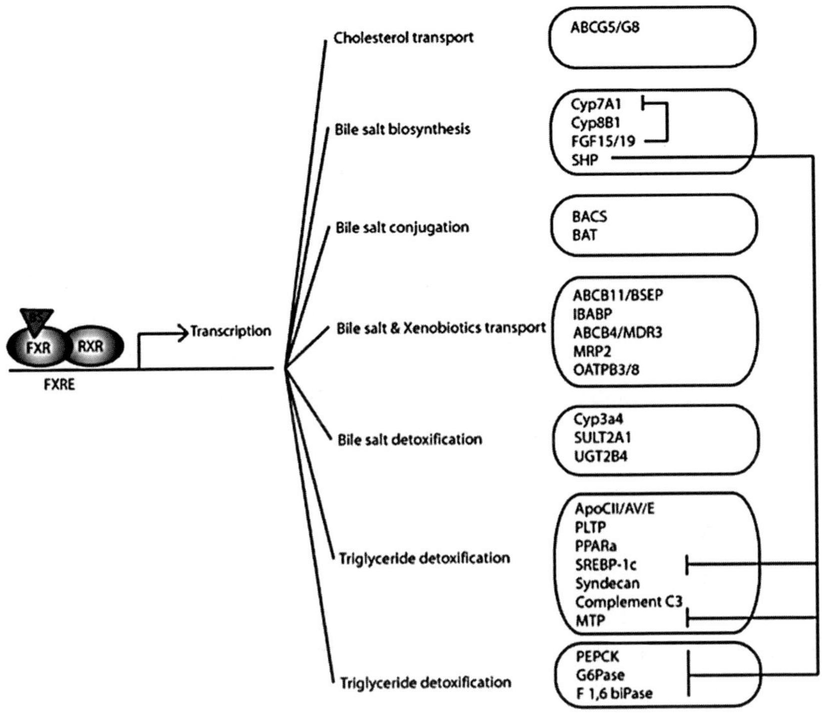

Fig. 1. Simplified gene regulation controlled by FXR. The main biological functions and genes regulated by FXR are indicated.

the retinoid X receptor (RXR, NR2B1) to DNA sequence motifs called FXR response elements (FXREs). Figure 1 shows the network of genes that are transcriptionally regulated by FXR. FXR regulates not only bile salt homeostasis, but also cholesterol, fat and glucose metabolism (Fig. 1).[1] More recently, the functions of FXR have been extended to regulation of inflammation in liver[2] and intestine,[3–5] liver growth/regeneration,[6,7] and carcinogenesis,[8–10] although the exact mechanisms remain elusive.

1.2 *The cholesterol sensor LXR*

The liver X receptors (LXRs) are nuclear receptors that are activated by endogenous oxysterols, oxidized derivatives of cholesterol. There are two isoforms of LXR, LXRα (NR1H3) and LXRβ (NR1H2). Both LXRα and LXRβ regulate gene expression by binding to DNA sequences associated with target genes as heterodimers RXR. LXRs act as cholesterol sensors: when cellular oxysterols accumulate as a result of increasing concentrations of cholesterol, LXR induces the transcription of genes that protect cells from cholesterol overload. The liver regulates *de novo* biosynthesis of cholesterol, the excretion of cholesterol into bile (directly or after conversion to bile acids), the secretion of cholesterol into blood as very low-density lipoproteins (VLDL), the modulation of receptor-mediated cellular cholesterol uptake, the formation of cholesteryl esters, and storage of cholesterol.

1.3 *The xenobiotic sensors PXR and CAR*

The Pregnane X Receptor (PXR, NR1I2 and the Constitutive Androgen Receptor (CAR, NR1I3) are xenobiotic receptors, regulating the expression of drug-metabolizing enzymes and transporters as adaptive responses to prevent the accumulation of toxic chemicals in the body. PXR and CAR share many target genes, such as Phase I metabolizing enzymes CYP3A, CYP2B and CYP2C; Phase II enzymes glutathione S-transferases, UDP-glucuronosyl-transferases (UGTs) and sulfotransferases (SULTs); and drug transporters, such as MDR1, MRP2 and OATP2. A variety of structurally diverse PXR ligands have been reported. These include pharmaceutical drugs, environmental pollutants, dietary supplements as well as as bile acids including LCA and ursodeoxycholic acid.[11] However, in contrast to PXR, CAR exhibits high constitutive activity and is thought to play a significant role in maintaining the basal expression of target genes. Transactivation mediated by CAR is thought to occur through multiple mechanisms. In addition to CAR ligands, such as phenobarbital and simvastatin, protein kinases, protein phosphatases and other factors have been implicated in the modulation of CAR nuclear translocation (reviewed in Meyer zu Schwabedissen).[12]

Both PXR and CAR are highly expressed in the liver and in the intestine, tissues that are important sites for the metabolism of both endogenous and exogenous chemicals.

1.4 *NRs regulate transcription of ABC transporters involved in bile formation*

ABC transporters have a crucial role in bile formation, because bile salts, phospholipids and cholesterol, which are the main components of bile, are transported from the hepatocyte into the bile canaliculus by ABC transporters. Secretion of bile salts by hepatocytes into the canalicular lumen is dependent upon the expression of the bile salt export protein (BSEP/ABCB11), whose expression is directly regulated by FXR.[13–15] In addition, the ABC transporter MDR3/ABCB4, responsible for transport of phospholipids, mainly phosphatidyl choline, from the canalicular membrane into bile, is also transcriptionally regulated by FXR.[16,17] The functional importance of ABCB11 and ABCB4 expression is underlined by pathophysiological consequences of mutations in both genes leading to Progressive Familial Intrahepatic Cholestasis (PFIC), Benign Recurrent Intrahepatic cholestasis (BRIC) and Intrahepatic Cholestasis of Pregnancy (ICP), type 2 and 3, respectively (See also Chapter 3).[18–22] Importantly, mutations and polymorphisms in FXR also predispose to ICP,[23] indicating that transcriptional regulation of ABCB11 and ABCB4 by FXR is crucial for normal bile formation.

Excretion of free cholesterol into the bile has been proposed to be executed by ABCG5 and ABCG8, which are two half transporters that form obligate heterodimers facilitating the biliary excretion of cholesterol into bile. These genes are arranged head-to-head and are co-ordinately expressed and induced following activation of LXR.[24,25] Mice lacking ABCG5 or ABCG8 exhibit profound reduction in biliary cholesterol levels and an accumulation of cholesterol in the liver after cholesterol feeding.[26] Mutations in the genes encoding either ABCG5 or ABCG8 result in β-sitosterolemia, an autosomal recessive disorder characterized by an increased risk of atherosclerosis and elevated plasma levels of phytosterols.[27,28]

ABCC2 (cMOAT/MRP2) plays a crucial role in the hepatobiliary transport of sulfate-, glucuronide-, and glutathione-conjugated metabolites as well as a variety of amphiphilic organic anions derived from hepatic metabolism. Recent studies have demonstrated that ABCC2 is regulated by PXR, FXR, LXR and the constitutive androstane receptor (CAR) and LXR.[17,29] The canalicular membrane also harbors the multidrug export pump MDR1/ABCB1 which transports bulky amphipathic organic cations (e.g., various drugs). PXR has been implicated in the induction of ABCB1 by different xenobiotics. In addition, physiological concentrations of bile salts induce ABCB1 *in vivo* via FXR, independently of PXR. However, CA-feeding in mice induced ABCB1 expression independently of PXR and FXR, illustrating the complexity of *in vivo* regulation of ABCB1.

In addition to these hepatobiliary excretion systems, the basolateral membrane of the hepatocyte also contains efflux pumps of the ABC transporter family. MRP3/ABCC3 and MRP4/ABCC4 are normally expressed at very low levels in hepatocytes, but are upregulated in situations of impaired bile flow, termed cholestasis, which may explain the shift towards renal excretion of bile acids in patients with chronic cholestasis.[30–32] Ligand-activated PXR was shown to induce MRP3 expression in human cell lines and *in vivo* in mice, effects which were not seen in the PXR–/–.[33,34] Induction of human and murine MRP4 can be achieved by treatment with CAR activators.[35] MRP3 is also upregulated in wild type but not Car null mice after treatment with various CAR activators,[36] indicating complex regulation of MRP3.

1.5 *Therapeutic potential of hepatobiliary nuclear receptors*

The apical hepatocyte ABC transporters determine the composition of bile. Pharmacological modulation of ABC transporters could potentially be beneficial in diseases in which expression or function of ABC transporters is affected. Although direct manipulation of the transport proteins is probably difficult, manipulating NRs is a promising alternative. Diseases that may be targeted by NR ligands include cholestatic liver disorders, including gall stone disease, primary biliary cirrhosis (PBC), primary sclerosing cholangitis (PSC), and intrahepatic cholestasis syndromes (PFIC, BRIC, ICP), but also fatty liver disease and chronic liver inflammation. In mice, genetic abrogation and pharmacological modulation of FXR, PXR and CAR influence cholestatic disease. Phase 1 and Phase 2 clinical trials of synthetic ligands that target biliary NRs are underway to intervene in cholestasis and metabolic syndromes.

Given the complexity of the regulatory networks and transcriptional regulation of genes by multiple NRs, there is a strong need for detailed understanding of human NR regulation networks. In addition, although 13% of all FDA-approved drugs target a NR,[37] serious side effects limit their utility and safety, probably due to general activation of all transcriptional actions of the NR. It may be necessary to develop selective ligands to target specific subsets of genes and separate "wanted" from "unwanted" effects. For rational design and evaluation of selective ligands, detailed insight into the molecular mechanisms by which NRs regulate gene transcription is required.

2. Targeting Regulation of ABC Transporters

Considering the expression in a variety of biological barriers, ABC transporters significantly affect the distribution of many drugs. Thereby, efflux transport can limit the penetration of a drug into the target tissue and its access to the target cells. Restricted delivery is considered as one major factor being responsible for poor pharmacotherapeutic response as well as for mere pharmacoresistance. Whereas basal expression levels can already constitute a limiting factor, the situation may be worsened when disease-associated or therapy-induced changes in efflux transporter expression further restrict distribution of the therapeutics.

Expression of efflux transporters is generally regulated in a highly dynamic manner allowing adaptation to changing requirements in detoxification and tissue protection. Therefore, it is obvious that drugs with cytotoxic effects which trigger cellular stress responses induce transporter expression as a mechanism of defence. In the treatment of cancer, induction of efflux transporter expression by chemotherapeutic drugs in tumor cells as well as cells of biological barriers (such as blood-brain barrier endothelial cells) can restrict drug concentrations at the target tumor cells.[38,39] The therapy-associated induction of efflux transporters can thereby critically contribute to therapeutic failure (Fig. 2).

2.1 *Pathophysiological induction of ABC proteins*

Pathophysiology-associated induction of efflux transporters and especially of ABCB1 (p-glycoprotein) has been repeatedly described in different disease states. It is particularly discussed as a major obstacle for pharmacotherapy of several central

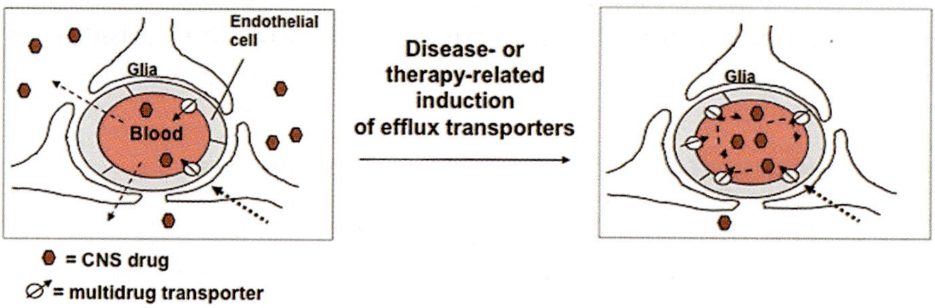

Fig. 2. Pathophysiology-associated mechanisms can induce expression of efflux transporters such as P-glycoprotein in response to seizure activity, ischemic brain damage, or HIV infection in brain capillary endothelial cells. Moreover, xenobiotics including drugs can drive transcriptional activation of the encoding genes. As a consequence efflux transporter over-expression limits brain penetration rates of CNS active drugs. Targeting the regulatory cascades that mediate the induction represents a promising strategy to control efflux transporter levels, and to optimize CNS therapy.

nervous system (CNS) diseases. In the epileptic brain an over-expression of ABCB1 and ABCC2 expression has been described in brain capillary endothelial cells and in selected parenchymal cells. Studies in rodent models of epilepsy revealed that over-expression occurs in a transient manner following seizure activity indicating that epileptic seizures are the major factor driving the induction of efflux transporters in the epileptic brain.[39,40] Ischemic brain damage modeled by middle cerebral artery occlusion in mice resulted in a comparable up-regulation of ABCB1.[41] As several antiepileptic drugs as well as neuroprotectants seem to represent ABCB1 substrates these data indicate that induction of ABCB1 and other efflux transporters may keep respective drug levels low following seizure activity or ischemic brain damage.

In the treatment of HIV infection, a major limitation in the efficacy of HIV protease inhibitors is the restricted access to the brain which implies the risk that the brain viral reservoir remains unaffected. ABCB1-mediated efflux seems to limit the brain penetration rates of different HIV protease inhibitors.[42–45] The up-regulation of ABCB1 in brain capillary endothelial cells triggered by the HIV-Tat protein[46] has been suggested to further reduce penetration and efficacy of the HIV protease inhibitors in long-term survivors of AIDS.

Awareness of the critical role of efflux transporters on the success rates of pharmacotherapy triggers many efforts to develop strategies to outwit transporter function. An obvious strategy is inhibition of transport by competitive or non-competitive transporter modulators. However, this strategy will affect transporter function throughout the body, thereby limiting the protective function of efflux transport and increasing the exposure of sensitive tissues to harmful xenobiotics. Taking these drawbacks into consideration, targeting the regulatory pathways that drive efflux transporter expression in disease states represents a promising alternate approach which would leave basal transporter expression and function unaffected. The strategy aims to prevent transcriptional activation by pathophysiological factors. The elucidation of the signaling cascades that result in transcription activation is the presupposition for the development of such preventive strategies. The mechanisms of expression regulation have so far been most intensely studied for ABCB1.

2.2 *Signaling pathways involved in ABCB1 induction in epilepsy*

Orphan nuclear receptors have been recognized as central regulators of drug-induced changes in the expression of ABCB1 and other members of the multidrug transporter families.[47] The orphan nuclear receptor PXR/SXR (termed pregnane X receptor [PXR] in rodents and steroid and xenobiotic receptor [SXR] in humans) proved to be expressed in biological barriers.[48] In brain capillary endothelial cells, PXR/SXR is discussed to function as a key xenobiotic sensor which mediates induction of ABCB1. Targeting these xenosensors by administration of antagonists such as ketoconazole has been suggested as a means to overcome drug-induced resistance mechanisms.[49,50] The proceeding molecular characterization of receptor structure bears implications for future design of antagonists that are more selective and potent than currently available molecules that target PXR/SXR.

The cellular mechanisms involved in seizure-induced overexpression of efflux transporters still require elucidation. However, relevant progress has been made recently with identification of key signaling factors that trigger ABCB1 in the epileptic brain.[51,52]

Candidate factors were initially selected based on available literature data for regulation of ABCB1 and for seizure-associated changes in expression profiles. Excessive glutamate release is a typical feature of epileptic seizures. In this context, it is of specific interest that glutamate has been described to affect ABCB1 expression.[53] The role of this excitatory neurotransmitter in regulation of ABCB1 expression was clearly substantiated by a combined *in-vitro*/*in-vivo* experimental approach.[51]

Consistent with our hypothesis, exposing isolated rodent brain capillaries to glutamate increased ABCB1 expression and transport activity. These increases were blocked by the NMDA receptor antagonist MK-801. In rats, intracerebral microinjection of glutamate at a nanomolar concentration, that did not cause behavioral or electrographic seizure activity, locally upregulated ABCB1 expression.

Moreover, it was possible to identify cyclooxygenase(COX)-2 as a central downstream factor of the glutamate-triggered and NMDA-receptor mediated signalling events.[52] In isolated brain capillaries both the nonselective COX inhibitor indomethacin as well as the COX-2 selective inhibitor celecoxib abolished ABCB1 induction by glutamate exposure. In contrast, co-incubation with a COX-2 inhibitor was without effect. Experiments with isolated brain capillaries from COX-2 knockout mice further affirmed the particular role of COX-2 as glutamate exposure was without impact on ABCB1 in these capillaries.

These data made us hypothesize that targeting of COX-2 can prevent seizure-induced ABCB1 overexpression in the epileptic brain. In line with this hypothesis, the COX-inhibitor indomethacin prevented the seizure-induced increase in capillary ABCB1 expression in the pilocarpine status epilepticus model in rats. These data are especially promising as a status epilepticus induced by the cholinomimetic pilocarpine represents an extremely strong trigger of ABCB1 overexpression. Overall the investigations identified a glutamate/NMDA-receptor/COX-2 signaling pathway in brain capillary endothelial cells that drives ABCB1 expression during seizures. Collectively, the data suggest that it might be possible to enhance brain uptake of antiepileptic drugs and to overcome transporter-mediated resistance by COX inhibition.[52] As enhanced glutamate release also represents a hallmark during ischemic brain damage, the induction of ABCB1 in the ischemic brain may also be related to glutamate release and subsequent activation of inflammatory events, and may be prevented using the same strategies as those already substantiated in an epilepsy model.

When considering these data as a departure point to design subsequent translational studies, one needs to bear in mind that the clinical use of COX-2 inhibitors can be associated with an enhanced risk for cardiovascular and cerebrovascular events. Moreover, contradictory data exist regarding an impact of COX-2 inhibition on seizure thresholds and seizure severity. These data range from the description of beneficial neuroprotective effects to aggravation of seizures, therefore raising further concerns regarding tolerability issues and demanding a thorough validation of safety and tolerability. For these reasons, it is of specific interest to elucidate the complete cascade that drives transcriptional activation of ABCB1 in response to excessive extracellular glutamate aiming to identify alternate targets.

Further elucidation of the general mechanisms involved in transporter regulation in different disease states may open additional avenues for new strategies to enhance delivery of therapeutics to target tissues. A variety of mechanisms that contribute to cellular stress responses, including phospholipase C, proteinkinase C, mitogen-activated protein kinase cascades, mobilization of intracellular Ca^{2+}, cytokines, nuclear factor kappa B, and heat shock factor 1, have been described to regulate multidrug transporter genes such as *ABCB1* (*MDR1*).[54–57] Using primary cultured rat brain endothelial cells to examine the effect of oxidative stress on expression of transporters, Felix and Barrand[58] found a stress-induced increase in ABCB1. Hartz *et al.*[59,66] defined a signaling pathway as part of the innate immune response through which ABCB1 activity is rapidly modulated. Their findings suggested that the inflammatory cytokine tumor necrosis factor (TNF)-alpha reduces ABCB1 activity via, TNF-R1 receptor activation, endothelin-1 release, and endothelin-B receptor signaling. In apparent contrast, long-term exposure of brain capillaries to TNF-alpha or endothelin-1 increased ABCB1 expression.[61] All these findings indicate a complex regulation of the major blood-brain barrier efflux transporter. Completing the view of involved regulatory mechanisms will inspire further research efforts for prevention of transporter-mediated resistance in different diseases.

2.3 *Targeted bench to patient translation*

The challenge for further translational development of the approach to target regulatory pathways will be the identification of patients which might benefit from this strategy. Considering putative side effects it would be advantageous to identify patients with efflux transporter overexpression. In this context, it is of specific interest that several studies are underway aiming to improve the possibilities for *in-vivo* imaging of transporter function.

The approaches are for instance based on the use of transporter substrates as PET radiotracers. Even more intriguing and probably more reliable is the approach to compare the brain distribution of a transporter substrate radiotracer in two subsequent PET scans with and without co-administration of a specific transporter modulator. A first study has recently been published on the combination of the PET tracer 11C-verapamil with the third generation ABCB1 modulator tariquidar.[62] A study using a combination of the PET tracer 18F MPPF with tariquidar has just been completed (La Fougere, unpublished data). Further development of these imaging techniques will render a basis for patient selection allowing a targeted application of strategies to overcome transporter-mediated resistance in the future.

3. Small-Molecule Corrector Therapy for Cystic Fibrosis

Cystic fibrosis (CF) is one of the most common hereditary diseases, with an incidence of approximately 1 in 2000 Caucasians. The clinical features of CF include chronic lung infection with progressive deterioration of lung function, pancreatic exocrine insufficiency, male infertility, and *meconium ileus* in the newborn. Lung disease is the principal cause of morbidity and death in CF, with median life expectancy of ~38 years in the U.S. CF is caused by mutations in the gene encoding the cystic fibrosis transmembrane conductance regulator (CFTR) protein,[63] which functions as a cAMP-activated Cl^- channel in the apical plasma membrane of epithelial cells in the airways, intestine, pancreas, sweat duct and other fluid-transporting tissues. CFTR is a member of the ATP-binding cassette (ABC) transporter family of proteins. Each CFTR molecule contains two membrane-spanning domains, two nucleotide binding domains that are involved in ATP binding/hydrolysis and a regulatory domain that becomes phosphorylated with channel opening.[64]

More than 1500 loss-of-function mutations in CFTR have been identified that cause CF, with ~90% of CF subjects having the ΔF508 mutation (deletion of phenylalanine at 508 position) as one or both CFTR gene alleles. The ΔF508 mutation produces relatively Cl^- impermeable epithelial cells because of defective cellular processing of the mutant protein leading to retention at the endoplasmic reticulum, accelerated degradation, and impaired Cl^- channel gating (Fig. 3A, top).[65–67] Other much less common mutations include mutations that cause defective channel gating, such as G551D, or premature CFTR truncation, such as W1282X.

Therapies for CF used at present do not correct the underlying CFTR defect, but are symptom-based, including antibiotics, mucolytics (such as DNase), anti-inflammatory agents, nebulized hypertonic saline, and lung transplantation. The mechanism by which defective CFTR causes CF lung disease is unknown, which has hampered the development of rational therapies for CF. As shown in Fig. 3A (bottom), there is evidence for reduced secretion of Cl^-, HCO_3^- and fluid by airway submucosal glands, resulting in a viscous, acidic airway surface liquid (ASL) that promotes bacterial colonization by impairment of mucociliary clearance and bacterial killing.[68] Another theory proposes that the major defect in CF is ENaC Na^+ channel hyperactivity in CF in which excessive Na^+ absorption by the airway surface in CF produces "ASL dehydration".[69] Based on these theories, various modulators of non-CF ion channels are in clinical trials, including activators of non-CF Cl^- channels and inhibitors ENaC Na^+ channels. However, a number of alternative theories have been proposed to relate CFTR dysfunction to CF disease, including

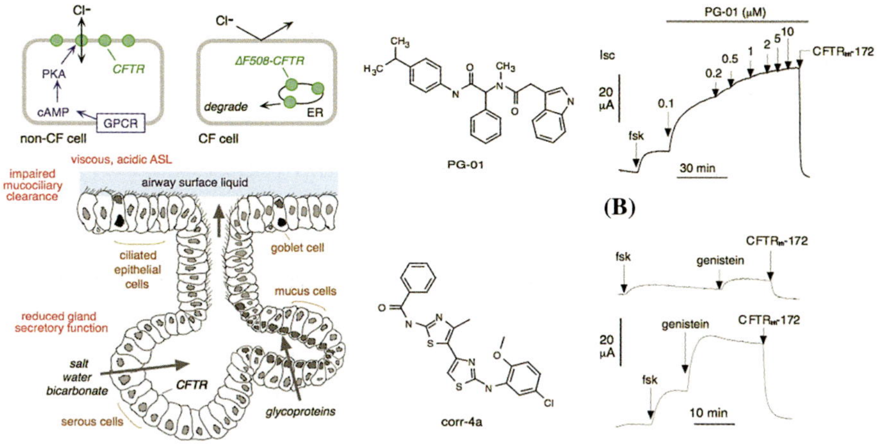

(A) (C)

Fig. 3. Small-molecule potentiators and correctors of ΔF508-CFTR, the most common CF-causing mutation. (**A**) (top) CFTR is a cAMP-activated Cl^- channel at the apical plasma membranes of epithelial cells. The most common CFTR mutation causing CF, ΔF508-CFTR, is misfolded, retained at the endoplasmic reticulum, and rapidly degraded. (bottom) Lung pathophysiology in CF showing reduced Cl^- and bicarbonate secretion by submucosal glands, producing a viscous, acidic airway surface liquid (ASL) that promotes bacterial colonization. (**B**) ΔF508-CFTR potentiator PG-10. Short-circuit current of ΔF508-CFTR-expressing epithelial cells (following low temperature rescue to permit ΔF508-CFTR plasma membrane targeting), showing that PG-01 corrects the defect in ΔF508-CFTR chloride channel gating. There is a small response to forskolin (fsk), followed by activation by PG-01 and inhibition by $CFTR_{inh}$-172. (**C**) ΔF508-CFTR corrector corr-4a. Short-circuit analysis of ΔF508-CFTR done on cells cultured for 24 h in the absence (top) or presence (bottom) of corr-4a, showing partial restoration of chloride current.

defective organellar function in airway epithelia and alveolar macrophages, excessive ceramide accumulation, high ASL salt concentration, and others.

3.1 *Correction and potentiation of mutant CFTR*

Compounds that target the mutant CFTR itself hold promise for treatment of CF, as restoring Cl^- permeability to CF cells is likely to correct the underlying cellular defect causing CF disease. Small-molecule therapies are under development for CF caused by several of the CFTR mutations, including mutations causing defective channel gating, intracellular processing, and biosynthesis. As the ΔF508 mutation causes defects in both channel gating and processing, efforts have been directed on two distinct types of compounds: compounds that increase ΔF508-CFTR open channel probability, which are referred to as "potentiators", and compounds that rescue defective ΔF508-CFTR cellular processing and promote plasma membrane expression, or "correctors". A single agent with both potentiator and corrector activities would be preferred. For CFTR mutations containing premature termination codons read-through therapies are under development. Gentamicin, which has read-through activity, was shown in human nasal potential difference measurements to partially restore CFTR function.[70] A newer candidate, PTC124, which does not have the renal and oto-toxicity of aminoglycosides, is in clinical trials for Duchenne muscular dystrophy. PTC124 produces ribosomal read-through of premature but not of normal stop codons, with phase 2 trials in CF showing early promise in nasal potential difference measurements.[71]

Many classes of compounds act as potentiators, including flavones, xanthines, benzimidazoles and dihydropyridines.[72] High-throughput screening efforts by two groups, each screening collections of ~150,000 diverse compounds, have yielded additional small-molecule potentiators.[73–75] The structure of the phenylglycine potentiator PG-01 is shown in Fig. 3B (left). Figure 3B (right) shows increased short-circuit current following addition of the cAMP agonist forskolin and PG-01 to epithelial cells in which ΔF508-CFTR has been targeted to the plasma membrane by a low temperature-rescue manoeuvre. The small response to a high concentration of forskolin represents the ΔF508-CFTR gating defect, as forskolin alone would fully activate wildtype CFTR. PG-01 strongly increases ΔF508-CFTR Cl^- conductance in a $CFTR_{inh}$-172-inhibitable manner. Patch-clamp analysis has indicated that PG-01 increases ΔF508-CFTR

Cl^- conductance with open probability similar to that of activated wildtype CFTR. In addition to correcting defective gating of ΔF508-CFTR, PG-01, and compound VX-770 (Vertex Pharmaceuticals), corrects defective gating of other CFTR mutants, including the less common G551D-CFTR. VX-770 is currently in phase 2 clinical trials in CF subjects having the G551D mutation. Unfortunately, it is unlikely that potentiator alone would be useful for treatment of CF patients having the common ΔF508 mutation because little if any ΔF508-CFTR protein is expressed at the plasma membrane.

Correctors have been identified that partially rescue ΔF508-CFTR misfolding and cellular misprocessing. Some neutraceuticals and drugs approved for other indications have been reported to have ΔF508-CFTR corrector activity, including curcumin,[76] miglustat[77] and sildenafil.[78] The possible use of curcumin, a component of the natural spice tumeric, generated considerable interest, though now many labs have found that curcumin does not have corrector activity,[79,80] contradicting the original report. Though more data are needed, miglustat and sildenafil appear to have little corrector activity as well. The compound 4-phenylbutyrate (Buphenyl), a drug approved for an erythrocyte urea cycle disorder, increases ΔF508-CFTR plasma membrane expression in cell culture models, probably by non-specific transcriptional enhancement. In early clinical trials, however, a large dose of 4-phenylbutyrate (20 grams three times a day) produced only small increases in apparent ΔF508-CFTR conductance in nasal potential difference measurements.[81]

Several classes of small-molecule ΔF508-CFTR correctors have been identified by high-throughput screening,[75,82] with the structure of the bisaminomethylbithiazole corr-4a shown in Fig. 3C (left). Figure 3C (right) shows increased short-circuit current in ΔF508-CFTR-expressing epithelial cells following 24 h incubation with corr-4a. The restoration of the Cl^- current is comparable to that conferred by low-temperature rescue. Original mechanism-of-action studies suggested that corr-4a improves ΔF508-CFTR folding, probably by a direct interaction, resulting in ER-to-Golgi trafficking and increased ΔF508-CFTR stability.[82] More recent studies suggested that small-molecule correctors bind directly to ΔF508-CFTR and improve its structure at the endoplasmic reticulum.[83–85] Studies with correctors in primary culture of human airway epithelial cells from ΔF508 homozygous subjects, which are the best surrogate available to predict efficacy in human clinical trials, show that maximum Cl^- current after stimulation by forskolin and a potentiator is less than 20% of that in airway cell cultures from non-CF subjects.[75,82] The percentage correction necessary to confer clinical benefit in CF is not known. Similarly, it is not known how well the data from cell culture models translate to efficacy in CF subjects. Clinical trials will be needed to establish efficacy of the various correctors under development.

4. Gene Therapy for Cystic Fibrosis

Cystic fibrosis (CF) is a single gene disorder with insufficient treatment options and a target organ, the lung, which can be non-invasively accessed. Thus, it is not surprising that CF has been at the forefront of gene therapy research. Since cloning of the CF gene (Cystic Fibrosis Transmembrane Conductance Regulator or CFTR) in 1989, 25 phase I/II clinical trials, involving approximately 420 CF patients, have been carried out using a variety of viral and non-viral gene transfer agents (GTAs).[86] Most early trials focused on the nasal epithelium, as a surrogate for the lung, to allow for easy access and sampling and, importantly, to ensure safety. Once an acceptable safety profile had been established GTAs were administered directly into the lung. Although many of these trials established proof-of-principle for gene transfer in the airways, a gene therapy-based treatment has not yet been developed. Not surprisingly airway epithelial cells, which have evolved to keep foreign particles out of the lung, are difficult to transfect. Here, we will summarize the key finding of these clinical studies and describe current pre-clinical and clinical research aimed at further developing gene therapy for CF.

4.1 *Clinical studies using viral vectors*

4.1.1 *Adenovirus*

Despite encouraging results in nasal and pulmonary tissues of pre-clinical models[87,88] and being well-tolerated at low to intermediate doses in humans,[89] adenovirus-mediated gene transfer in the absence of epithelial damage has been inefficient in CF patients.[90] This is mainly due to the absence of the coxsackie-adenovirus receptor (CAR) on the apical surface of the majority of human airway epithelial cells (AECs)[91] and highlights the important differences in receptor distribution in animal models and man. In addition to problems with low transduction efficiency the use of adenovirus for a chronic disease like CF is limited by effective cellular and humoral immune responses against the virus, which prevents re-administration of this transient vector. Thus, Harvey *et al.* delivered three doses of Ad-CFTR to the lungs of CF patients three months apart and demonstrated that after the third administration vector-specific CFTR mRNA was no longer detectable.[89] Suitable strategies that prevent immune responses to the virus and are compatible with use in CF patients have so far not been identified.

4.1.2 *Adeno-associated-virus*

Adenoviruses were superseded by first generation adeno-associated (AAV) vectors due to their good safety profile, broad tissue tropism, long duration of expression, and suggestion of superior escape from immune system surveillance compared with other viruses. Targeted Genetics, Inc led the development of AAV2 vectors for CF gene therapy.[92] Several clinical trials have been carried out in the nose, sinus and lungs of CF patients, all using TgAAV-CFTR which uses the weak AAV long terminal repeat (LTR) promoter to drive expression. A phase II trial assessing the safety and efficacy of repeated administration (three doses one month apart) of AAV2 has been carried out.[92] Viral shedding and increases in neutralizing antibodies were noted, but virus administration was safe. Importantly, a significant reduction in sputum interleukin 8 (a pro-inflammatory cytokine) and a small improvement in lung function (FEV1) was seen after the first, but not second or third administrations. On the basis of these studies Targeted Genetics Corp. initiated a large repeat-administration study (100 subjects), sufficiently powered to detect significant changes in lung function. Whilst TgAAVCF was well tolerated, the study did not meet its primary efficacy end point of statistically significant improvement in lung function[93] and the CF program at Targeted Genetics was discontinued in 2005. There may be several reasons for these disappointing results: (1) AAV2 is too inefficient in transducing airway epithelial cells via the apical membrane, (2) the LTR promoter used to drive CFTR expression is too weak, (3) repeat administration of AAV2 to the lung is not possible due to the development of an anti-viral immune response. Research aimed at addressing these potential limitations of AAV is actively being pursued (see below).

4.2 *Clinical studies using non-viral vectors*

Nine clinical trials have evaluated non-viral gene transfer to nasal or lung epithelium.[86] The majority of these studies have shown partial correction of the chloride transport defect in nasal epithelium. For example, we were able to demonstrate approximately 25% correction of the chloride transport defect in the lung.[94] In contrast to viral GTAs, non-viral formulations are more likely to be repeatedly administrable and proof-of-principle for this has been demonstrated in man.[95] Despite these promising studies it is unknown, if non-viral gene transfer agents, which are less efficient than viral vectors, will be able to correct clinically relevant endpoints such as infection and inflammation in the CF lung. To assess this questions clinical trials have to run over extended period of times (maybe 12 months), thus involving repeated administration of the gene transfer complex.

In a large scale screening program the UK CF Gene Therapy Consortium (http://www.cfgenetherapy.org.uk) has identified the currently most efficient non-viral gene transfer agent for airway transfection and will, over the next three years assess, if repeated administration of this complex can alter clinically relevant end-points in CF. The first phase of this program has recently started with CF patients receiving a single dose of pGM169, a plasmid carrying the CFTR cDNA under control of a hybrid elongation factor 1a promoter (hCEFI), complexed to the cationic lipid GL67A. Although we and others have previously used GL67A in CF patients the plasmid DNA is novel and is, therefore, considered "first-in-man" by the regulatory agencies. pGM169 is a fourth generation plasmid, which is completely depleted of pro-inflammatory nucleotide sequences (CpG islands) and is importantly regulated by a promoter, that at least in mouse models, leads to prolonged (>4 weeks expression) after a single-dose.[96] The single-dose trial is focused on safety, but will also assess duration of gene expression in man. This information will inform the dosing interval in the subsequent repeat-dose clinical trial which will run over 12 months in about 100 patients.

4.3 *Recent pre-clinical developments*

4.3.1 *Viral vectors-adenovirus*

As mentioned above adenoviral vectors are severely hampered by low transduction efficiency of human airway epithelium and the induction of strong immune responses. Over the year various attempts have been made to overcome these problems including administration of immunosuppressants and corticosteroids,[97] as well as the opening of tight-junctions to gain access to the CAR receptor on the basolateral membrane.[98] A controversial issue is whether these strategies are compatible with clinical use given the heavy bacterial colonization present in the CF lung.

Helper-dependent ("gutted") adenoviral vectors, which are depleted of all viral genes, may be less immuno-stimulatory and have improved safety profiles compared to first- and second-generation viruses, which have only a subset of viral genes deleted,[99] but currently do not prevent induction of immune responses. Despite the current limitations for human use, adenoviral vectors have proven useful for pre-clinical proof-of-principle studies.[100,101]

4.3.2 *Viral vectors-AAV*

Research into the optimal AAV-vectors for airway transduction is heavily pursued. AAV serotype 2 was the first AAV-based vector assessed for lung transfection is being actively pursued. More recently AAV2 with capsids from serotypes 1, 5, 6 and 9 (AAV2 genome plus AAV1, 5, 6 or 9 capsids) have been evaluated for lung gene therapy and appear to be more efficient in transducing airway epithelial cells than AAV2.[102] The atomic structure of AAV2 has also been identified, which may enable rational engineering of vector capsids for specific cell targeting.[103] Molecular evolution studies based on DNA shuffling and related recombination techniques are currently utilized to generate new AAV vectors,[104,105] which may prove useful for the transduction of airway epithelium.

Strategies to overcome the AAV packaging problem, thereby allowing incorporation of stronger promoters in addition to the large CFTR gene, have been developed, including approaches based on trans-splicing[106] and homologous recombination.[107] The principle of these techniques is to split the therapeutic cDNA and required promoter elements, and package them into two viruses, which when transfecting the same cell may recombine and generate a full-length therapeutic gene. One would speculate that both of these strategies would lead to reduced transfection efficiency, when compared with administration of one intact virus to the lung. However, surprisingly, Halbert *et al.* have demonstrated that AAV2/6 (ITR from AAV2 and capsid from AAV6) recombination-dependent vectors transduced lung cells in mice almost as efficiently as intact vector, with 10% of AECs being positive.[108]

Shortening the CFTR cDNA is an alternative approach that has been developed to overcome the AAV packaging problems. For example Ostedgaard *et al.* have shown that a CFTR cDNA in which a portion of the R-domain has been deleted leads to partial correction of chloride transport *in vitro*[108] and Fischer *et al.* established proof-of-principle that AAV2/5 carrying a truncated CFTR cDNA leads to vector-specific mRNA and protein expression in non-human primates.[109] In addition to modifications in vector serotype and the cDNA, Halbert *et al.* have shown that selection of the promoter/enhancer elements is important for transduction efficiency. In the context of AAV2/5 the CAG hybrid promoter consisting of a CMV enhancer, a beta-actin promoter and splice donor, and a beta-globin splice acceptor was most effective and lead to >90% transduction efficiency in mouse airways.[110]

It has been postulated that AAV may not infect antigen-presenting dendritic cells and thereby, avoid activation of the host immune system. However, the feasibility of repeated AAV administration is still unresolved. Results have varied greatly, and may depend on the host, delivery route and AAV serotype tested.[111–113] A clinical trial assessing repeat administration of an AAV vector known to efficiently transduce the human airway epithelium is the only way to address this question reliably. Interestingly Lui *et al.* have recently shown that AAV vectors may be able to transduce progenitor cells in the mouse lung,[114] which if efficient may help to overcome repeat administration problems.

4.3.3 *Viral vectors-Lentivirus*

Lentiviral vectors transduce non-dividing cells and integrate into the genome, thereby, theoretically expressing the transgene for the life-time of the cell. Several groups, including our own, have attempted to generate lentiviral vectors for persistent gene expression in the airway epithelium. Whilst it has been challenging to identify appropriate viral envelope proteins, the baculovirus GP64[115] protein and the Sendai-virus derived F and

HN envelope proteins[116] transduce airway epithelial cells efficiently. Importantly, gene expression *in vivo* persists in mouse airway epithelium for >17 months (our data). It is currently unclear, if prolonged expression is due to vector integration into pulmonary stem or progenitor cells, or due to the long life-expectancy of airway epithelial cells, which as recently reported, may have a half-life of up to 17 months.[117] Importantly, we and others, have shown that lentiviruses can be repeatedly administered to the airways of mice.[115,118] It is currently unclear, how the virus evades the immune system, but this feature as well as the prolonged expression profile make the virus an attractive candidate for CF gene therapy, and the UK CF Gene Therapy Consortium, in collaboration with the CF Trust and DNAVEC Corp (Japan) is intensively developing this vectors. A major bottleneck is currently the scale up of vector production for large scale pre-clinical *in vivo* studies.

4.3.4 *Non-viral vectors*

Over the last few years there have not been many convincing advances in improving non-viral gene vectors for the airway gene transfer. Attempts at improving non-viral gene transfer through application of physical energy, such as ultrasound or exposure to magnetic fields, have been moderately successful in the murine lung,[119,120] but currently don't merit progression into clinical trials. We have shown that adjuncts, such as doxorubicin, increase lipid-mediated gene transfer in mice, but the mechanism is unknown and this treatment is not suitable for human.[121] Some progress has been made in improving polyethylenimine (PEI)-based vectors. Davies *et al.* have improved the PEI/DNA complexes through incorporation of a simple filtration step, allowing preparation of more concentrated formulations, which lead to more efficient gene transfer in the lung.[122] Improvements in the plasmids used for non-viral gene transfer are at least as important as improving the lipid or polymer vector. A good example is the generation of plasmids completely devoid of CpG sequences and, thereby, less inflammatory (mentioned above). The hCEFI promoter leads to longer lasting expression than the commonly used CMV promoter,[96] but the search for promoters suitable for airway epithelial cell gene transfer is continuing. The optimal combination of non-viral vector and plasmid DNA will ultimately determine efficiency and safety.

4.4 *Other hot topics*

4.4.1 *Cell therapy*

Bone marrow-derived stem cells (BMSC) may have the capacity to trans-differentiate and repopulate various organs. Various groups have attempted to assess, whether BMSC can transdifferentiate into airway epithelium.[123] In our view systemic administration of BMSC currently does not hold great promise for CF for a number of reasons: (a) data on trans-differentiation in the lung is conflicting; (b) myeloablation and extensive damage of the lung with bleomycin or detergents appear to be pre-requisites for engraftment, neither of which is likely to be clinically applicable; and (c) in reports that provided evidence of engraftment and differentiation, the frequency in airways is very low ($<1\%$).

Interestingly, systemic administration of stem cells has been superseded by studies assessing topical administration of various types of stem cells, which may be a more efficient route to deliver cell therapy to the airways. This was supported in studies by Wong *et al.*[124] and Rejman *et al.*[125] However, even after topical administration, the detection of transplanted cells in the recipient lung remained a rare event. Duchesneau *et al.*[126] extended these studies and showed that destruction of endogenous Clara cells with Naphtalene 2 days prior to transplantation, in combination with Busulfan-induced myoablation 1 day prior to transplantation, achieved maximal retention in the lung. However, the retention frequency of transplanted cells remained low and the differentiation status largely unresolved.

4.4.2 *Animal models*

Various types of CF knockout mice have been available for almost 20 years. Unfortunately these mice do not develop CF lung disease[127] and have, therefore, been of limited value for the development of gene therapy and other treatments. More recently, homozygote CF pigs and ferrets have been generated through somatic cell nuclear transfer of targeted fetal fibroblasts.[128,129] Importantly, both CF pigs and ferrets develop CF-like disease[130,131] and will be valuable research tools for understanding disease pathophysiology and for the development of new treatments.

5. The Role of ABC Multidrug Transporters in Normal and Cancer Stem Cells

5.1 *Stem cells and the ABC multidrug transporter network*

Stem cells are special cells capable of continuous self-renewal that provide a large number of differentiated cell types belonging to various lineages. Early embryonic stem cells have the potential to differentiate into all of our cell types, while tissue-derived ("adult") stem cells are assumed to have narrower differentiation potential. Adult stem cells are present throughout life in tissue niches, and are activated under conditions when cell renewal is necessary. A relatively recent hypothesis gaining increasing acceptance is that the heterologous population of cancer cells also contains stem cell-like components, and the high self-renewal properties of these so-called cancer stem cells is the driving force responsible for tumorigenesis.[132–134]

Cancer stem cells and normal stem cells have similar gene expression profiles and are controlled by the same set of survival factors. As a result of mutations in key oncogenes and/or tumor suppressor genes, cancer stem cells escape from normal regulation and differentiation, yielding the expansion of malignant cell clones. In cancer stem cells the activation of stem-cell like cell-signaling pathways (e.g., notch, sonic hedgehog, wnt-catenin, and FGF-2) were observed,[135] together with high proliferative capacity, increased telomerase activity, and increased expression of anti-apoptotic genes. Such cancer stem cells have been suggested to be present in malignancies as diverse as brain, breast, colon cancer, and melanoma.[132,136–144] Transplantation of a single cancer stem cell can give rise to tumors in animal models.[145] In situ, cancer stem cells show a low division rate, and significantly, they endure the most unfavorable physiological conditions, as well as treatment with anticancer agents. As a result, cancer stem cells can be re-activated even after years of silent, slow reproduction and despite continuous chemotherapy. Since cancer stem cells are believed to make up as little as 0.1–1% of the cells in a tumor, it is difficult to study their properties. However, a common suggested feature of these cells is their multidrug resistant (MDR) phenotype that allows them to survive traditional cancer chemotherapy and radiation therapy. In addition to having high levels of DNA repair and a low ability to enter apoptosis, cancer stem cells express high levels of multidrug resistance ABC drug transporters, which altogether seem to result in elevated drug resistance.

ABC multidrug transporters play a key role in the defense of cells and tissues against various toxic chemicals, especially of hydrophobic and amphipathic nature. These ABC proteins, which include members of the ABCB subfamily (ABCB1, and potentially ABCB3, ABCB5 and ABCB6), those of the ABCC subfamily (ABCC1-C6), and the ABCG subfamily (ABCG2, and potentially ABCG1, G4 and G5+G8), form a network of transporters, which we postulated to participate in a "chemoimmunity system", resembling in many aspects the classical immune system. Together with the key drug metabolizing enzymes and other transporters, the task of this chemoimmunity network is to prevent cellular and tissue damage afflicted by toxic metabolites or chemicals that would easily enter cells as a result of their rapid penetration through cell membranes.[146]

The chemoimmunity network is similar to the immune system in its basic (innate) presence and also in its potential for rapid activation or up-regulation (adaptive response), leading to the elimination of foreign agents. However, in contrast to the classical immune system, a general feature of the relatively small number of participants of this protein network is their "promiscuity", that is the recognition of a wide variety of foreign molecules based on rather general characteristics, such as molecular size, hydrophobicity, nature of hydrogen bondings, etc. This is true for the relevant key phase I-II metabolizing enzymes (such as the CYPs and/or the conjugating enzymes), but also for the transporter proteins (phase 0 and phase III metabolism).[147]

Features of the relevant ABC transporters are discussed in detail throughout this book. In the following sections we describe how the ABC multidrug transporter network may provide a key defense mechanism in normal and cancer stem cells, and how this network may be an important component in the design of efficient cancer chemotherapy.

5.2 *Stem cell markers and ABC transporters*

Abundant expression of MDR-ABC transporters in various progenitor cell types has been widely described. It was first observed in hematopoietic progenitor cell preparations that a so called "side population" (SP) of cells was less efficiently stained by the fluorescent Hoechst 33342 dye.[148–151] The Hoechst-dim SP cells were found to be the most potent bone marrow repopulating early hematopoietic stem cells, and several reports have confirmed that SP cells are multipotent stem cells in a variety of tissue sources[152–157] and even in ES cell cultures.[158] The basis of this SP feature is high level expression of the ABCG2 protein (and in some cases probably also the ABCB1/MDR1 protein) and their active Hoechst dye extrusion function.[149]

In fact, the expression of ABCG2 in the stem cell-like SP has been suggested to be useful as a selective marker for the early progenitors. The isolation of a discrete and highly enriched population of neural stem cells from adult- and embryonic-derived neural progenitor cells (neurospheres) have been reported using the Hoechst dye staining method.[159] However, in this study, variable levels of the ABCG2 and/or ABCB1 messengers and proteins were detected in the embryonic or adult type SP cells. Another recent study failed to find any relationship between Hoechst dye extruding SP cells and ABCG2 expression levels in human umbilical cord blood.[160]

In murine hematopoietic and fetal liver stem cells, the SP cells were also found in greatly variable quantities, and drug or cytokine treatment differently affected Rhodamine-123 and Hoechst 33342 efflux activities.[161] Stem cells from *mdr-1a/1b-/-* mice showed dramatic decrease in Rhodamine-123 efflux, which is most probably performed by the Mdr1 protein only. The authors suggested that MDR-ABC protein activities fluctuated in the same fashion as a number of other stem cell markers, suggesting regulation by common control mechanisms. Still, the regenerative function of stem cells was independent of ABC transporter expression.

In many related dye extrusion studies, verapamil is applied as a general inhibitor of the MDR-ABC transporters. However, although verapamil is a powerful inhibitor of the MDR1-ABCB1 protein, in spite of a common belief, emphasized in many publications, it is only a poor and partial inhibitor of the human ABCG2.[162] Thus, confusion of the relevant ABC transporters may arise in such experiments. Since there are powerful, selective inhibitors for ABCG2 (e.g., FTC or Ko143) and for MDR1 as well (e.g., Reversin 121 or LY335979, Zosuquidar), the specific role of the MDR-ABC transporters in selected stem cell populations should be studied using these agents.

Although ample evidence indicated the role of MDR-ABC transporters in tissue-derived stem and progenitor cells, there were only scattered data, based mostly on RT-PCR studies,[163] regarding their expression in human embryonic stem (HuES) cells. In our laboratory we studied the expression and regulation of MDR-ABC transporters in several human embryonic stem cell lines,[164] performed comparative expression studies for MDR1, MRP1 and ABCG2, and looked for transcriptional regulation of the ABCG2 protein.

Selective, high level functional ABCG2 expression was observed in undifferentiated HuES cells, and expression of this protein significantly decreased during early cell differentiation. ABCG2 in HuES cells provided protection against mitoxantrone toxicity, with a drug-stimulated overexpression of the transporter. No significant expression of ABCB1/ABCC1 was found either in undifferentiated or partially differentiated HuES cells, even if exposed to low drug concentrations. Examination of the ABCG2 mRNA in HuES cells indicated the use of selected promoter sites and a truncated 3′ untranslated region, suggesting functionally distinct regulation of this transporter at the transcriptional level in undifferentiated stem cells.

The important cell protective function of Abcg2 in murine ES cells has also been documented.[165] However, recent study suggested that ABCG2 expression level was low in certain undifferentiated human ES cells, and only cell differentiation induced the expression of this transporter.[166]

The abundance of ABC transporters in various stem cells suggests a dedicated physiological function, linked to maintenance of pluripotency and self renewal. To address this exciting hypothesis, investigators turned to knockout animal models, which provided extensive information on the physiological role of ABC transporters in the past. To investigate the potential role of ABC transporters in SP cells, Jonker *et al.* generated an *Abcg2/Mdr1a/1b* triple knockout mouse.[167] In bone marrow, Abcg2 (Bcrp1) was almost exclusively responsible for the SP, and in the mammary gland both Mdr1 and Abcg2 contributed to the SP phenotype. Interestingly, the bone marrow of the *Mdr1a/1b-/-* double knockout mice displayed an elevated SP, suggesting that Abcg2 may compensate for the loss of Mdr1a/1b in bone marrow.

The physiological role of ABC multidrug transporters in the progenitor/stem cell function is still unclear. The knockout animals are healthy and viable. Loss of the relevant ABC transporter function is not accompanied by reduced stem cell number or function.[150] Furthermore, overexpression of ABCG2 in monkey hematopoietic stem cells does not interfere with hematopoietic repopulation and cell maturation.[168] Although ABCG2 expression was suggested to enhance the growth potential of hematopoietic stem cells,[169] these results suggest that, in laboratory conditions, ABC transporters are dispensable for stem cell function. Rather, in line with earlier conclusions, ABC transporters are present to protect the valuable stem cell sanctuary. Indeed, multidrug transporters may significantly contribute to stem cell protection against endo- or xenobiotics. Since both normal and cancer stem cells have to survive and proliferate under relatively hypoxic and unfavorable metabolic conditions, the potential role of MDR-ABC transporters may be compensation for toxic cell metabolites.

ABCG2 expression is an important protective factor under hypoxic conditions.[170] Progenitor cells obtained from Abcg2 knock-out mice show reduced ability to form colonies under hypoxia. Moreover, blocking of Abcg2 function in normal progenitor cells reduced survival under hypoxic conditions. The primary cause of this hypoxic susceptibility was accumulation of cellular haem and/or porphyrins. Haem is specifically bound by ABCG2, and drug transport by ABCG2 is significantly modulated by haem. This study clearly showed that, during hypoxia, ABCG2 expression was upregulated through an interaction of HIF-1 with the ABCG2 promoter region. The results suggest that ABCG2 permits enhanced cell survival in oxygen-poor environments by reducing accumulation of toxic heme metabolites, and hypoxia increases ABCG2 expression in stem cells.[171–173]

A recent study in murine ES cells demonstrated that during colony expansion, the level of ABCG2 greatly increased, whereas cellular protoporphyrin IX levels decreased.[165] Inhibition of ABCG2 expression or function arrested cell division, reduced expression of pluripotency markers and resulted in elevated intracellular levels of protoporphyrin IX. The authors concluded that ABCG2 protects ES cells from toxic endobiotic accumulation during colony expansion, and that regulation of the ABCG2-dependent defense mechanism is important in the self-renewal of ES cells.

Thus, ABCG2 seems to play a role in preventing toxic effects caused by hypoxia or increased levels of cellular free radicals.

Accordingly, the expression level of this multidrug transporter is significantly regulated by these toxic effects. In addition to transcription-level regulation, data in the literature also indicate that ABCG2 is under complex post-translational modulation, affecting its functional presence in the cell membrane.[174,175] This rapid and efficient regulation may explain some of the controversial findings in ABCG2 expression levels in various stem-cell derived cellular fractions.

As to the role of other MDR-ABC transporters in stem cells, we have already discussed the variable expression and function of ABCB1/MDR1, and we will mention the potential involvement of a newly recognized MDR1-like protein, ABCB5 in melanocyte progenitors and melanoma stem cells. According to a recent study,[176] a member of the ABCB subfamily, the mitochondrial "half-transporter" ABCB6 may also be involved in regulation of cellular haem and porphyrin balance, thus expression of this transporter may contribute to progenitor cell survival under hypoxic conditions.

5.3 *Cancer stem cell ABC and cancer drug resistance*

As described above, the ABCG2 and ABCB1/MDR1 genes are believed to be expressed in the normal tissue-derived and embryonic stem cells, and they may actually characterize stem cell populations. These MDR-ABC transporters were also suggested to be present in most tumor stem cells. Hoechst dye extruding SP cells have also been identified in many tumor cell lines and these SP cells were shown, in several cases, to be highly tumorigenic.[139,151,177–179] Down-regulation of ABCG2 expression significantly suppresses cancer stem cell proliferation,[180] and ABCG2 expression, due to its transporter activity, may contribute to tumor stem cell protection. In prostate cancer, it has been suggested that ABCG2 protects the putative tumor stem cells from androgen deprivation, hypoxia, or adjuvant chemotherapy, and these cells provide a nucleus for a recurrent prostate cancer.[181] A recent report[182] indicates that, even in an established cancer cell line (MCF7), the SP fraction and relevant high ABCG2 expression label a highly tumorigenic fraction and both are down-regulated during induced differentiation or epithelial-mesenchymal transition.

Still, the main question is that whether SP cells in the tumors bear a relationship to cancer stem cells? The overlapping properties of the relatively ill-defined SP cells (their only common feature is low Hoechst dye uptake) and the even more ill-defined cancer stem cells (by using specific, e.g., surface markers, a definitive cancer stem cell phenotype has yet to be conclusively demonstrated)[183] are difficult to assess and convincingly establish.

Hematopoietic malignancies are well characterized and cell populations are relatively easy to separate and examine for protein expression. The role of SP fraction and the ABC transporters in cancer stem cells is widely suggested. In CML, both ABCB1 and ABCG2 were more highly elevated in a tumorigenic, primitive subset of the leukemia cells.[184,185] In AML de Grouw *et al.*[186] demonstrated that ABC transporters were differentially expressed in the stem-cell like populations than in the more differentiated leukemia cells, whereas ABCB1, ABCG2 and ABCC1 were variably expressed in normal and early AML cells. Still, leukemia stem cells were found to reside in the SP cells, and such SP-like cells were found in the bone marrow of more than 80% of the AML patients.[187]

As a summary, we are still uncertain how tumor stem cells relate to the ABC-transporter-dependent Hoechst dye-extruding populations, and the how closely drug resistance often found in hiding and repopulating cancer stem cells is connected to MDR-ABC mediated drug efflux. Hypoxia may be an important factor to up-regulate ABCG2 (and potentially ABCB6) in a niche of cancer stem cells and this may, in turn, provide increased protection of cancer stem cells against chemotherapy. As mentioned above, the functional expression of ABCG2 occurs via numerous pathways, and a recent paper describes regulation of ABCG2 transcript stability via a putative miRNA in a colon cancer cell line.[188]

In the past few years the significance of MDR-ABC dependent drug resistance seems to be better appreciated in the targeted therapy of leukemia by protein kinase inhibitors (TKI). The application of imatinib mesylate (Glivec, STI 571), an inhibitor of the Bcr-Abl hyperactivity brought a breakthrough in treating CML. While comparisons with other treatments demonstrate improved overall long term survival, some patients become resistant or intolerant to imatinib therapy, and relapse is frequent in cases of interruption of the treatment (refractory disease). Several mechanisms mediating imatinib-resistance have been identified (such as the amplification of the *bcr-abl* gene, the overexpression of the BCR-ABL protein, compensatory activation of the Src kinases), and, most importantly, the appearance of point mutations in the kinase domain of Abl. Since imatinib, as any other small molecule protein kinase inhibitor with an intracellular target, has to accumulate in tumor cells, it must overcome pharmacological barriers restricting passage into diverse body compartments and cancer cells. Penetration into these compartments is modulated by ABC transporters that promote transport of a broad variety of drugs. Overexpressed in cancer cells, drug transporters can reduce intracellular drug concentrations, which may allow evolution of BCR-ABL point mutations or BCR-ABL independent pathways leading to TKI drug resistance.[189–192]

The existence of primitive CML stem cells has been well documented.[190] These cells are refractory to TKI therapy and therefore provide a sanctuary for the Philadelphia chromosome. It has become clear that eradication of CML requires a therapeutic modality that can eliminate CML stem cells. Although the differential expression of ABC transporters in hematopoietic and committed cells suggest that the ABC-transporter-mediated protection against xenobiotics (and cytotoxic agents) is conferred by simultaneous activity of several family members, the definitive role of ABC transporters in the protection of CML stem cells remains to be determined.

Imatinib (similarly to some of the new, more efficient kinase inhibitors) is both a high affinity transported substrate and — at higher concentrations — an effective inhibitor of

ABCG2, thus resistance by this transporter is provided only in a given drug concentration range.[193] Increased expression of ABCB1 and ABCG2 predicts resistance in CML stem cells to several therapeutic agents that are effluxed by these transporters, and examining drug-transporter interactions may be an important tool to achieve a good therapeutic response. Despite the established role of MDR-ABC transporters in *in vitro* resistance of cells against various TKIs,[189,194,195] the influence of the ABC transporter network on the clinical response to TKI treatment is unclear. Whether ABC transporters should be considered clinically important mediators of TKI resistance should be investigated by correlative studies evaluating the effect of transporter expression on treatment efficacy and patient survival.

Interestingly, more and more studies indicate that a previously un-explored, MDR1/Pgp like transporter, ABCB5, may have an important role in certain tumor stem cells. This ubiquitously expressed and variably processed protein[196] is highly expressed in melanocyte progenitor cells, causes drug resistance against numerous chemotherapeutic agents[197,198] and may have a special role in melanoma stem cell populations.[199–202] In a stem cell like sub-population of melanoma cells, the ABCB5 protein was found in variable, but relatively high frequency. Expression was suggested to be associated with disease progression and mark a small population of melanoma cells with the capability of initiating tumors in mice.[203] A recent paper documents that by drug-selection, ABCB5 can be up-regulated in CML type (K562) cells, and that ABCB5-expressing cells represent a drug-resistant, stem cell like subpopulation.[204]

In summary, it is strongly suggested that in order to eradicate a tumor, therapy should target cancer stem cells. We therefore emphasize, for successful chemotherapy, the relevant ABC multidrug transporters should also be examined. According to a recent review,[205] drugs that inhibit ATP-binding cassette transporter proteins may "reveal the Holy Grail" of cancer therapy.

Current observations suggest a major role of ABCG2 (and potentially ABCB5) in cancer stem cell like populations, and question the relevance of further ABC transporters that are frequently overexpressed (such as ABCB1 and ABCC1, which are mostly found in differentiated tissues or tumor cell types). Targeting of these latter transporters in clinical trials has brought little or no success in clinical outcome.[206–208] One explanation may be that these proteins are less relevant in the hidden cancer stem cells than in the widely examined, partially differentiated cancer cell offspring. As we continue to investigate the causes of cancer multidrug resistance, we may still be following the bad example of looking for our lost keys where the light is better, instead of looking for them where we actually lost them.

References

1. Lefebvre, P., Cariou, B., Lien, F., Kuipers, F., and Staels, B. (2009) *Physiol. Rev.* **89**, 147.
2. Wang, Y.D., Chen, W.D., Wang, M., Yu, D., Forman, B.M., and Huang, W. (2008) *Hepatology* **48**, 1632.
3. Vavassori, P., Mencarelli, A., Renga, B., Distrutti, E., and Fiorucci, S. (2009) *J. Immunol.* **183**, 6251.
4. Inagaki, T., Moschetta, A., Lee, Y.K., Peng, L., Zhao, G., Downes, M., Yu, R.T., Shelton, J.M., Richardson, M.A., Repa, J.J., Magelsdorf, D.J., and Kliewer, S.A. (2006) *Proc. Natl. Acad. Sci. U.S.A.* **103**, 3920.
5. Gadaleta, R.M., van Erpecum, K.J., Oldenburg, B., Willemsen, E.C., Renooij, W., Murzilli, S., Klomp, L.W., Siersema, P.D., Schipper, M.E., Danese, S., Penna, G., Laverny, G., Adorini, L., Moschetta, A. and Mil, S.W. (2011) *Gut* **60**(4), 463–472.
6. Milona, A., Owen, B.M., van, M.S., Dormann, D., Mataki, C., Boudjelal, M., Cairns, W., Schoonjans, K., Milligan, S., Parker, M., White, R., and Williamson, C. (2010) *Am. J. Physiol. Gastrointest. Liver Physiol.* **298**, G151.
7. Huang, W., Ma, K., Zhang, J., Qatanani, M., Cuvillier, J., Liu, J., Dong, B., Huang, X., and Moore, D.D. (2006) *Science* **312**, 233.
8. Kim, I., Morimura, K., Shah, Y., Yang, Q., Ward, J.M., and Gonzalez, F.J. (2007) *Carcinogenesis* **28**, 940.

9. Maran, R.R., Thomas, A., Roth, M., Sheng, Z., Esterly, N., Pinson, D., Gao, X., Zhang, Y., Ganapathy, V., Gonzalez, F.J., and Guo, G.L. (2009) *J. Pharmacol. Exp. Ther.* **328**, 469.
10. Modica, S., Murzilli, S., Salvatore, L., Schmidt, D.R., and Moschetta, A. (2008) *Cancer Res.* **68**, 9589.
11. Kakizaki, S., Yamazaki, Y., Takizawa, D., and Negishi, M. (2008) *Curr. Drug Metab.* **9**, 614.
12. Meyer zu Schwabedissen, H.E. and Kim, R.B. (2009) *Mol. Pharm.* **6**, 1644.
13. Ananthanarayanan, M., Balasubramanian, N., Makishima, M., Mangelsdorf, D.J., and Suchy, F.J. (2001) *J. Biol. Chem.* **276**, 28857.
14. Gerloff, T., Geier, A., Roots, I., Meier, P.J., and Gartung, C. (2002) *Eur. J. Biochem.* **269**, 3495.
15. Plass, J.R., Mol., O., Heegsma, J., Geuken, M., Faber, K.N., Jansen, P.L., and Muller, M. (2002) *Hepatology* **35**, 589.
16. Huang, L., Zhao, A., Lew, J.L., Zhang, T., Hrywna, Y., Thompson, J.R., De, P.N., Royo, I., Blevins, R.A., Pelaez, F., Wright, S.D., and Cui, J. (2003) *J. Biol. Chem.* **278**, 51085.
17. Kast, H.R., Goodwin, B., Tarr, P.T., Jones, S.A., Anisfeld, A.M., Stoltz, C.M., Tontonoz, P., Kliewer, S., Willson, T.M., and Edwards, P.A. (2002) *J. Biol. Chem.* **277**, 2908.
18. Dixon, P.H., Weerasekera, N., Linton, K.J., Donaldson, O., Chambers, J., Egginton, E., Weaver, J., Nelson-Piercy, C., de, S.M., Warnes, G., Elias, E., Higgins, C.R., Jonston, D.G., McCarthy, M.I., and Williamson, C. (2000) *Hum. Mol. Genet.* **9**, 1209.
19. Dixon, P.H., van Mil, S.W., Chambers, J., Strautnieks, S., Thompson, R.J., Lammert, F., Kubitz, R., Keitel, V., Glantz, A., Mattsson, L.A., Marschall, H.U., Molokhia, M., Moore, G.E., Linton, K.J., and Williamson, C. (2009) *Gut* **58**, 537.
20. Jacquemin, E. (2001) *Semin. Liver Dis.* **21**, 551.
21. Mullerbach, R., Linton, K.J., Wiltshire, S., Weerasekera, N., Chambers, J., Elias, E., Higgins, C.F., Johnston, D.G., McCarthy, M.I., and Williamson, C. (2003) *J. Med. Genet.* **40**, e70.
22. Strautnieks, S.S., Bull, L.N., Kinsely, A.S., Kocoshis, S.A., Dahl, N., Arnell, H., Sokal, E., Dahan, K., Childs, S., Loing, V., Tanner, M.S., Kagalwalla, A.F., Nemeth, A., Pawlowska, J., Baker, A., Mieli-Vergani, G., Freimer, N.B., Gardiner, R.M., and Thompson, R.J. (1998) *Nat. Genet.* **20**, 233.

23. van Mil, S.W., Milona, A., Dixon, P.H., Mullenbach, R., Geenes, V.L., Chambers, J., Shevchuk, V., Moore, G.E., Lammert, F., Glantz, A.G., Mattsson, L.A., Whittaker, J., Parker, M.G., White, R., and Williamson, C. (2007) *Gastroenterology* **133**, 507.
24. Berge, K.E., Tian, H., Graf, G.A., Yu, L., Grishin, N.V., Schultz, J., Kwiterovich, P., Shan, B., Barnes, R., Hobbs, H.H. (2000) *Science* **290**, 1771.
25. Repa, J.J., Berge, K.E., Pomajzl, C., Richardson, J.A., Hobbs, H., and Mangelsdorf, D.J. (2002) *J. Biol. Chem.* **277**, 18793.
26. Yu, L., Gupta, S., Xu, F., Liverman, A.D., Moschetta, A., Mangelsdorf, D.J., Repa, J.J., Hobbs, H.H., and Cohen, J.C. (2005) *J. Biol. Chem.* **280**, 8742.
27. Lee, M.H., Lu, K., and Patel, S.B. (2001) *Curr. Opin. Lipidol.* **12**, 141.
28. Lu, K., Lee, M.H., Hazard, S., Brooks-Wilson, A., Hidaka, H., Kojima, H., Ose, L., Stalenhoef, A.F., Mietinnen, T., Bjorkhem, I., Bruckert, E., Pandya, A., Brewer, H.B., Jr., Salen, G., Dean, M., Srivastava, A., and Patel, S.B. (2001) *Am. J. Hum. Genet.* **69**, 278.
29. Adachi, T., Nakagawa, H., Hagiya, Y., Yasuoka, T., and Ishikawa, T. (2009) *Mol. Pharm.* **6**, 1678.
30. Donner, M.G., Warskulat, U., Saha, N., and Haussinger, D. (2004) *Biol. Chem.* **385**, 331.
31. Soroka, C.J., Lee, J.M., Azzaroli, F., and Boyer, J.L. (2001) *Hepatology* **33**, 783.
32. Zelcer, N., Reid, G., Wielinga, P., Kuil, A., van der Heijden, I., Schuetz, J.D., and Borst, P. (2003) *Biochem. J.* **371**, 361.
33. Teng, S., Jekerle, V., and Piquette-Miller, M. (2003) *Drug Metab. Dispos.* **31**, 1296.
34. Teng, S. and Piquette-Miller, M. (2007) *Br. J. Pharmacol.* **151**, 367.
35. Wagner, M., Halilbasic, E., Marschall, H.U., Zollner, G., Fickert, P., Langer, C., Zatloukal, K., Denk, H., and Trauner, M. (2005) *Hepatology* **42**, 420.
36. Zhang, J., Huang, W., Qatanani, M., Evans, R.M., and Moore, D.D. (2004) *J. Biol. Chem.* **279**, 49517.
37. Overington, J.P., Al-Lazikani, B., and Hopkins, A.L. (2006) *Nat. Rev. Drug Discov.* **5**, 993.
38. Lee, G. and Bendayan, R. (2004) *Pharm. Res.* **21**, 1313.

39. Loscher, W. and Potschka, H. (2005) *Nat. Rev. Neurosci.* **6**, 591.
40. Sisodiya, S.M. (2003) *Curr. Opin. Neurol.* **16**, 197.
41. Spudich, A., Kilic, E., Xing, H., Kilic, U., Rentsch, K.M., Wunderli-Allenspach, H., Bassetti, C.L., and Hermann, D.M., (2006) *Nat. Neurosci.* **9**, 487.
42. Kim, R.B., Fromm, M.F., Wandel, C., Leake, B., Wood, A.J., Roden, D.M., and Wilkinson, G.R. (1998) *J. Clin. Invest.* **101**, 289.
43. Washington, C.B., Wiltshire, H.R., Man, M., Moy, T., Harris, S.R., Worth, E., Weigl, P., Liang, Z., Hall, D., Marriot, L., and Blaschke, T.F. (2000) *Drug Metab. Dispos.* **28**, 1058.
44. Edwards, J.E., Brouwer, K.R., and McNamara, P.J. (2002) *Antimicrob. Agents Chemother.* **46**, 2284.
45. Banks, W.A., Ercal, N., and Price, T.O. (2006) *Curr. HIV Res.* **4**, 259.
46. Hayashi, K., Pu, H., Tian, J., Andras, I.E., Lee, Y.W., Henning, B., and Toborek, M. (2005) *J. Neurochem.* **93**, 1231.
47. Masuyama, H., Suwaki, N., Tateishi, Y., Nakatsukasa, H., Segawa, T., and Hiramatsu, Y. (2005) *Mol. Endocrinol.* **19**, 1170.
48. Bauer, B., Hartz, A.M., Fricker, G., and Miller, D.S. (2004) *Mol. Pharmacol.* **66**, 413.
49. Ekins, S., Ecker, G.F., Chiba, P., and Swaan, P.W. (2007) *Xenobiotica* **37**, 1152.
50. Huang, H., Wang, H., Sinz, M., Zoeckler, M., Staudinger, J., Redinbo, M.R. Teotico, D.G., Locker, J., Kalpana, G.V., and Mani, S. (2007) *Oncogene* **26**, 258.
51. Bauer, B., Hartz, A.M., Pekcec, A., Toellner, K., Miller, D.S., and Potschka, H. (2008) *Mol. Pharmacol.* **73**, 329.
52. Bauer, B., Hartz, A.M., Pekcec, A., Toellner, K., Miller, D.S., and Potschka, H. (2008) *Mol. Pharmacol.* **73**, 1444.
53. Zhu, H.J. and Liu, G.Q. (2004) *Life Sci.* **75**, 1313.
54. McRae, M.P., Brouwer, K.L., and Kashuba, A.D. (2003) *Drug Metab. Rev.* **35**, 19.
55. Shtil, A.A. and Azare, J. (2005) *Int. Rev. Cytol.* **246**, 1.
56. Tchenio, T., Havard, M., Martinez, L.A., and Dautry, F. (2006) *Mol. Cell. Biol.* **26**, 580.
57. Ho, E.A. and Piquette-Miller, M. (2006) *Curr. Cancer Drug Targets* **6**, 195.
58. Felix, R.A. and Barrand, M.A. (2002) *J. Neurochem.* **80**, 64.

59. Hartz, A.M., Bauer, B., Fricker, G., and Miller, D.S. (2004) *Mol. Pharmacol.* **66**, 387.
60. Hartz, A.M., Bauer, B., Fricker, G., and Miller, D.S. (2006) *Mol. Pharmacol.* **69**, 462.
61. Bauer, B., Hartz, A.M., and Miller, D.S. (2007) *Mol. Pharmacol.* **71**, 667.
62. Bankstahl, J.P., Kuntner, C., Abrahim, A., Karch, R., Stanek, J., Wanek, T., Wadsak, W., Kletter, K., Muller, M., Loscher, W., and Langer, O. (2008) *J. Nucl. Med.* **49**, 1328.
63. Rommens, J.M., *et al.* (1989) *Science* **245**, 1059.
64. Sheppard, D. and Welsh, M. (1999) *Physiol. Rev.* **79**, S23.
65. Denning, G.M., *et al.* (1992) *Nature* **358**, 761.
66. Du, K., Sharma, M., and Lukacs, G.L. (2005) *Nat. Struct. Mol. Biol.* **12**, 17.
67. Dalemans, W., *et al.* (1991) *Nature* **354**, 526.
68. Verkman, A.S., Song, Y., and Thiagarajah, J. (2005) *Am. J. Physiol.* **284**, C2.
69. Boucher, R.C. (2007) *J. Intern. Med.* **261**, 5.
70. Wilschanski, M., *et al.* (2003) *N. Engl. J. Med.* **349**, 1433.
71. Kerem, E., *et al.* (2008) *Lancet* **372**, 719.
72. Amaral, M.D. and Kunzelmann, K. (2007) *Trends Pharmacol. Sci.* **28**, 334.
73. Yang, H., *et al.* (2003) *J. Biol. Chem.* **278**, 35079.
74. Pedemonte, N., *et al.* (2005) *Mol. Pharmacol.* **67**, 1797.
75. Van Goor, F., *et al.* (2006) *Am. J. Physiol. Lung Cell. Mol. Physiol.* **290**, L1117.
76. Egan, M.E., *et al.* (2004) *Science* **304**, 600.
77. Norez, C., *et al.* (2006) *FEBS Lett.* **580**, 2081.
78. Dormer, R.L., *et al.* (2005) *Thorax* **60**, 55.
79. Song, Y., *et al.* (2004) *J. Biol. Chem.* **279**, 40629.
80. Grubb, B., *et al.* (2006) *Am. J. Respir. Cell Mol. Biol.* **34**, 355.
81. Rubenstein, R.C., Egan, M.E., and Zeitlin, P.L. (1997) *J. Clin. Invest.* **100**, 2457.
82. Pedemonte, N., *et al.* (2005) *J. Clin. Invest.* **115**, 2564.
83. Wang, Y., Loo, T.W., Bartlett, M.C., and Clarke, D.M. (2007) *J. Biol. Chem.* **282**, 33247.
84. Loo, T.W., Bartlett, M.C., and Clarke, D.M. (2008) *Biochem. J.* **413**, 29.
85. Wellhauser, L., Kim Chiaw, P., Pasyk, S., Li, C., Ramjeesingh, M., and Bear, C. (2009) *Mol. Pharmacol.* **75**, 1430.

86. Griesenbach, U. and Alton, E.W. (2009) *Adv. Drug Deliv. Rev.* **61**, 128.
87. Katkin, J.P., Gilbert, B.E., Langston, C., French, K., and Beaudet, A.L. (1995) *Hum. Gene Ther.* **6**, 985.
88. Scaria, A., St George, J.A., Jiang, C., Kaplan, J.M., Wadsworth, S.C., and Gregory, R.J. (1998) *J. Virol.* **72**, 7302.
89. Harvey, B.G., Leopold, P.L., Hackett, N.R., Grasso, T.M., Williams, P.M., Tucker, A.L., Kaner, R.J., Ferris, B., Gonda, I., Sweeney, T.D., Ramalingam, R., Kovesdi, I., Shak, S., and Crystal, R.G. (1999) *J. Clin. Invest.* **104**, 1245.
90. Joseph, P.M., O'Sullivan, B.P., Lapey, A., Dorkin, H., Oren, J., Balfour, R., Perricone, M.A., Rosenberg, M., Wadsworth, S.C., Smith, A.E., St. George, J.A., and Meeker, D.P. (2001) *Hum. Gene. Ther.* **12**, 1369.
91. Walters, R.W., Grunst, T., Begelson, J.M., Finberg, R.W., Welsh, M.J., and Zabner, J. (1999) *J. Biol. Chem.* **274**, 10219.
92. Moss, R.B., Rodman, D., Spencer, L.T., Aitken, M.L., Zeitlin, P.L., Waltz, D., Milla, C., Brody, A.S., Clancy, J.P., Ramsey, B., Hamblett, N., and Heald, A.E. (2004) *Chest* **125**, 509.
93. Moss, R.B., Milla, C., Colombo, J., Accurso, F., Zeitlin, P.L., Clancy, J.P., Spencer, L.T., Pilewski, J., Waltz, D.A., and Heald, A.E. (2007) *Hum. Gene. Ther.* **18**, 726.
94. Alton, E.W., Stern, M., Farley, R., Jaffe, A., Chadwick, S.L., Phillips, J., Davies, J., Smith, S.N., Browning, J., Davies, M.G., Hodson, M.E., Durham, S.R., Li, D., Jeffrey, P.K., Scallan, M., Balfour, R., Eastman, S.J., Cheng, S.H., Smith, A.E., Meeker, D., and Geddes, D.M. (1999) *Lancet* **353**, 947.
95. Hyde, S.C., Southern, K.W., Gileadi, U., Fitzjohn, E.M., Mofford, K.A., Waddell, B.E., Gooi, H.C., Goddard, C.A., Hannavy, K., Smyth, S.E., Egan, J.J., Sorgi, F.L., Huang, L., Cuthbert, A.W., Evans, M.J., Colledge, W.H., Higgins, C.F., Webb, A.K., and Gill, D.R. (2000) *Gene Ther.* **7**, 1156.
96. Hyde, S.C., Pringle, I.A., Abdullah, S., Lawton, A.E., Davies, L.A., Varathalingam, A., Nunez-Alonso, G., Green, A.M., Bazzani, R.P., Sumner-Jones, S.G., Chan, M., Li, H., Yew, N.S., Cheng, S.H., Christopher, B.A., Davies, J.C., Griesenback, U., Porteous, D.J., Sheppard, D.N., Munkonge, F.M., Alton, E.W., and Gill, D.R. (2008) *Nat. Biotechnol.* **26**, 549.
97. Price, A.R., Limberis, M.P., Wilson, J.M., and Diamond, S.L. (2007) *Gene. Ther.* **14**, 1594.

98. Koehler, D.R., Frndova, H., Leung, K., Louca, E., Palmer, D., Ng, P., McKerlie, C., Cox, P., Coates, A.L., and Hu, J. (2005) *J. Gene. Med.* **7**, 1409.
99. Brunetti-Pierri, N. and Ng, P. (2008) *Gene. Ther.* **15**, 553.
100. Koehler, D.R., Sajjan, U., Chow, Y.H., Martin, B., Kent, G., Tanswell, A.K., McKerlie, C., Forstner, J.F., and Hu, J. (2003) *Proc. Natl. Acad. Sci. U.S.A.* **100**, 15364.
101. van Heeckeren, A.M., Schluchter, M.D., Xue, W., and Davis, P.B. (2006) *Am. J. Respir. Crit. Care Med.* **173**, 288.
102. Limberis, M.P., Vandenberghe, L.H., Zhang, L., Pickles, R.J., and Wilson, J.M. (2009) *Mol. Ther.* **17**, 294.
103. Xie, Q., Bu, W., Bhatia, S., Hare, J., Somasundaram, T., Azzi, A., and Chapman, M.S. (2002) *Proc. Natl. Acad. Sci. U.S.A.* **99**, 10405.
104. Maheshri, N., Koerber, J.T., Kaspar, B.K., and Schaffer, D.V. (2006) *Nat. Biotechnol.* **24**, 198.
105. Excoffon, K.J., Koerber, J.T., Dickey, D.D., Murtha, M., Keshavjee, S., Kaspar, B.K., Zabner, J., and Schaffer, D.V. (2009) *Proc. Natl. Acad. Sci. U.S.A.* **106**, 3865.
106. Duan, D., Yue, Y., and Engelhardt, J.F. (2001) *Mol. Ther.* **4**, 383.
107. Halbert, C.L., Allen, J.M., and Miller, A.D. (2002) *Nat. Biotechnol.* **20**, 697.
108. Ostedgaard, L.S., Rokhlina, T., Karp, P.H., Lashmit, P., Afione, S., Schmidt, M., Zabner, J., Stinski, M.F., Chiorini, J.A., and Welsh, M.J. (2005) *Proc. Natl. Acad. Sci. U.S.A.* **102**, 2952.
109. Fischer, A.C., Smith, C.L., Cebotaru, L., Zhang, X., Askin, F.B., Wright, J., Guggino, S.E., Adams, R.J., Flotte, T., and Guggino, W.B. (2007) *Mol. Ther.* **15**, 756.
110. Halbert, C.L., Lam, S.L., and Miller, A.D. (2007) *Hum. Gene. Ther.* **18**, 344.
111. Beck, S.E., Jones, L.A., Chestnut, K., Walsh, S.M., Reynolds, T.C., Carter, B.J., Askin, F.B., Flotte, T.R., and Guggino, W.B. (1999) *J. Virol.* **73**, 9446.
112. Halbert, C.L., Standaert, T.A., Aitken, M.L., Alexander, I.E., Russel, D.W., and Miller, A.D. (1997) *J. Virol.* **71**, 5932.
113. Halbert, C.L., Rutledge, E.A., Allen, J.M., Russell, D.W., and Miller, A.D. (2000) *J. Virol.* **74**, 1524.
114. Liu, X., Luo, M., Guo, C., Yan, Z., Wang, Y., Lei-Butters, D.C., and Engelhardt, J.F. (2009) *Mol. Ther.* **17**, 285.

115. Sinn, P.L., Arias, A.C. Brogden, K.A., and McCray, P.B., Jr. (2008) *J. Virol.* **82**, 10684.
116. Kobasyashi, M., lida, A., Ueda, Y., and Hasegawa, M. (2003) *J. Virol.* **77**, 2607.
117. Rawlins, E.L. and Hogan, B.L. (2008) *Am. J. Physiol. Lung Cell Physiol.* **295**, L231.
118. Griesenbach, U., Mitomo, K., Inoue, M., Meng, C., Tabata, T., Yasuji, U., Farley, R., Hasegawa, M., and Alton, E.W. (2008) *Mol. Ther.* **16** (Abstract No. 754).
119. Xenariou, S., Griesenbach, U., Ferrari, S., Dean, P., Scheule, R.K., Cheng, S.H., Geddes, D.M., Plank, C., and Alton, E.W. (2006) *Gene. Ther.* **13**, 1545.
120. Xenariou, S., Griesenbach, U., Liang, H.D., Zhu, J., Farley, R., Somerton, L., Singh, C., Jeffery, P.K., Ferrari, S., Scheule, R.L., Cheng, S.H., Geddes, D.M., Blomley, M., and Alton, E.W. (2007) *Gene. Ther.* **14**, 768.
121. Griesenbach, U., Meng, C., Farley, R., Gardner, A., Brake, M.A., Frankel, G.M., Gruenert, D.C., Cheng, S.H., Scheule, R.K., and Alton, E.W. (2009) *Biomaterials* **30**, 1971.
122. Davies, L.A., McLachlan, G., Sumner-Jones, S.G., Ferguson, D., Baker, A., Tennant, P., Gordon, C., Vrettou, C., Baker, E., Zhu, J., Alton, E.W., Collie, D.D., Porteous, D.J., Hyde, S.C., and Gill, D.R. (2008) *Mol. Ther.* **16**, 1283.
123. Wong, A.P., Keating, A., Lu, W.Y., Duchesneau, P., Wang, X., Sacher, A., Hu, J., and Waddell, T.K. (2009) *J. Clin. Invest.* **119**, 336.
124. Wong, A.P., Keating, A., Lu, W.Y., Duchesneau, P., Wang, X., Sacher, A., *et al.* (2009) *J. Clin. Invest.* **119**(2), 336.
125. Rejman, J., Colombo, C., and Conese, M. (2009) *Mol. Ther.* **17**(7), 1257.
126. Duchesneau, P., Wong, A.P., and Waddell, T.K. (2010) *Mol. Ther.* **18**(10), 1830.
127. Dorin, J.R. (2006) Animal models, in: Bush, A., Alton, E.W., Davies, J.C., Griesenbach, U., Jaffe, A. (eds.) *Progress in Respiratory Research: Cystic Fibrosis in the 21st Century*. Karger, pp. 84–92.
128. Rogers, C.S., Hao, Y., Rokhlina, T., Samuel, M., Stoltz, D.A., Li, Y., Petroff, E., Vermeer, D.W., Kable, A.C., Yan, Z., Spate, L., Wax, D., Murphy, C.N., Rieke, A., Whitworth, K., Linville, M.L., Korte, S.W., Engelhardt, J.F., Welsh, M.J., and Prather, R.S. (2008) *J. Clin. Invest.* **118**, 1571.

129. Sun, X., Yan, Z., Yi, Y., Li, Z., Lei, D., Rogers, C.S., Chen, J., Zhang, T., Welsh, M.J., Leno, G.H., and Engelhardt, J.F. (2008) *J. Clin. Invest.* **118**, 1578.
130. Ostedgaard, L.S., Meyerholz, D.K., Chen, J.H., Pezzulo, A.A., Karp, P.H., Rokhlina, T., *et al.* (2011) *Sci. Transl. Med.* **3**(74), 74ra24.
131. Sun, X., Sui, H., Fisher, J.T., Yan, Z., Liu, X., Cho, H.J., *et al.* (2010) *J. Clin. Invest.* **120**(9), 3149.
132. Reya, T., *et al.* (2001) *Nature* **414**(6859), 105.
133. Ben-Porath, I., *et al.* (2008) *Nat. Genet.* **40**(5), 499.
134. Wong, D.J., *et al.* (2008) *Cell Stem Cell* **2**(4), 333.
135. Dvorak, P., Dvorakova, D., Hampl, A. (2006) *FEBS Lett.* **580**, 2869.
136. Monk, M. and Holding, C. (2001) *Oncogene* **20**(56), 8085.
137. Pardal, R., Clarke, M.F., and Morrison, S.J. (2003) *Nat. Rev. Cancer* **3**(12), 895.
138. Clarke, M.F., *et al.* (2006) *Cancer Res.* **66**(19), 9339.
139. Kruger, J.A., *et al.* (2006) *Blood* **108**(12), 3906.
140. Li, L. and Neaves, W.B. (2006) *Cancer Res.* **66**(9), 4553.
141. Tan, B.T., *et al.* (2006) *Lab. Invest.* **86**(12), 1203.
142. Buzzeo, M.P., Scott, E.W., and Cogle, C.R. (2007) *Leukemia* **21**(8), 1619.
143. Cho, R.W. and Clarke, M.F. (2008) *Curr. Opin. Genet. Dev.* **18**(1), 48.
144. Tomasson, M.H. (2009) *J. Cell. Biochem.* **106**(5), 745.
145. Quintana, E., *et al.* (2008) *Nature* **456**(7222), 593.
146. Sarkadi, B., *et al.* (2006) *Physiol. Rev.* **86**(4), 1179.
147. Szakacs, G., *et al.* (2008) *Drug Discov. Today* **13**(9–10), 379.
148. Goodell, M.A., *et al.* (1996) *J. Exp. Med.* **183**(4), 1797.
149. Zhou, S., *et al.* (2001) *Nat. Med.* **7**(9), 1028.
150. Zhou, S., *et al.* (2002) *Proc. Natl. Acad. Sci. U.S.A.* **99**(19), 12339.
151. Scharenber, C.W., Harkey, M.A., and Torok-Storb, B. (2002) *Blood* **99**(2), 507.
152. Asakura, A., *et al.* (2002) *J. Cell. Biol.* **159**(1), 123.
153. Asakura, A. and Rudnicki, M.A. (2002) *Exp. Hematol.* **30**(11), 1339.
154. Bunting, K.D. (2002) *Stem Cells* **20**(1), 11.
155. Summer, R., *et al.* (2003) *Am. J. Physiol. Lung Cell. Mol. Physiol.* **285**(1), L97.
156. Meissner, K., *et al.* (2006) *J. Histochem. Cytochem.* **54**(2), 215.
157. Honda, M.J., *et al.* (2007) *Int. Endod. J.* **40**(12), 949.
158. Vieyra, D.S., Rosen, A., and Goodell, M.A. (2009) *Stem Cells. Dev.* **18**, 1155.

159. Kim, M. and Morshead, C.M. (2003) *J. Neurosci.* **23**(33), 10703.
160. Alt, R., Wilhelm, F., Pelz-Ackermann, O., Egger, D., Niederwieser, D., and Cross, M. (2009) *Exp. Hematol.* **37**(2), 294.
161. Uchida, N., *et al.* (2004) *Blood* **103**(120), 4487.
162. Sarkadi, B., *et al.* (2004) *FEBS Lett.* **567**(1), 116.
163. Bhattacharya, B., *et al.* (2004) *Blood* **103**(8), 2956.
164. Apati, A., *et al.* (2008) *Biochim. Biophys. Acta.* **1778**(12), 2700.
165. Susanto, J., *et al.* (2008) *PLoS One* **3**(12), e4023.
166. Zeng, H., *et al.* (2009) *Stem Cells* **27**, 2435.
167. Jonker, J.W., *et al.* (2005) *Stem Cells* **23**(8), 1059.
168. Bozorgmehr, F., *et al.* (2007) *Stem Cells* **25**(10), 2610.
169. Ahmed, F., *et al.* (2008) *Stem Cells* **26**(3), 810.
170. Krishnamurthy, P., *et al.* (2004) *J. Biol. Chem.* **279**(23), 24218.
171. Krishnamurthy, P. and Schuetz, J.D. (2005) *Biometals* **18**(4), 349.
172. Krishnamurthy, P., *et al.* (2008) *Cancer Res.* **68**(13), 4983.
173. Krishnamurthy, P., Xie, T., and Schuetz, J.D. (2007) *Pharmacol. Ther.* **114**(3), 345.
174. Meyer zu Schwabedissen, H.E., *et al.* (2006) *Drug Metab. Dispos.* **34**(4), 524.
175. Wang, H., *et al.* (2006) *Am. J. Physiol. Endocrinol. Metab.* **290**(5), E798.
176. Lynch, J., *et al.* (2009) *Cancer Res.* **69**(13), 5560.
177. Hirschmann-Jax, C. *et al.* (2004) *Proc. Natl. Acad. Sci. U.S.A.* **101**(39), 14228.
178. Kondo, T., Setoguchi, T., and Taga, T. (2004) *Proc. Natl. Acad. Sci. U.S.A.* **101**(3), 781.
179. Ho, M.M., *et al.* (2007) *Cancer Res.* **67**(10), 4827.
180. Chen, Z., *et al.* (2009) *Int. J. Cancer* **126**, 841.
181. Xiong, A., *et al.* (2008) *Tissue Eng. Part A* **14**(6), 995.
182. Yin, L., Castagnino, P., and Assoian, R.K. (2008) *Cancer Res.* **68**(3), 800.
183. Shipitsin, M. and Polyak, K. (2008) *Lab. Invest.* **88**(5), 459.
184. Wulf, G.G., *et al.* (2001) *Blood* **98**(4), 1166.
185. Jordanides, N.E., *et al.* (2006) *Blood* **108**(4), 1370.
186. De Grouw, E.P., *et al.* (2006) *Leukemia* **20**(4), 750.
187. De Jonge-Peeters, S.D., *et al.* (2007) *Crit. Rev. Oncol. Hermatol.* **62**(3), 214.
188. To, K.K., *et al.* (2008) *Mol. Cancer. Res.* **6**(1), 151.
189. Cervenak, J., *et al.* (2006) *Cancer Lett.* **234**(1), 62.

190. Turk, D. and Szakacs, G. (2009) *Curr. Opin. Drug Discov. Devel.* **12**(2), 246.
191. Hegedus, C., *et al.* (2009) *Adv. Drug Deliv. Rev.* **61**(1), 47.
192. Hegedus, C., *et al.* (2009) *Curr. Cancer Drug Targets* **9**(3), 252.
193. Thomas, J., *et al.* (2004) *Blood* **104**(12), 3739.
194. Ozvegy-Laczka, C., *et al.* (2005) *Drug Resist. Updat.* **8**(1–2), 15.
195. Ozvegy-Laczka, C., *et al.* (2004) *Mol. Pharmacol.* **65**(6), 1485.
196. Chen, K.G., *et al.* (2005) *Pigment Cell. Res.* **18**(2), 102.
197. Frank, N.Y., *et al.* (2005) *Cancer Res.* **65**(10), 4320.
198. Frank, N.Y., *et al.* (2003) *J. Biol. Chem.* **278**(47), 47156.
199. Fang, D., *et al.* (2005) *Cancer Res.* **65**(20), 9328.
200. Heimerl, S., *et al.* (2007) *Melanoma Res.* **17**(5), 265.
201. Bhoumik, A. and Ronai, Z. (2008) *Pigment Cell Melanoma Res.* **21**(2), 111.
202. Zabierowski, S.E. and Herlyn, M. (2008) *Cancer Cell* **13**(3), 185.
203. Schatton, T., *et al.* (2008) *Nature* **451**(7176), 345.
204. Frank, N.Y. and Frank, M.H. (2009) *Leuk. Res.* **33**(10), 1303.
205. Misaghian, N., *et al.* (2009) *Leukemia* **23**(1), 25.
206. Fisher, G.A. and SIkic, B.I. (1995) *Hematol. Oncol. Clin. North Am.* **9**(2), 363.
207. Lepper, E.R., *et al.* (2005) *Pharmacogenomics* **6**(2), 115.
208. Szakacs, G., *et al.* (2006) *Nat. Rev.* **5**(3), 219.

Index